ADVANCED CONTROL SYSTEM TECHNOLOGY

C. J. Chesmond
P. A. Wilson
and M. R. Le Pla

Edward Arnold
A division of Hodder & Stoughton
LONDON   MELBOURNE   AUCKLAND

© 1991 Yenesta Pty Ltd, P. A. Wilson and M. R. le Pla

First published in Great Britain 1991

British Library Cataloguing in Publication Data

Chesmond, C. J.
  Advanced control system technology.
  I. Title   II. Wilson, P. A.   III. Pla, M. R. Le
  629.8

  ISBN 0-340-54550-X

Typeset in 11/12 Times by MS Filmsetting Limited, Frome, Somerset
Printed in Great Britain for Edward Arnold,
a division of Hodder and Stoughton Limited,
Mill Road, Dunton Green, Sevenoaks, Kent TN13 2YA,
by St Edmundsbury Press Limited, Bury St Edmunds, Suffolk
and bound by Hartnolls Limited, Bodmin, Cornwall

Preface

In the preface to the companion volume – *Basic Control System Technology* – which was first published in 1990, the release of the present volume was foreshadowed. It was also pointed out that the two monographs were to be treated as complementary; this approach has indeed been followed.

For the reader who is entering the Automatic Control field for the first time, it is advisable to have read the *Basic* volume before embarking on the present one. In fact, it has been expedient in a few places to cross-refer the reader of the *Advanced* volume to information presented in the other.

Technology is moving fast, with computers now dominating the role of Controller in feedback loops. Of necessity, much of this book concentrates on those aspects of computer hardware and software which are relevant to automatic control systems.

Moreover, the rapid overlapping of such fields as Data Communications, Feedback Control, Computer Engineering and Software Engineering has been reflected intentionally in this *Advanced* volume. This represents the situation as it now stands, with the technology. Thus, the competent Control Engineer now needs to have a good working knowledge of all of these fields, particularly in the way in which they inter-relate when controlling plant automatically. The Control Engineer, now more than ever before, needs to embrace the *holistic* systems approach to which reference was made in the preface of the *Basic* volume.

The present work has been produced by three authors, so as to draw on as wide a range of expertise as possible. In addition, we should particularly like to thank our good friend Ian Brown for the help given in preparing the chapter on realtime operating systems.

Acknowledgements would not be complete without including the steadfast support of our wives during the long hours needed to compile this material.

Contents

1

Signal and Data Conversion

1.1 Introduction

It is commonplace for the output signal of one element in a control loop to be incompatible with the input signal requirements of the element which it is intended to drive. It therefore becomes necessary to interpose special hardware, in order to convert the characteristics of the driving signal to the appropriate form. Examples of configurations in which this would be necessary are the following:

- a transducer with a pneumatic output signal required to drive into an electronic controller
- a digital transducer required to drive into an analog electronic controller
- a synchro control transformer required to drive a DC servoamplifier.

Many other examples are easily conceived. Hardware which is capable of providing signal conversion of this nature will obviously be very varied in nature, depending largely upon the forms of the input and output signals of concern, and, for this reason, a survey of the alternative elements available must necessarily be limited in extent in this particular volume. (Note, for example, that complete volumes have been devoted just to analog-digital and digital-analog conversion techniques.)

The range of hardware available for signal conversion is summarised in Table 1.1., which should be interpreted by making reference to the Key. It should be noted that, if necessary, signal converters may be cascaded: thus, for example, an AC voltage signal may be converted into a set of signals representing parallel digital data by cascading together a demodulator and an analog-digital converter having parallel digital outputs. The type of hardware to which reference is made in the key of Table 1.1 will now be described in detail, with the exception of those previously described in *Basic Control System Technology*.

1.2 Voltage–to–current converters

Section 8.5.3 in *Basic Control System Technology* described how negative feedback may be used in order to create a voltage-controlled current source. A specific application of this technique is shown in Fig. 1.1.

It is readily shown for this configuration that, provided the amplifier output voltage V_o is not saturating and the amplifier can be regarded as 'ideal',

$$\frac{V_i}{R_1} + \frac{I_L R_D}{R_2} = 0$$

Solving for I_L yields $I_L = -\dfrac{R_2}{R_1 R_D} V_i$

Table 1.1   Signal conversion hardware

TO \ FROM	DC voltage	DC current	AC voltage (single)	AC synchro voltage pattern	AC resolver voltage pattern	Serial digital data	Parallel digital data	Pneumatic pressure (3 to 15 psi control air)
DC voltage	–	E	G				O	P	X
DC current	A	–							Y
AC voltage (single)	B		–	H	L	Q	R	
AC synchro voltage pattern				–	I	S	T	
AC resolver voltage pattern				I	–	U	V	
Serial digital data	C			J	M	–	W	
Parallel digital data	D			K	N	W	–	
Pneumatic pressure (3 to 15 psi control air)	F							–

Note   (1) A blank entry signifies that no component exists to effect a conversion directly: conversion may be achieved by cascading two or more of the listed devices

(2) H, L and X, the section references given in the Key are to be found in the companion volume *Basic Control System Technology*.

Key:

A	current feedback around a high gain amplifier	:	refer to Section 1.2
B	modulator; phase-sensitive, where necessary	:	refer to Section 1.4
C	analog-digital converter (ADC) with serial output	:	refer to Section 1.6
D	analog-digital converter (ADC) with parallel output	:	refer to Section 1.6
E	resistor; with bufferred load, where necessary	:	refer to Section 1.3
F	current–to–air converter (transducer)	:	refer to Section 1.19
G	demodulator (rectifier); phase-sensitive, where necessary	:	refer to Section 1.5
	or RMS-to-DC converter	:	refer to Section 1.13
H	synchro control transformer with locked rotor, or use one line-to-line voltage	:	refer to Section 3.2.5
I	Scott-T connected transformer pair	:	refer to Section 1.10
J	synchro-digital converter with serial output	:	refer to Section 1.11
K	synchro-digital converter with parallel output	:	refer to Section 1.11
L	resolver control transformer with locked rotor, or use one line-to-line voltage	:	refer to Section 3.2.6
M	resolver-digital converter with serial output	:	refer to Section 1.8
N	resolver-digital converter with parallel output	:	refer to Section 1.8
O	digital-analog converter (DAC) with serial input	:	refer to Section 1.7
P	digital-analog converter (DAC) with parallel input	:	refer to Section 1.7
Q	multiplying digital-analog converter (MDAC), with sinewave reference and serial input	:	refer to Section 1.7
R	multiplying digital-analog converter (MDAC), with sinewave reference and parallel input	:	refer to Section 1.'7
S	digital-synchro converter with serial input	:	refer to Section 1.12
T	digital-synchro converter with parallel input	:	refer to Section 1.12
U	digital-resolver converter with serial input	:	refer to Section 1.9
V	digital-resolver converter with parallel input	:	refer to Section 1.9
W	shift register or counter	:	refer to Section 1.14
X	gauge pressure transducer with voltage output	:	refer to Section 5.4
Y	air-to-current converter (transducer)	:	refer to Section 1.18

References to Sections 3.2.5, 3.2.6 and 5.4 above are to be found in the companion volume, *Basic Control System Technology.*

Sections 1.15, 1.16 and 1.17 contain descriptions of code converters, frequency-voltage converters and voltage-frequency converters, respectively, which cannot logically be entered into Table 1.1 but which sometimes feature in data conversion.

The limiting value for I_L is related to the saturation value V_{osat} of the amplifier output voltage by $I_{Lmax} = \dfrac{V_{osat}}{R_D + R_L}$ and V_{osat} will usually be 1.5 to 2 volts (V) in magnitude less than the voltage of the amplifier supply rails (which is typically 12 V, 15 V, or 18 V).

The principal disadvantage with the network of Fig. 1.1 lies in the fact that the load must 'float', because the network requires the dropping resistor to be tied to signal common. This problem can be eliminated by interchanging R_D and R_L and feeding back the voltage drop $I_L R_D$ through a differential amplifier stage, as shown in Fig. 1.2.

The operation of the differential amplifier stage is described in Section 12.2.3 of *Basic Control System Technology* and the version shown here has a voltage gain of unity. The formulae quoted above for I_L therefore still apply.

Where the range of I_L is required to be offset (by, for example, 4 milliamperes (mA)), this can be effected by injecting an appropriate bias current into the negative input terminal of the inverting amplifier.

Figure 1.3 shows a simpler version (which only uses one operational amplifier) of the network of Figure 1.2. Analysis shows that, provided the relationship $R_2 R_3 = R_1 R_D$ is satisfied and the amplifier is not saturated, the output current I_L is related to the input voltage V_i by the expression

$$I_L = \frac{-V_i R_2}{R_1 R_D}; \text{ moreover, } I_L = -V_i/R_D \text{ if } R_1 = R_2.$$

The limiting value of I_L is given by

$$I_{L_{max}} = \frac{V_{osat}}{R_D\left(1 + \frac{R_L}{R_3}\right) + R_L}; \text{ if } R_L \ll R_3,$$

then their approximates to $I_{L_{max}} \cong \frac{V_{osat}}{(R_D + R_L)}.$

Figure 1.4 shows a non-inverting alternative to the network of Figure 1.1. For this arrangement, $I_L R_D = V_i$, so that $I_L = V_i/R_D$. Again, R_L and R_D may be interchanged if a differential amplifier stage is inserted into the feedback path. If required, I_L may be offset by adding a suitable bias to the input of the non-inverting amplifier, using the type of input resistor network described in Section 12.2.2. of *Basic Control System Technology*.

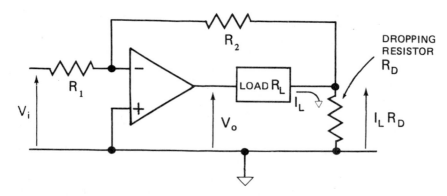

Fig. 1.1   Basic sign-inverting operational amplifier network for a voltage-controlled current source with a floating load

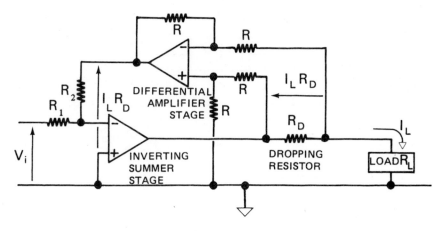

Fig. 1.2 Sign-inverting operational amplifier network for a voltage-controlled current source, with a grounded load

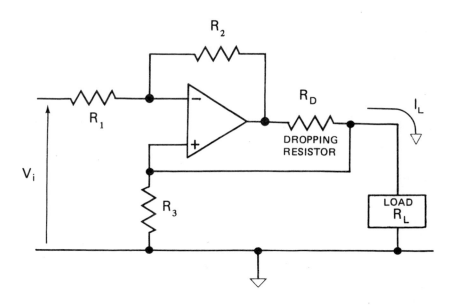

Fig. 1.3 Alternative and simpler sign-inverting operational amplifier network for a voltage-controlled current source, with a grounded load

The configurations shown here will be suitable for converting either DC or AC signals; power to the operational amplifier must be supplied for bipolar operation in the latter case.

Where the source of V_i has a low internal resistance and sufficient current drive, it may be possible to drive the load R_L directly from the source (yielding $I_L = V_i/R_L$), rendering the amplifier stages unnecessary.

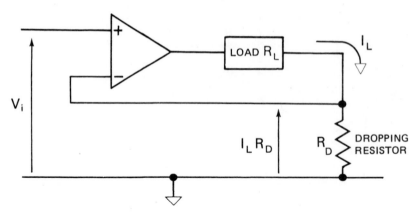

Fig. 1.4   Simple non-inverting operational amplifier network for a voltage-controlled current source, with a grounded load

1.3   Current–to–voltage converters

The usual method for converting a current signal into a voltage signal is by use of an appropriate dropping resistor. Figure 1.5 shows one possible configuration, which assumes that it is appropriate to connect one side of the resistor R_D to signal common.

   The buffer amplifier must be included where the load resistance R_L is likely to vary and is sufficiently low as to create a significant shunting effect on R_D. The amplifier may be inverting or non-inverting, as required, typically using the configurations discussed in Section 12.2 of *Basic Control System Technology*: referring to these networks, $R_D I_i$ becomes the voltage source V_1 and the other source V_2 is redundant unless required as a bias source to offset V_o.

   Where the dropping resistor cannot be tied to signal common, for whatever reason, it may be floated and allowance made for this by use of a

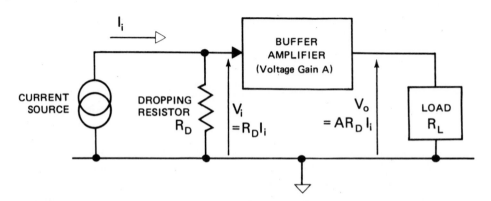

Fig. 1.5   Simple configuration producing a current-controlled voltage source

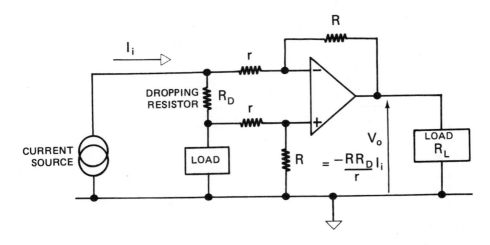

Fig. 1.6 Use of a differential amplifier in a current-controlled voltage source configuration

differential amplifier network of the type described in Section 12.2.3 of *Basic Control System Technology*: this is shown in Fig. 1.6.

The configurations shown here will be suitable for converting either DC or AC signals: in the latter case, the operational amplifier must be supplied for bipolar operation.

The process industries are currently tending to standardise on 1 to 5 V DC for the voltage signal range, in contrast to 4 to 20 mA DC for the most common current signal range: in this particular case, the dropping resistor would require to have a value of 250 ohm, assuming the amplifier stage to have a voltage gain of unity.

1.4 Modulators

The function of a modulator, in the control system context, is usually to convert a DC voltage into an AC voltage of fixed (carrier) frequency, such that the magnitude (defined in terms of either peak on RMS value) of the AC is proportional to the magnitude of the DC. This process, in a Telecommunications context, is referred to as 'suppressed – carrier amplitude modulation'.

Wherever the DC voltage is going to reverse in polarity, corresponding to a reversal in the sense of the data being represented by the DC signal, it is usual to cause the AC voltage to reverse its phase relative to the phase relationship pertaining before the occurrence of the DC sign reversal. The type of modulator which can achieve this phase reversal is said to be 'phase-sensitive'. Figure 1.7 shows typical waveforms for various input signal conditions and assumes that the output waveform is required to be sinusoidal.

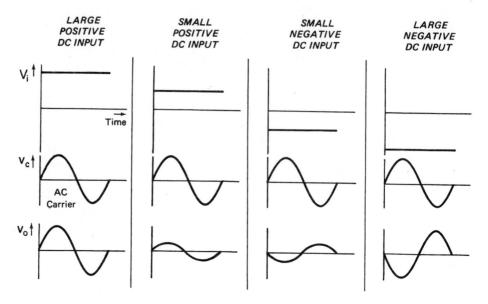

Fig. 1.7   Representative waveforms for a phase-sensitive modulator

Figure 1.8 represents the static characteristic required for the Phase-Sensitive Modulator (PSM).

The negative RMS output voltage values are to be interpreted as a reversal in the relative phase of the output voltage waveform.

The most common type of PSM network currently uses a monolithic analog multiplier: such a device is capable of multiplying together two analog voltages, as indicated in Fig. 1.9. Not shown in this diagram are

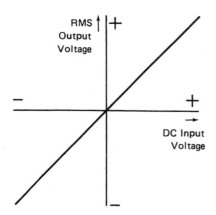

Fig. 1.8   Representative static characteristic of a phase-sensitive modulator

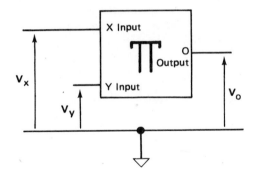

Fig. 1.9 Symbolic representation of an analog multiplier

external components usually necessary for setting output nulls, sensitivity, and for providing output power boosting: refer to Section 2.4.2 for further details.

The law for such a device is typically of the form $v_o = \dfrac{v_x v_y}{10}$, where v_x, v_y and v_o are in volts (V). Thus, $v_o = 10\,V$ when $v_x = v_y = 10\,V$. It is commonplace for the excursions of v_x, v_y and v_o to be limited to the vicinity of $10\,V$, so that external attenuation or amplification of signals may be necessary, depending on the particular application. In PSM applications, the multiplier is required to handle all four possible alternative combinations of input signal polarity (i.e. $++$, $+-$, $-+$, and $--$), so that it needs to be configured for 'four-quadrant' operation. Figure 1.10 shows such a multiplier being used as a PSM: it behaves according to the formula $v_o = kV_i V_{ref_m} \sin \omega_c t$, where k is a constant.

The phase-sensitive relationship may be expressed more clearly by rewriting the equation in the form:

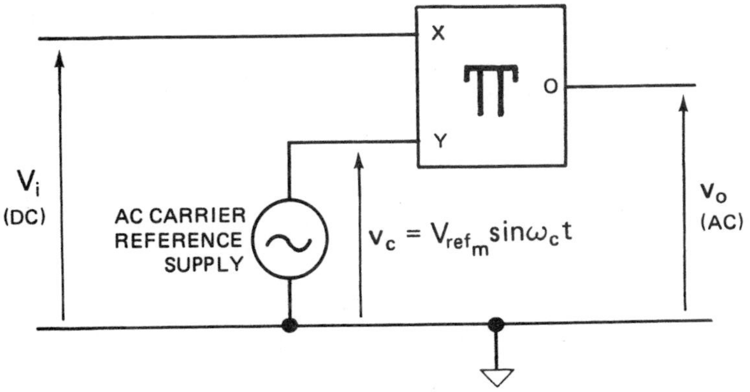

Fig. 1.10 Use of an analog multiplier as a phase-sensitive modulator

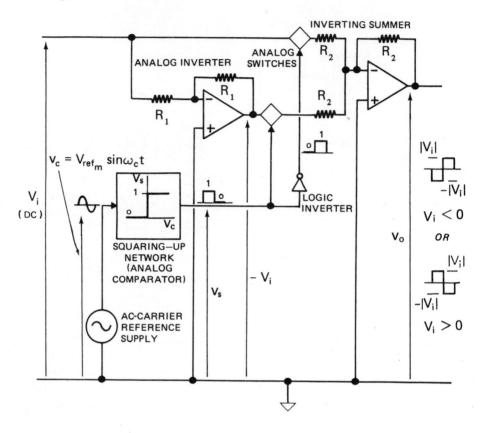

Fig. 1.11   Use of digitally-controlled analog switches to produce a squarewave phase-sensitive modulator

$$v_o = k|V_i|V_{ref_m} \sin \omega_c t, \qquad\qquad V_i > 0$$

and

$$v_o = k|V_i|V_{ref_m} \sin (\omega_c t + \pi), \qquad V_i < 0$$

Occasionally, a square output waveform is an acceptable alternative to the sinusoidal waveform. In this situation, modulation may be effected by causing the carrier, after being converted to a squarewave, to 'gate' digitally-controlled analog switches. A suitable network is exemplified by Fig. 1.11: the analog switches are assumed to provide a low resistance analog path when the digital signal v_s is in the 1 state and an open circuit when the digital signal v_s is in the 0 state.

$$v_o = + V_i, \qquad v_c > 0$$
$$v_o = - V_i, \qquad v_c < 0;$$

in other words, $v_o = V_i \operatorname{sgn}(v_c)$.

Whilst a squarewave modulator can be more accurate (in terms of linearity, offset nulls, etc.) than a sinewave modulator, it should be borne in

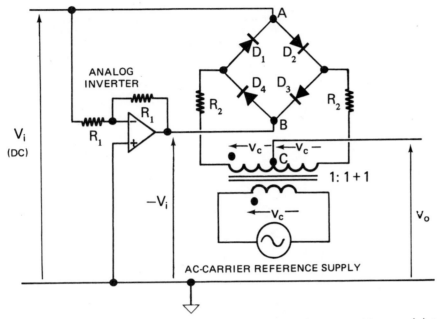

Fig. 1.12 Circuit for a diode ring squarewave phase-sensitive modulator

mind that the harmonic content of the squarewave may be detrimental to the control elements being fed by the modulator. It is conceivable that some of this harmonic content could be reduced by the application of a low-pass output filter (see Section 3.2), bearing in mind that normally the carrier frequency ω_c is constant.

Squarewave modulators have been constructed using special diode bridges, but these are no longer commonplace in modern equipment, due to the inaccuracies introduced by the forward characteristics of the diodes. Figure 1.12 shows a typical diode-ring modulator.

For the half cycle of the reference supply when v_c is positive, diodes D_1 and D_2 conduct and behave (ideally) as short-circuits, so that nodes A and C are at the same potential, with the result that $v_o = V_i$. For the alternate half cycle when v_c is negative, diodes D_3 and D_4 conduct and behave (ideally) as short-circuits, so that nodes B and C are at the same potential and $v_o = -V_i$.

Thus, $\quad v_o \cong + V_i, \qquad v_c > 0$

and $\quad\;\; v_o \cong - V_i, \qquad v_c < 0$

Investigation will show that the network ceases to function as described should V_i exceed one half the peak value of v_c.

1.5 Demodulators

The function of a demodulator, in the control system context, is the reverse of that of a modulator: that is, usually to convert an AC voltage of fixed

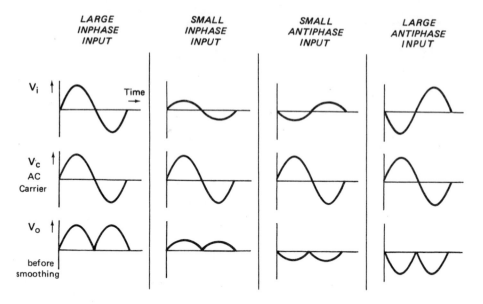

Fig. 1.13   Representative waveforms for a phase-sensitive demodulator

(carrier) frequency into a DC voltage, such that the magnitude of the DC is proportional to the magnitude (defined in terms of either peak or RMS value) of the AC. Wherever the AC voltage is going to reverse in phase (relative to the carrier reference), corresponding to a reversal in the sense of the data being represented by the AC signal, it is usual to cause the DC voltage to reverse polarity.

The type of demodulator which can achieve this phase reversal is said to be 'phase-sensitive'. Figure 1.13 shows typical waveforms for various input signal conditions and assumes that the input sinewave is precisely inphase or antiphase, relative to the AC-carrier reference.

Figure 1.14 represents the static characteristic required for the Phase-Sensitive Demodulator (PSD), which is alternatively known as a Phase-Sensitive Rectifier (PSR).

Interpretation of Fig. 1.13 indicates that, before v_o is smoothed, the PSD is required to implement the relationship

$$v_o = + v_i, \qquad v_c > 0$$
$$v_o = - v_i, \qquad v_c < 0; \qquad \text{in other words,}$$
$$v_o = v_i \, \text{sgn} \, (v_c).$$

A diode-ring PSD can be constructed along the lines of the diode-ring PSM, but with the AC and DC signals interchanged, as shown in Fig. 1.15.

It will be left to the reader to derive the principle of operation of this network. It can also be shown that the network will cease to function as required should the peak value of v_i exceed one half the peak value of v_c.

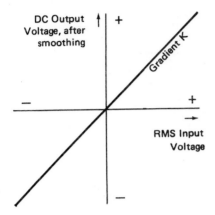

Fig. 1.14 Representative static characteristic of a phase-sensitive demodulator

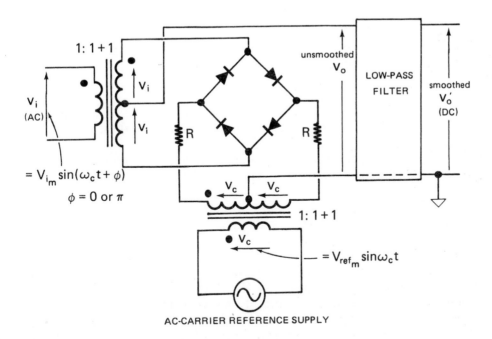

Fig. 1.15 Circuit for a diode ring phase-sensitive demodulator

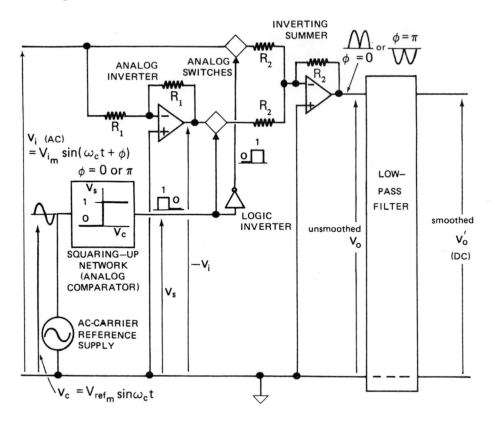

Fig. 1.16   Use of digitally-controlled analog switches to produce a phase-sensitive demodulator

This type of PSD is no longer commonplace, because of inaccuracies introduced by the forward characteristics of the diodes and the availability of superior circuits.

A more accurate PSD can be created using digitally-controlled analog switches, similar to the network of Fig. 1.11. This type of PSD is shown in Fig. 1.16.

This circuit operates in the same way as the Fig. 1.11 network, so that, before smoothing by the low-pass filter, $v_o = + v_i$, $v_c > 0$ and $v_o = - v_i$, $v_c < 0$.

It is quite usual for the source of the AC input voltage v_i to introduce a parasitic phase shift ε so that, in practice, $\phi = \varepsilon$ or $(\varepsilon + \pi)$. This has the effect of diminishing the sensitivity K, which is represented by the gradient of the characteristic of Fig. 1.14, according to the relationship $K = K_m \cos \varepsilon$, where K_m is the 'ideal sensitivity', obtained when $\varepsilon = 0$. Thus, when $\varepsilon = \pi/2$, the output from the PSD is zero for all input signal levels and the network becomes unusable: parasitic phase shifts must therefore be kept to a minimum or corrected for (by, for example, using power-factor correction techniques).

1.6 Analog–digital converters (ADCs)

Most contemporary ADCs are either monolithic (that is, achieved within single integrated circuit chips) or are supplied as plastic or metal encapsulated plug-in circuit modules. This is not to say that each device is complete in itself: to enable it to function, separate sample-hold amplifier, clock-pulse generator, precision voltage reference source, control logic, and (possibly) output logic may need to be added externally to the basic device.

Several alternative conversion techniques, which will be described, are employed by ADC manufacturers. However, suppliers often do not identify the particular principle employed in a specific converter, although, in fact, this principle can often be inferred from the specification for the converter. It is commonplace to discover an ADC described in terms such as 'high-speed', 'high-accuracy', 'low-cost', etc. In practice, it is rarely necessary to have a knowledge of the principle of operation in order to be able to incorporate a particular ADC into a system: it is usually adequate to have a reasonably detailed specification (including waveforms) for the device.

The following are considered to be the most important terms in a specification for an ADC.

- *Input Voltage Range*. Typical bipolar input ranges are $-5\,V$ to $+5\,V$, $-10\,V$ to $+10\,V$, and $-10.24\,V$ to $+10.24\,V$; typical unipolar input ranges are $0\,V$ to $+5\,V$, $0\,V$ to $+10\,V$, and $0\,V$ to $+10.24\,V$. It is important to appreciate that not all devices can be configured in all ranges: in some cases, a particular ADC may be suitable only for unipolar operation. In any case, the input range will bear a close relationship to the voltage value required for the external precision voltage reference source.

- *Conversion Time*. This will be closely related to the principle of operation employed, the clock-pulse frequency, and the word length (see below). 'Conversion time' is defined as the time required for the ADC to generate a complete digital word representative, in code, of the value of an input voltage sample. Typical conversion times range from about 100 nanoseconds to 50 milliseconds. As a general rule for any particular type of converter, the conversion time will increase when the word length is increased: that is, an 8-bit converter can be expected to be faster than a 12-bit converter of the same type, for example.

- *Word Length*. This term denotes the number of bits in the digital output word which represents, in code, the value of the input voltage sample. The word length is closely related to the term 'resolution', which is the smallest change in input voltage required to result in a change in the output code. Assuming a weighted type of binary code (see below), then, if the word length is represented by n and the nominal maximum input voltage which can theoretically be converted is represented by $V_{i_{max}}$, the resolution will be given by $V_{i_{max}}/2^n$ in the case of a unipolar converter, and by $V_{i_{max}}/2^{n-1}$ in the case of a bipolar converter. Resolution should not be confused with the term 'accuracy', which is the degree of precision to which the digital word represents the analog input value: accuracy depends on the degree of precision to which the analog components of the ADC have been manufactured.

It is possible for an ADC to achieve 12-bit resolution but with 10-bit accuracy, for example. If this converter is rated for 0 V to $- 10$ V input range, then it can sense a change of $10/2^{12}\,V = 2.4\,mV$ in input value but the output word produced will only represent the input value to an accuracy of $10/2^{10}\,V = 9.8\,mV$.

Typical word lengths are 8, 10, 12, 14, and 16 bits, although a number of ADCs with greater lengths are manufactured. In addition to conversion time, price will increase with word length.

- *Output Code.* This is the relationship between the output bit pattern and the equivalent input voltage value. In the majority of cases, the code will be a form of weighted binary code: the term 'weighted' means that a given bit in the output word is equivalent to the same value of input voltage increment, irrespective of the states of the other bits in the word. Coding of unipolar ADCs is relatively straightforward, because no (sign) bit needs to be allocated to represent the state (+ or −) of the input voltage polarity; in the case of bipolar ADCs, it will be necessary to allocate one output bit to indicate sign, so that the number of bits available for representation of voltage magnitude is reduced by one. Table 1.2 lists a number of alternative codes for a converter having an 8-bit word length.

It will be seen from Table 1.2 that many different output codes are available. Alternative codes not shown in the table include Binary-Coded Decimal (BCD) codes although, in general, these would more often be used for display purposes rather than in control loops. The need to represent 0 V input can result in a slight asymmetry in the orientation of the code, as can be seen with some examples in the table.

1.6.1   Single slope ADC

The internal organisation of a single slope ADC is shown in Fig. 1.17.

The principle of operation of this network can be demonstrated by referring to the waveforms of Fig. 1.18.

Immediately prior to each reset pulse, the count size stored in the counter will equal the number of clock pulses counted during the previous period when the comparator was in the 1 state: this will therefore be proportional to the analog input voltage level. The accuracy of this ADC depends on the precision of the sawtooth gradient. The speed of this ADC will be relatively low, depending on the frequency of the clock and on the word length.

Table 1.2 Typical ADC output codes

Nominal Input Value	ADC type						
	Unipolar			Bipolar			
	Natural (Straight) Binary	Complementary Straight Binary	Sign & Natural Binary Magnitude	Offset Binary	Complementary Offset Binary	Two's Complement Binary	Complementary Two's Complement Binary
$+V_{imax}$	11111111	00000000	01111111				10000000
$+(V_{imax} - V_{LSB})$	11111110	00000001	01111110	11111111	00000000	01111111	10000001
....	....	....	....	....	....	....	....
$+V_{LSB}$	00000001	11111110	00000001	10000001	01111110	00000001	11111110
0	00000000	11111111	00000000	10000000	01111111	00000000	11111111
$-V_{LSB}$			10000001	01111111	10000000	11111111	00000000
....			....	....	....	....	....
$-(V_{imax} - V_{LSB})$			11111110	00000001	11111110	10000001	01111111
$-V_{imax}$			11111111	00000000	11111111	10000000	

Note $V_{LSB} = V_{imax}/2^8$ volts for an 8-bit unipolar converter
$V_{LSB} = V_{imax}/2^7$ volts for an 8-bit bipolar converter

1.6.2 Dual slope ADC

This is a variant of the ramp type of ADC (above) which obviates the effect in the latter of variations in the sawtooth gradient. Using a system based on that of Fig. 1.19, V_i is applied to the integrator, the output of which now ramps up, until the counter registers a full count.

At this instant, the input to the integrator is transferred to the precision negative voltage reference $-V_{ref}$ and the counter is reset to zero. The output of the integrator ramps down and, on reaching zero volts, causes the counter to stop: at this juncture, the count size will be representative of the value of V_i. Figure 1.20 shows waveforms which demonstrate the principle of operation.

If X represents the peak value of the integrator output voltage, then

$$X = KV_i T_1 = KV_{ref} T_2$$

where K is the integrator sensitivity, in second^{-1}

T_1 is the duration of the up ramp, the time to register full count

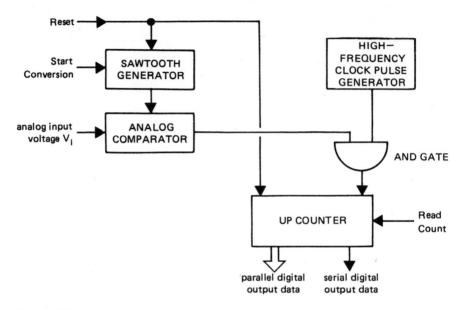

Fig. 1.17   Internal organisation of a single slope analog-digital converter

T$_2$ is the duration of the down ramp

Thus, $T_2 = \dfrac{V_i}{V_{ref}} T_1$. Since T_1 will be proportional to the full count size, it follows that T_2 will be proportional to the count size registered during read-out. Thus, the numerical value of the digital output $= \dfrac{V_i}{V_{ref}} \times$ (the numerical value of the full count), $0 \leqslant \dfrac{V_i}{V_{ref}} \leqslant 1$.

Since K does not feature in the final equations, it follows that, provided the integrator is stable, the absolute value of K is not critical.

1.6.3   Feedback ADC

This arrangement employs a digital-analog converter (DAC), to be described in Section 1.7, in a feedback loop as shown in Fig. 1.21.

This type of converter can be very fast because of the inherently high speed of the DAC and the ability to track upward and downward variations in V_i. Basically, V_f is a DC voltage measure of the current count size, which is caused by the comparator output state to change in such a way as to enable V_f to track variations in V_i. The speed of this ADC will be limited by the clock frequency, which must be chosen to suit the maximum switching rates of the counter and the DAC.

A simplified version of this ADC employs only an up-counter; the counter must then be reset periodically and the count size read only when counting has ceased.

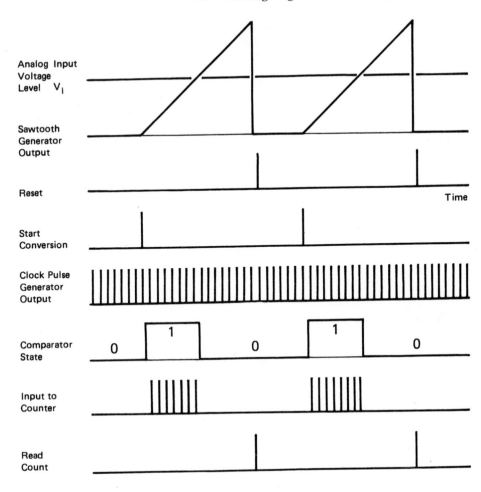

Fig. 1.18 Representative internal waveforms for the single slope analog-digital converter of **Fig. 1.17**

1.6.4 Successive approximations ADC

This is one of the faster types of ADC, because the speed depends only upon the switching times of the analog switches, comparators, and analog inverting and summing amplifiers. The first two stages of such a converter are shown in Fig. 1.22.

The operation of this network can best be explained by means of the flow chart in Fig. 1.23. In effecting the comparison XX (for example), the hardware in fact compares twice the balance with $V_{ref}/2$, so that the same reference voltage source can be used for all stages in the converter. Similarly, the subtraction of $V_{ref}/4$ from the balance, at ZZ, is effected by causing $V_{ref}/2$ to be subtracted from twice the balance. All stages in the converter thus use identical hardware and the same reference voltage signal and, as a result, considerably enhanced accuracy can be achieved.

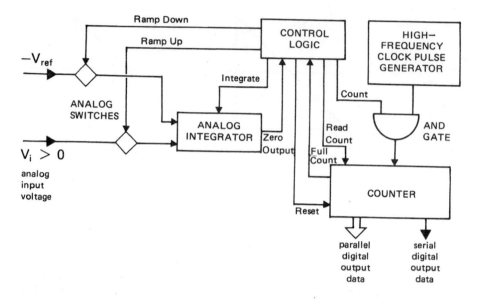

Fig. 1.19   Internal organisation of a dual slope analog–digital converter

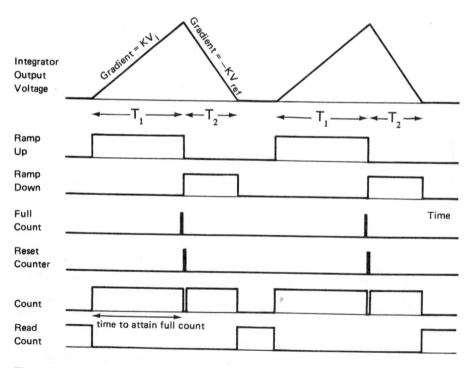

Fig. 1.20   Representative internal waveforms for the dual slope analog-digital converter of **Fig. 1.19**

1.6.5 Voltage/frequency ADC

With this type of ADC, the input voltage V_i is converted into a pulse train by a voltage/frequency converter (voltage-controlled oscillator), so that the frequency of the pulse train is proportional to V_i. The pulse train is gated into a counter for a controlled period of time so that the final count size, which becomes the digital output, is proportional to the pulse frequency and hence is a representation of the value of V_i.

1.6.6 Flash ADC

This is by far the fastest type of ADC and involves the large-scale use of analog comparators, as shown in Fig. 1.24. The comparators are enabled momentarily by a pulse on the Clock line. All of those comparators referenced to a voltage less than V_{in} will (for example) register a HIGH output state, whilst the remainder will register a LOW state, as shown in Fig. 1.25.

All Exclusive–ORs (XORs) except one will have a LOW output state; the odd-one-out will be that associated with those two (adjacent) comparators with voltage reference levels on either side of the level of V_{in}. The encoder consists of high-speed combinational logic which converts 1-out-of-m code to n-bit binary code, where $m = 2^n$. Quite clearly, if a great word length n is required, the number of comparators and XORs needs to be considerable and the device will necessarily use custom-designed large-scale integrated circuits and be expensive. The conversion time will be determined by the turn-on time of the comparators and the speed of the logic.

1.6.7 Typical ADC external circuitry

Figure 1.26 shows the additional circuit elements which will typically be required before an ADC can be used within a control system.

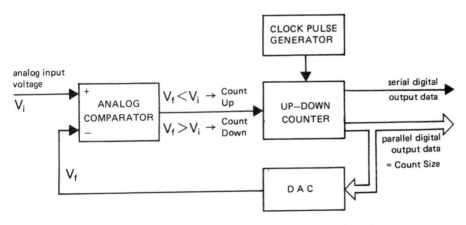

Fig. 1.21   Internal organisation of a feedback analog-digital converter

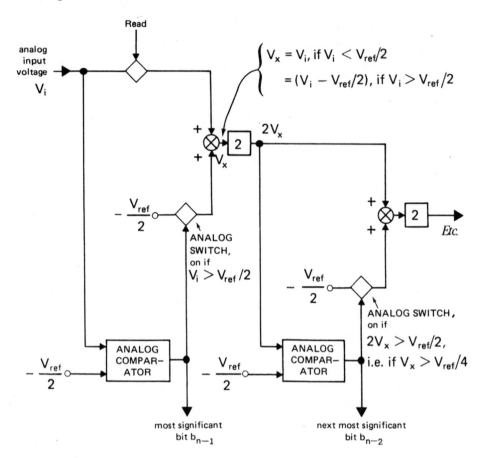

Fig. 1.22   First (most significant) two stages of a successive approximations analog-digital converter

The sample-hold amplifier is included so that the ADC is presented with a constant voltage for the duration of one conversion. If the converter is fast and V_i is changing only slowly, then this amplifier could be superfluous. Sample-hold amplifiers usually take the form of an integrated circuit requiring a high quality capacitor as the sole external component.

The bistables are often necessary when the ADC operates in such a way that its digital output states change sequentially rather than concurrently. The bistables provide temporary digital storage, so that all the output bits will remain constant during any conversion and will be updated concurrently whenever a conversion is completed.

The need (or otherwise) for the remaining elements of Fig. 1.26 has already been explained for the particular types of ADC. In some cases, one or more of these elements may be incorporated into the ADC by the manufacturer.

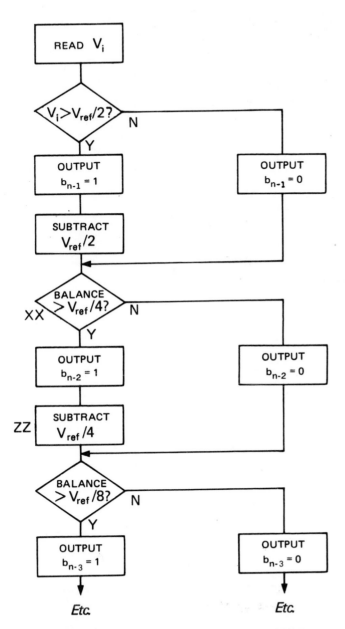

Fig. 1.23   Flow chart for the first (most significant) three stages of a successive approximations analog-digital converter

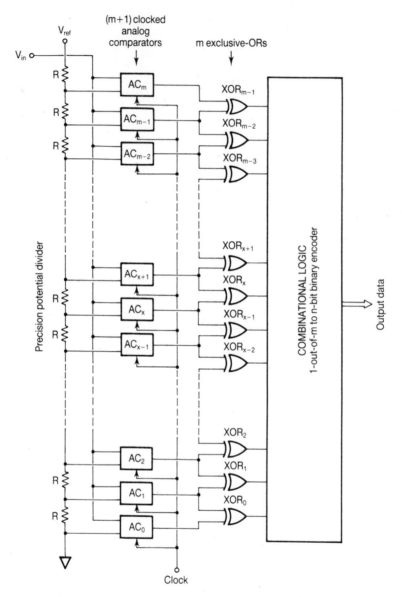

Fig. 1.24   Internal organisation of a flash ADC

1.7   Digital–analog converters (DACs) and multiplying digital–analog converters (MDACs)

The principal difference between the DAC and the MDAC is that the former is associated with a fixed voltage reference source, whilst the latter is associated with a varying voltage reference. In the DAC case, the reference source may be incorporated into the device. In terms of the principle of operation, the DAC and MDAC circuits behave identically, so that the

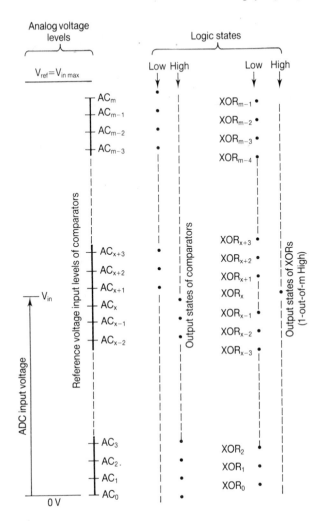

Fig. 1.25   Pattern of analog voltage levels and logic states within the flash ADC of **Fig. 1.24**

following description, which will refer to DACs, will be equally applicable to MDACs.

Most contemporary DACs are either monolithic (that is, achieved with single integrated-circuit chips) or are supplied as plastic or metal encapsulated plug-in circuit modules. Only one conversion technique is used and this involves the switching of resistor networks using solid state digitally-controlled analog switches.

The following are considered to be the most important terms in a specification for a DAC.

- *Output Voltage Range.* Typical bipolar output ranges are $-5\,\text{V}$ to $+5\,\text{V}$, $-10\,\text{V}$ to $+10\,\text{V}$, and $-10.24\,\text{V}$ to $+10.24\,\text{V}$; typical unipolar output ranges are $0\,\text{V}$ to $+5\,\text{V}$, $0\,\text{V}$ to $+10\,\text{V}$, and $0\,\text{V}$ to

+ 10.24 V. It is important to appreciate that not all devices can be configured in all ranges: in some cases, a particular DAC may be suitable only for unipolar operation. In any case, the output range will bear a close relationship to the voltage value required for the precision voltage reference source.

- *Settling Time.* This is the time required, following a change in digital input data, for the output voltage to change and settle to a value different from the prescribed value by an amount no greater than a voltage equivalent to one half of one least-significant bit. Because of the nature of the circuitry, settling times are very short, typically being in the range from 15 nanoseconds to 40 microseconds.

- *Word Length.* This term denotes the number of bits in the digital input word, which represents, in code, the value required for the output voltage. The word length is closely related to the term 'resolution', which is the smallest change in output voltage which can result from a change in the input code. Assuming a weighted type of binary code, then, if the word length is represented by n and the nominal maximum output voltage which can theoretically be generated is represented by $V_{0_{max}}$, the resolution will be given by $V_{0_{max}}/2^n$ in the case of a unipolar converter, and by $V_{0_{max}}/2^{n-1}$ in the case of a bipolar converter. Typical word lengths are 8, 10, 12, 14 and 16 bits, although a number of DACs with greater lengths are manufactured: price will increase with word length.

- *Input Code.* This is the relationship between the input bit pattern and the equivalent output voltage value, and, in the majority of cases, it will be a form of weighted binary code. Coding of unipolar DACs is relatively straightforward because no (sign) bit needs to be allocated to represent the state (+ or −) of the output voltage polarity; in the case

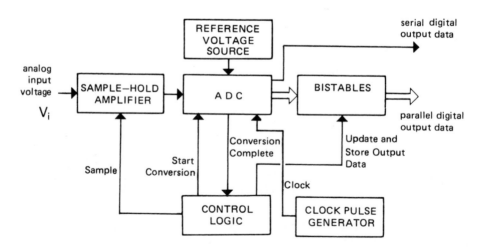

Fig. 1.26 Typical configuration of components for a complete analog–digital conversion process

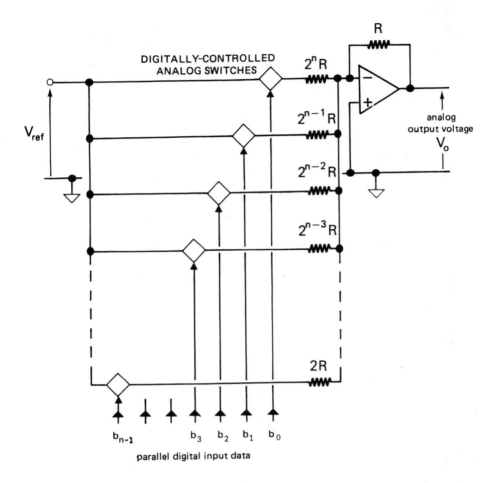

Fig. 1.27   A configuration of resistors and switches to represent the principle of the digital–analog conversion process

of bipolar DACs, it will be necessary to allocate one input bit to indicate sign, so that the number of bits available for representation of voltage magnitude is reduced by one. The codes listed in Table 1.2 (Section 1.6) for ADCs are equally applicable to DACs: the entries in the first column are now to be interpreted as 'Nominal Output Value'.

The principle of operation of a DAC is represented theoretically by Fig. 1.27.

The law for this configuration is:

$$V_o = - V_{ref}[2^{-1}b_{n-1} + \ldots + 2^{-(n-2)}b_2 + 2^{-(n-1)}b_1 + 2^{-n}b_0]$$

where bits $b_0, \ldots, b_{n-1} = 0$ or 1, depending on the state of the input word. Thus, when $b_o = b_1 = \ldots = b_{n-1} = 1$, $V_o = - V_{ref}(1 - 2^{-n})$, so that the nominal maximum value of V_o is slightly less than V_{ref} in magnitude.

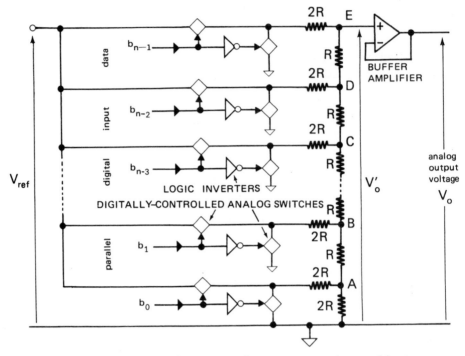

Fig. 1.28 A practical configuration of resistors, switches and logic to achieve accurate digital–analog conversion

Although feasible, this method of construction is not practicable, because high accuracy is precluded by the large range of resistance values which would usually be required.

For high accuracy, the resistors are designed to have comparable values, often precisely determined by laser trimming techniques. The resulting 'ladder network' is sometimes contained in a package separate from the DAC although, currently, it is more common to incorporate the network in the DAC device. Figure 1.28 shows a suitable ladder network arrangement, in which there are only two different resistor values, related by a 2:1 ratio.

Provided that the V_{ref} voltage source has a low output resistance, it can be shown that the resistance measured from every node A, B, C, D, E to ground is virtually R ohms, irrespective of the states of the analog switches. However, the magnitude of V'_o will depend on the states of the switches.

Increasing the number of bits does not change the weighting of each bit, starting with the most-significant, but it does increase the degree of resolution, since the least-significant bit is reduced in weighting by a factor of 2 for each bit added to the word length.

It can be shown that

$$V_o = V'_o = V_{ref}[2^{-1}b_{n-1} + \ldots + 2^{-(n-2)}b_2 + 2^{-(n-1)}b_1 + 2^{-n}b_0]$$

Additional switches may be incorporated in order to make the converter bipolar: for example, the state of a sign bit could be used to reverse the sign of V_{ref}. Alternatively, the polarity of V_{ref} could be fixed, and the output voltage offset by $- V_{ref}/2$, using external circuitry.

Figure 1.29 shows the additional circuit elements which typically will be required before a DAC can be used within a control system.

The operational amplifiers are normally external to the DAC and the quantity (1 or 2) will depend on whether the swing of the output voltage is unipolar or bipolar, respectively.

In the case of the MDAC, the V_{ref} input signal will be derived from a variable (analog) DC voltage source so that, for this case also,

$$V_o = V_{ref} \times \text{(analog value represented by the digital input word)}$$

1.8 Resolver–digital converters

These converters are used for direct conversion of resolver AC voltages to digital data (representing a shaft angular displacement) and also often form the bases of Synchro-Digital Converters (see Section 1.11). Because of their nature, appropriate versions may be used as Inductosyn-Digital Converters. When several angular displacement transducers are required to be interfaced to a digital controller, it becomes economically advantageous to use resolvers or synchros, together with a time-shared converter of the appropriate type, in preference to using a set of digital transducers. Resolver-Digital Converters can assume two alternative forms: the tracking type and the sampling type.

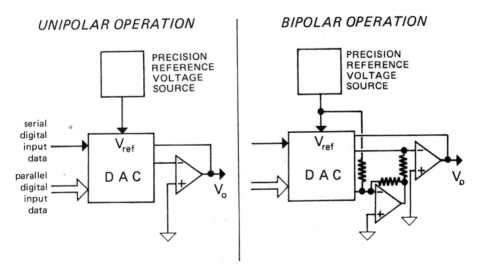

Fig. 1.29 Typical arrangement of components for complete digital–analog conversion processes

1.8.1   Tracking resolver–digital converter

The two output voltages V_1 and V_2 from a resolver transmitter can be represented by $V_1 = kV_{ref} \sin \theta$, and $V_2 = kV_{ref} \cos \theta$, where V_1, V_2 and V_{ref} are RMS values, θ is shaft angle, and k is the transformation ratio. Thus,

$$\theta = \tan^{-1}(V_1/V_2)$$

The tracking converter employs what is known as an 'electronic servo', which is represented by Fig. 1.30.

The up-down counter increments an implied shaft angle ϕ and electronically adjusts the divider ratios according to the relationships $k_1 = \cos \phi$ and $k_2 = \sin \phi$.

Thus, when the null state is achieved (error voltage $= 0$), $k_1V_1 - k_2V_2 = 0$ so that $kV_{ref} \sin \theta \cos \phi = kV_{ref} \cos \theta \sin \phi$; whence, $\tan \phi = \tan \theta$ and $\phi = \theta$.

At any instant, the count size represents, in digital form, the current value of ϕ and hence of θ. Practical accuracies are in the region of 10

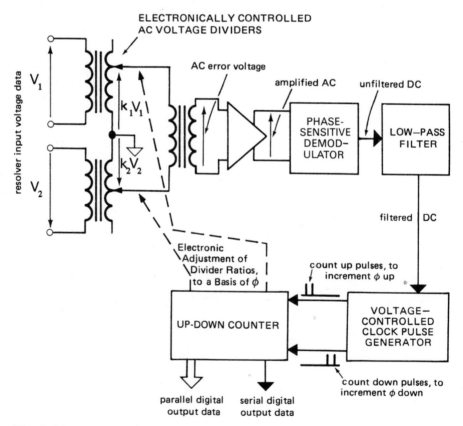

Fig. 1.30   Symbolic representation of a tracking resolver–digital conversion process

seconds of arc. The tracking converter is relatively immune to parasitic phase shifts in the resolver signals and to harmonics and noise; however, it is not particularly suitable for multiplexing (time-sharing) duties, due to limitations in maximum slew rate.

1.8.2   Sampling resolver–digital converters

These converters usually sample the peak values of the resolver voltages and then manipulate these sample values. Figure 1.31 shows schematically the internal arrangement.

Successive approximation techniques, similar to those indicated in Fig. 1.23, are used to adjust the resistive divider ratios according to the relationships $k_1 = \cos\phi$ and $k_2 = \sin\phi$, where the value of ϕ is implied by the state of the logic.

A null is achieved when $k_1 V_{1m} = k_2 V_{2m}$, so that

$$kV_{ref_m}\sin\theta\cos\phi = kV_{ref_m}\cos\theta\sin\phi,\text{ whence}$$
$$\tan\phi = \tan\theta \text{ and } \phi = \theta$$

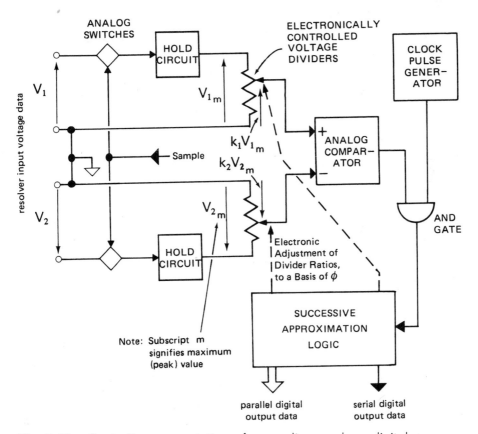

Fig. 1.31   Symbolic representation of a sampling resolver–digital conversion process

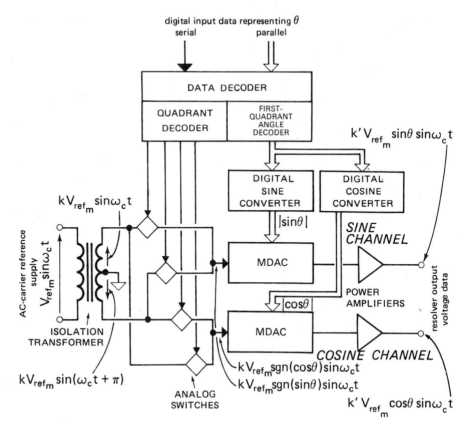

Fig. 1.32 Symbolic representation of a digital—resolver conversion process

Sampling converters are less accurate than their tracking converter counterparts and are sensitive to parasitic phase shifts, harmonics, and noise in the resolver signals, which may therefore necessitate accurate pre-filtering. However, sampling converters are suitable for multiplexing duty, being capable of sampling at up to twice the carrier frequency (the frequency of the reference supply).

1.9 Digital—resolver converters

These converters are used for direct conversion of digital data (representing shaft angular displacement) and also often form the bases of Digital-Synchro Converters (see Section 1.12). These converters are, in effect, solid-state replacements for resolver transmitters, with the value of the input angle represented by a digital word.

Figure 1.32 shows a typical arrangement for a Digital-Resolver Converter. The decoder interprets the most-significant bits to sense the quadrant in which the angle lies: the quadrant decoding selects either inphase or antiphase AC reference voltage, as appropriate, for the sine and cosine channels. The decoder also converts the data to the equivalent first-quadrant value and this angular value is input to digital sine and cosine converters. The quadrant decoder thus selects the sign of the AC reference for the two channels, by gating the appropriate analog switches, according to the relationships $++$, $+-$, $--$, $-+$ for the four quadrants. In each case, the two signs represent the required sign for the sine and cosine functions, respectively. The data representing equivalent first-quadrant angle are then converted to sine and cosine data, using hardware implementing the sine and cosine algorithms, so that the output data from these converters now represent the magnitudes of the sine and cosine functions.

The sine channel MDAC (see Section 1.7) multiplies the AC analog input signal $k.V_{refm} \cdot sgn\,(\sin\theta) \cdot \sin\omega_c t$ by the digital input representing $|\sin\theta|$, to yield an AC analog output signal $k'.V_{refm} \cdot \sin\theta \cdot \sin\omega_c t$ which is then input to a power amplifier: this, in turn, outputs $k'.V_{refm} \cdot \sin\theta \cdot \sin\omega_c t$. The MDAC in the cosine channel causes the associated power amplifier, by the same reasoning, to output $k'.V_{refm} \cdot \cos\theta \cdot \sin\omega_c t$.

The two AC output voltages thus represent a pair of resolver voltages, so that the converter is, in effect, a solid-state Resolver Transmitter. Because of its nature, it can also be arranged to function as a Digital-Inductosyn Converter.

1.10 Scott–T transformers

A pair of suitable signal transformers may be connected to enable 4-wire resolver data to be converted to 3-wire synchro data, and vice-versa. Such a configuration is traditionally known as a 'Scott-T' connected pair of transformers, the need for which originated in 2-phase to 3-phase power conversion, and vice-versa.

The operation of the configuration will be explained in relation to synchro-resolver conversion, as indicated by Fig. 1.33.

All voltages are RMS values. By the application of simple trigonometry, it can be shown that

$$V_{2CT} = \frac{\sqrt{3}}{2}kV_{ref}\cos\theta \qquad V_{3CT} = \frac{1}{2}kV_{ref}\sin\theta$$

and $\;V_{1CT} = -\dfrac{1}{2}kV_{ref}\sin\theta$

where θ is the angular displacement of the source synchro.

By giving transformer T_1 a turns ratio of $\dfrac{\sqrt{3}}{2}{:}1$ and T_2 a ratio of $\dfrac{1}{2}+\dfrac{1}{2}{:}1$, it is readily seen that the output voltages will be given by

$$V_{o_1} = kV_{ref}\cos\theta$$
and $\quad V_{o_2} = kV_{ref}\sin\theta$

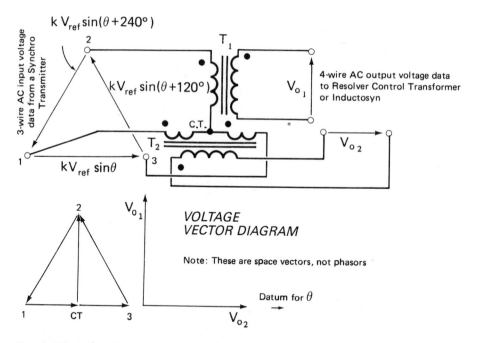

Fig. 1.33   Configuration of Scott–T connected transformers to convert 3-wire synchro voltage data to 4-wire resolver voltage data

The maximum values of these variable RMS voltages are equal but separated by 90° vectorially, and therefore the transformers provide a suitable data source for a resolver transmission system. Equally readily, it may be shown that the transformer pair will provide suitable resolver-synchro transformation, if the direction of input-output transmission is reversed.

Scott-T transformer pairs are commonly used in Synchro-Digital and Digital-Synchro Converters, which are the subject of the next two sections.

1.11   Synchro–digital converters (SDCs)

To generate equivalent digital angular data it is usual to convert synchro data to resolver data, and subsequently to convert the latter to digital angular data in the construction of an SDC, because it is relatively easier to process two AC voltages rather than three.

Figure 1.34 shows the general arrangement, with a Scott-T transformer pair (refer to Section 1.10) performing the synchro-resolver conversion, followed by one of the types of Resolver-Digital Converter described in Section 1.8. Because of the low power levels involved, the transformers may be sufficiently small (and thus called 'microtransformers') to enable them to be encapsulated with the Resolver-Digital Converter in the same package.

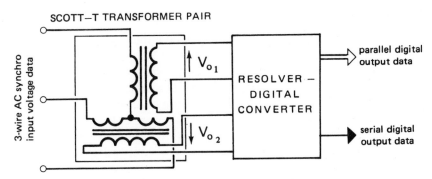

Fig. 1.34   Internal configuration of a synchro–digital converter

1.12   Digital–synchro converters (DSCs)

It is usual to convert digital angular data to resolver data and subsequently to convert the latter to synchro data in the construction of a DSC, because it is relatively easier to generate two AC voltages rather than three.

Figure 1.35 shows the general arrangement, with a Digital-Resolver Converter (described in Section 1.9) being followed by a Scott-T transformer pair (refer to Section 1.10). In some cases, the transformers may be sufficiently small (and thus called 'microtransformers') to be encapsulated with the Digital-Resolver Converter in the same package. However, because of the possible load demands, it may be necessary for the transformers and even the power amplifiers deriving V_{i_1} and V_{i_2} to be packaged separately. The Digital–Synchro Converter may be thought of as a solid-state Synchro Torque Transmitter (TX) or Control Transmitter (CX); in addition, these techniques have been extended to produce solid-state Differential Transmitters (TDXs and CDXs) and Control Transformers (CTs).

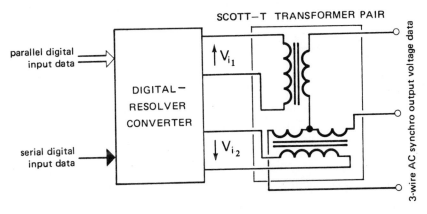

Fig. 1.35   Internal configuration of a digital–synchro converter

1.13   RMS–to–DC converters

These converters are usually monolithic integrated circuits generating a DC output voltage which is proportional to the true-RMS value of an applied AC input voltage. Because they do not sense the phase of the AC voltage, these converters can only be used either for instrumentation purposes (to enable a DC instrument to display the RMS value of an AC voltage) or as non-phase-sensitive demodulators in those AC-carrier control systems in which phase reversals are sensed by some alternative means. Figure 1.36 shows how an RMS-to-DC Converter could be used as part of a phase-sensitive detector network.

The most common type of converter network mechanises the relationship $V_{RMS} = \overline{v_i^2}/V_{RMS}$, using the type of circuitry implied by Fig. 1.37, where the bar over v_i^2 signifies time-averaging.

Provided that the period of v_i is long in relation to the time constant of the low-pass filter, it is readily seen that V_o will be a good approximation to V_{RMS}: for a given filter time constant, there will be finite limits to the frequency range of v_i.

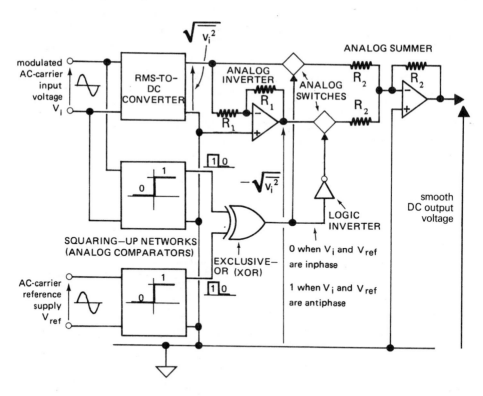

Fig. 1.36   Application of a RMS–to–DC converter in the construction of a phase-sensitive demodulator

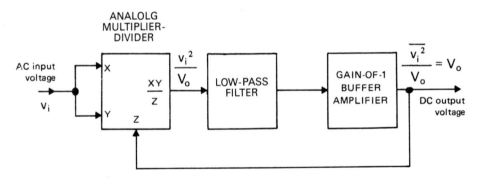

Fig. 1.37 RMS–to–DC conversion process using an analog multiplier–divider

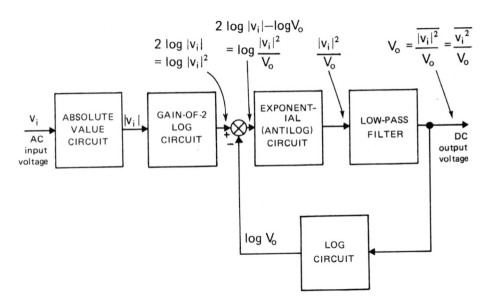

Fig. 1.38 RMS–to–DC conversion process using analog log and antilog circuits

An alternative type of converter uses Log and Antilog circuits to implement the relationship $\log \dfrac{|v_i|^2}{V_o} = 2\log|v_i| - \log V_o$, as shown in Fig. 1.38.

1.14 Shift registers and counters

A shift register is a set of logic 'cells', each of which is capable of sustaining a binary 1 or a binary 0 state. A digital network or circuit which can be used to create a cell is known as a 'Bistable', 'Bistable Multivibrator', or 'Flip-Flop'. A bistable can be arranged to change its output state whenever its input signals change in an appropriate manner, in which case it is known as an 'asynchronous' element, or it can be arranged to change its output state whenever its input signals are in an appropriate state and a pulse from a clock pulse generator is present, in which case it is known as a 'synchronous' element.

Figure 1.39 shows an 8-bit shift register constructed from eight synchronous bistables. When any input bit $b_{0_i} \ldots b_{7_i}$ is set to a 1, the corresponding output bit $b_{0_o} \ldots b_{7_o}$ will assume a 1 state on the next clock pulse and will maintain this state until the next clock pulse occurs. Thus, the register can store an 8-bit digital word for the duration of one clock interval. The register has been configured for parallel input and parallel output.

If Q_8 is linked to J_7, Q_7 to J_6, and so on to Q_2 linked to J_1, then a 1 bit applied to b_{7_i} will cause b_{7_o} to be set to 1 on the next clock pulse, b_{6_o} to be set to 1 on the following clock pulse, and so on until b_{0_o} is set to 1 after eight clock pulses have occurred. Thus, data can be caused to ripple through the register and it has been configured for serial input and parallel output, if all outputs Q_1 to Q_8 are used. If only output Q_1 is used and the clock pulse train is maintained then, after eight clock pulses have occurred, the sequence of 1s and 0s emerging from output Q_1 will be a replica of the sequence of 1s and 0s which were input to J_8, and the register has been configured for serial input and serial output with a time delay separating the data streams.

If a full 8-bit input is word loaded in through $b_{0_i} \ldots b_{7_i}$, before the outputs are effectively cross-connected to adjacent Set inputs, then the time sequence of bits emerging from Q_1 will be a replica of the bit pattern loaded

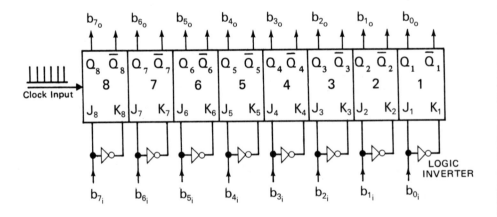

J = Set Input; K = Reset Input, Q = Principal Output; Q̄ = Inverse Output

Fig. 1.39   Construction of an 8-bit shift register from eight bistable elements

into the register initially, so that it has now been configured for parallel input and serial output.

Thus, a set of bistables, together with appropriate combinational logic, can be organised for storage of data, for temporarily delaying data, for conversion of data from serial to parallel format, and for conversion of data from parallel to serial format. Figure 1.40 summarises these various configurations.

If the outputs of a shift register are fed back through appropriate combinational logic, in order to set and reset the shift register inputs, then the network can be configured to behave as a counter: for example, the output states of the register could be arranged to represent a count, in an appropriate binary code, of the number of pulses occurring on a serial data input line, commencing at a specific instant in time. The counter could be reset to zero, for example, by causing a logic 1 to be applied to all bistable Reset inputs simultaneously. Certain types of counter (for example, natural binary and binary-coded-decimal) are available already pre-configured, as monolithic integrated circuits.

Counters can also be arranged to convert the frequency of a periodic waveform into a digital word, and the result is a frequency-to-digital converter, which normally is referred to as a 'Frequency Counter'. Such an element is useful, for example, for converting the frequency of the pulse train produced by a digital velocity transducer or a turbine type of flowmeter into a parallel digital word.

Figure 1.41 shows two alternative arrangements, catering for both high and low input signal frequencies. In the former case, the number of input rising edges (say) occurring within one (fixed) period of the clock waveform is counted, and the count value is stored by the shift register. In the latter case, the period of the input waveform is measured initially, by counting the number of clock pulses occurring within one period of the input waveform, the count size being stored by the shift register; finally, the value of the period may be inverted, to yield frequency data, by means of digital hardware mechanising a reciprocal law algorithm (see Section 2.7).

Frequently, the input signal will need to be preconditioned, by suitable circuitry, to standardise the voltage levels and to 'square-up' the rising and/ or falling edges of the waveform: this may require amplifiers, attenuators, analog comparators, Schmitt triggers, etc.

1.15   Code converters

A code converter network would typically be used where the output code from a digital transducer is incompatible with the input code of the digital controller to which the transducer is to supply data. For example, a Gray code shaft encoder might have to be interfaced to a digital controller configured to accept natural binary code, as indicated in Table 1.3.

One technique for genrating the translation process is to devise a Boolean equation for each output E...H, each in terms of the set of input states A...D. Thus, in this simple example,

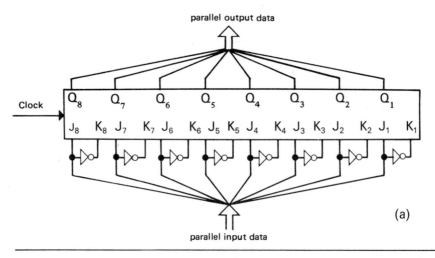

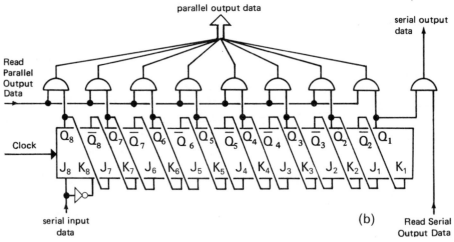

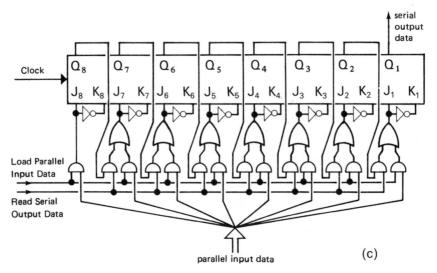

Fig. 1.40 The addition of combinational logic to a shift register to effect: (a) parallel input–parallel output with delay; (b) serial input–serial/parallel output with delay; (c) parallel input–serial output with delay

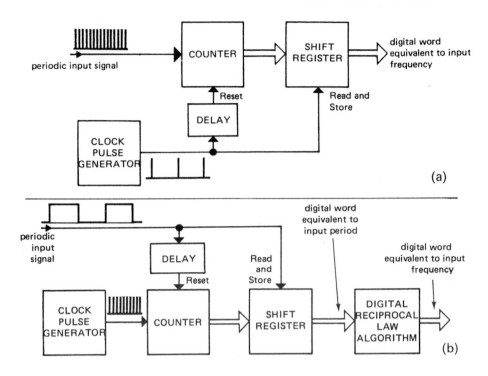

Fig. 1.41   Internal configurations of frequency counters using shift registers for:
(a) high signal frequencies; (b) low signal frequencies

$$E = A$$
$$F = \bar{A}.B$$
$$G = \bar{B}.C. + \bar{A}.B.\bar{C}$$
$$H = A.D + \bar{B}.\bar{C}.D + \bar{B}.C.\bar{D} + \bar{A}.B.C.D + \bar{A}.B.\bar{C}.\bar{D}$$

In these formulae, . represents a logical AND, + a logical OR, and the bar ‾ a logical inversion (a NOT operation).

Provided that the number of input signals does not exceed preferably four or, at most, five, the process of formulating Boolean equations can be assisted by the construction of a Karnaugh map for each of the output signals. This is a particularly useful aid to minimising the quantity of hardware required, especially in those cases where not all combinations of input state can arise: for example, those states corresponding to 10 to 15 decimal, inclusive, in the above illustrative case.

Once the equations have been minimised, they can be mechanised by AND, OR, NAND, NOR, and inverter gates, using a suitable logic series. However, this simple approach does not take into account the fact that it might be possible to simplify the network further using one or more of the following techniques:

Table 1.3   Gray and natural binary codes

Decimal value	Network input states (Gray code)				Network output states (natural binary code)			
	A	B	C	D	E	F	G	H
0	0	0	0	0	0	0	0	0
1	0	0	0	1	0	0	0	1
2	0	0	1	1	0	0	1	0
3	0	0	1	0	0	0	1	1
4	0	1	1	0	0	1	0	0
5	0	1	1	1	0	1	0	1
6	0	1	0	1	0	1	1	0
7	0	1	0	0	0	1	1	1
8	1	1	0	0	1	0	0	0
9	1	1	0	1	1	0	0	1

- rewriting the equations in terms of the other outputs, in addition to the inputs, on the right-hand side
- sharing hardware between the logic channels implementing the equations
- using other types of gate: for instance, the exclusive-OR.

It is possible to buy certain code converters (for example, binary-coded-decimal to natural binary converters) pre-configured in monolithic integrated circuits, although the choice is limited. In some cases, these ICs can be cascaded, in order to extend the word length. Code conversion logic may also be programmed into Programmable Logic Array (PLA) elements: these are integrated circuits containing a large number of identical 'cells', each consisting of a small number of logic gates. Interconnections between gates and cells are programmed into the device using techniques similar to those used for programming PROMs (Programmable Read-Only Memories).

One potential problem with combinational logic code converters arises from the variation in propagation delays through the different channels, which can result in short-lived (that is, of nanosecond duration) spurious output states occurring during input transitions. The circuitry being fed by the converter must be designed so that it cannot respond to these parasitic states.

An alternative technique for code conversion, which can be used when counters are available to operate in the two codes of interest, is illustrated symbolically in Fig. 1.42.

The two counters are supplied with precisely the same number of pulses to count (up or down, as the case may be), so that they both register exactly

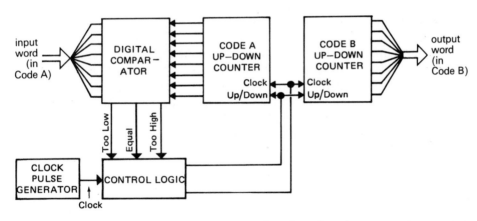

Fig. 1.42   Use of synchronised counters for digital code conversion

the same count size, in terms of their respective codes. The state of the digital comparator senses any disparity between the input word and the count value in the Code A counter, causing the control logic to require the two counters to increase or decrease their count size, according to requirements. Modern IC counters can count at frequencies up to 1 GHz, which means that conversions can be very fast.

It must be stressed that both combinational and sequential logic design can require considerable skill and experience. Many books are available to deal with these complex topics. Note that logic networks can be constructed either by hardwiring logic elements or by programming Programmable Logic Arrays.

A third technique can achieve extremely fast code conversion if modern semi-conductor memory is used. The input word is interpreted as the address of a location in read-only memory (ROM) and stored in this location, as a data word, is the corresponding output word required. All possible input states must be accommodated by a corresponding number of locations and the full set of corresponding output states must be stored, in the form of a look-up table, with the word length of each location equal to the output word length required. The conversion time will be equal to the access time of the memory, and this can be less than 350 nanoseconds.

1.16   Frequency–voltage converters

These are usually encapsulated circuits or monolithic integrated circuits. The purpose of the converter is to generate a DC voltage proportional in value to the frequency of an applied periodic input waveform. Typically, these circuits will operate over a frequency range from DC to the region of 100 kHz and will function with a variety of input waveforms, including sinewave, squarewave, triangularwave and pulse train. In control systems, the principal application would be the conversion of a pulse train from a digital velocity transducer or the sinewave from a signal alternator or power alternator into a DC feedback voltage, for use in speed or frequency control.

The principle of operation is shown symbolically in Fig. 1.43. The capacitor C is charged from a precision voltage reference source, during the positive half-cycle of the input waveform, and discharges into the active low-pass filter, during the negative half-cycle. The capacitor is arranged to accumulate a fixed quantity of charge, during any one charging interval, so that the average value of the current flowing during the discharge interval will be proportional to the input frequency.

1.17   Voltage–frequency converters

These are usually encapsulated circuits or monolithic integrated circuits. The purpose of the converter is to generate a periodic output waveform having a frequency proportional to the magnitude of an applied (unipolar) DC input voltage. Since a 'Voltage-Controlled Oscillator' (VCO) performs the same function, the two terms are synonymous.

Most VCOs are based on a triangularware oscillator using a precision analog integrator which alternately integrates positive and negative precision reference voltages: the gain of the integrator is made to be proportional to the DC input voltage, so that the integrator output voltage ramps up and down (with a gradient having a magnitude proportional to the DC input voltage), depending on whether the positive or the negative reference is being applied at that instant in time. Analog comparator and amplitude shaping circuits enable the triangularwave to be translated into a sinewave, squarewave, sawtooth and pulse train.

The VCO will require external components, including a precision capacitor for the oscillator integrator: the value of this capacitor will determine the frequency range, which is unlikely to exceed three decades. By providing means to enable the integrator to have a different gain for the

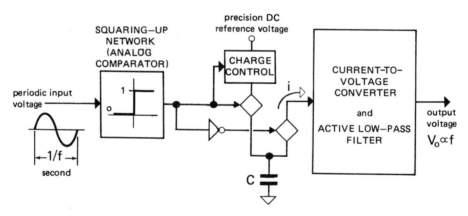

Fig. 1.43   Typical internal configuration of a frequency–voltage converter

two different directions of charging, it becomes possible to modify the triangular waveshape in the direction of a sawtooth waveshape, to vary the mark/space ratio of the squarewave and the pulse width of the pulse train, and to reduce the level of possible harmonic distortion of the sinewave.

Typically, the converter would have control inputs to provide some or all of the following facilities:

- output frequency control
- output amplitude control
- output null offset adjustment
- mark/space ratio control
- sinewave distortion control.

These converters have a wide application in the instrumentation field but their use in the control field is limited largely to tracking filters employing Phaselock techniques. However, many modern oscillators will employ the type of circuitry described above but configured to have preset values of output amplitude and frequency: a common application for such an oscillator would be an AC voltage reference source for AC-carrier transducers and other control elements.

1.18   Air–to–current converters

The usual purpose of these converters is to enable a 3 to 15 psi (20 to 100 kPa) pneumatic signal source to be interfaced to an electronic process controller. The converter typically generates an output signal in the range of 4 to 20 mA DC.

Figure 1.44 shows a typical arrangement for one of these converters, which are sometimes referred to as 'P/I Converters'. It is almost identical in operation to the electronic force balance transmitter described in Section 5.2 of *Basic Control System Technology*, the principal difference being that it responds to a single input (pneumatic) pressure: no further explanation should therefore be necessary.

1.19   Current–to–air converters

The usual purpose of these converters is either to enable a 4 to 20 mA DC electronic signal source to be interfaced to a pneumatic process controller or to enable an electronic process controller to drive a pneumatically-actuated final control element. One of these converters would enable an electronic function generator to be used as a signal source for testing pneumatic control systems.

Figure 1.45 shows a typical arrangement for one of these converters, which are sometimes referred to as 'I/P Converters'. It is almost identical in operation to the pneumatic force balance transmitter described in Section 5.2 in *Basic Control System Technology*, the principal difference being that

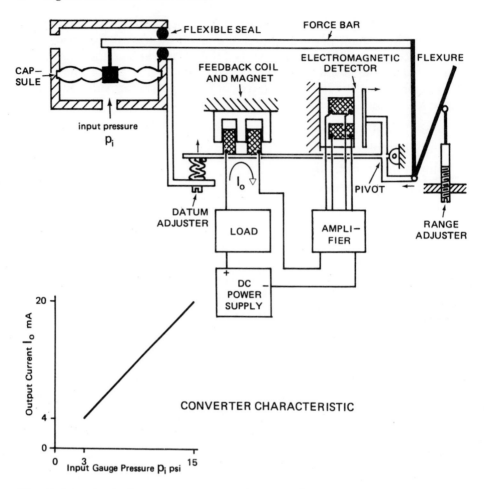

Fig. 1.44   Symbolic representation of a typical air–to–current converter configuration, with its static characteristic

it responds to a force established by an applied DC input current: no further explanation should therefore be necessary.

1.20   Analog input channels for digital processors

Analog input hardware for digital computer interfaces to plant is often divided into two categories: high level, in which the signal magnitudes range up to 10 V DC, and low level, in which the signal magnitudes range up to 1 V DC. Subdivision is necessary because the method of preamplification of the signal, prior to analog–digital conversion, is different for the two categories. The level of input signal experienced will depend upon the nature of the analog transducer which provides the signal source.

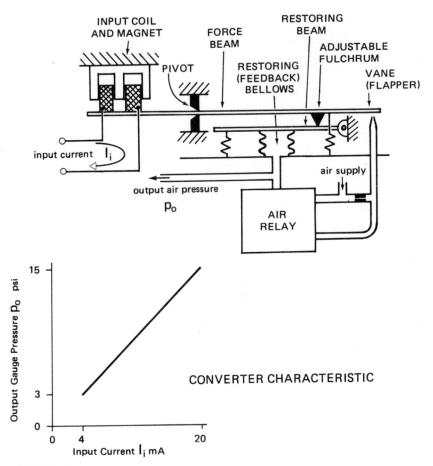

Fig. 1.45   Symbolic representation of a typical current—to—air converter configuration, with its static characteristic

1.20.1   Signal conditioning

Frequently, the analog signal from a transducer may be required to be 'conditioned' before it can be processed by the ADC. The need for conditioning can arise because the signal level is inappropriate and/or because protection of the interface is required against excessive voltage or current transients. Conditioning of a particular signal can necessitate one or more of the following:

- attenuation
- amplification
- air-to-current conversion
- current-to-voltage conversion
- noise filtering

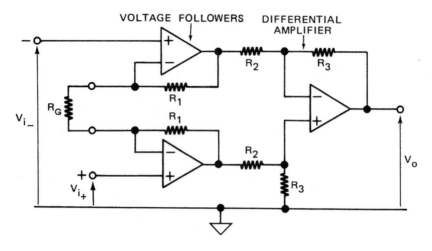

Fig. 1.46 Typical internal circuit for an instrumentation amplifier

- compensation for nonlinearities
- electrical isolation
- electrical protection.

Signal amplification and common-mode noise rejection is often effected by use of high-quality special-purpose encapsulated 'Instrumentation Amplifiers'. A typical internal organisation of one of these amplifiers is shown in Figure 1.46. For this arrangement, it can be shown that

$$V_o = \frac{R_3}{R_2}\left(1 + \frac{2R_1}{R_G}\right)(V_{i_+} - V_{i_-}),$$

provided that the amplifier imperfections are ignored, resistors are perfectly matched, and the amplifiers are unsaturated. The external (or sometimes internal) resistor R_G enables the overall voltage gain to be preset at the input terminals. The differential input network enables common-mode noise components to cancel, and the voltage follower arrangement at the input enables a very high input impedance to be presented to the signal source.

More elaborate versions of these amplifiers are known as 'Isolation Amplifiers', and these effect electrical isolation up to a level of (say) 8 kV between the input and output networks. One type of isolation amplifier uses transformers to decouple the networks, and a typical arrangement is shown in Fig. 1.47.

Several variations on this arrangement are marketed commercially as encapsulated modular devices, some of which use opto-couplers as alternatives to the transformers shown in Fig. 1.47.

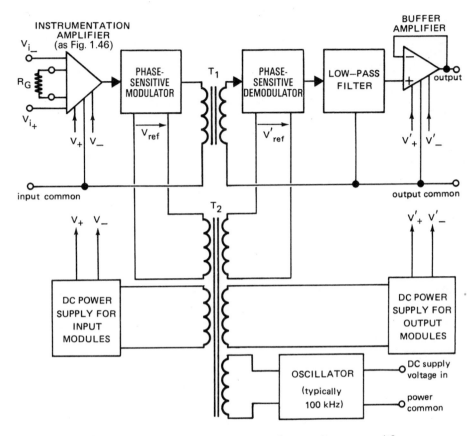

Fig. 1.47   Typical internal configuration of an isolation amplifier

Figure 1.48 shows a typical protection arrangement for a double-ended voltage input signal, involving a fuse, zener diodes, and gas-discharge suppressors.

1.20.2   Analog multiplexers

A Multiplexer (MUX) is a selection switch which determines which analog input signal is to be connected to the ADC at any particular moment in

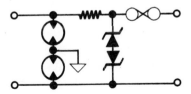

Fig. 1.48   Typical protection circuit for a voltage channel

time. The computer is programmed to address a specified input signal and this address is sent (via the address bus) to the multiplexer, which decodes the address and closes the appropriate switch. The speed at which the MUX can switch determines the maximum scanning rate (number of points/second) of the analog input system. Most MUXs are composed of either electromechanical relays or metal oxide semiconductor field-effect transistors (MOSFETs).

Typically, the relay types used are either dry reed or mercury wetted. Because of the inherent time constant of the coil circuit, scannning rates are relatively low: typically, 300 samples/second; allowance must also be made for contact bounce, when dry reeds are used. Relays are electrically rugged and they can accept a high common-mode voltage: that is, a high common voltage on both inputs, when the voltage source is double-ended, the voltage usually being specified relative to earth potential. Relay MUX systems are more expensive than equivalent MOSFET MUX systems.

The MOSFET systems can switch with high scannning rates: typically, up to several thousand samples/second. These MUXs need to be protected from parasitic voltage spikes and they can tolerate common-mode voltages only up to approximately 10 V, with double-ended voltage sources.

MUX systems can also be classified in terms of the type of configuration used for the connections. Figure 1.49 shows typical configurations, in which the relay symbol should be replaced by a solid state switch symbol if MOSFET switches are being used. Single-ended configurations involve a minimum number of connections but give relatively poor noise rejection; double-ended, differential configurations yield improved noise rejection, because common-mode noise will virtually cancel out. However, differential flying capacitor configurations give the best noise rejection.

1.20.3   Analog input systems

The performance of an analog input system is a function of all of the components in that system, so that it is important to consider these components in relation to each other. In general, each ADC must be preceded by either a buffer amplifier or a sample-hold amplifier, in order to minimise the loading effect of the ADC input and to provide the ADC with a stationary signal during each conversion process: the sample-hold may form an integral part of the ADC. For low-level signals (in the mV range), a high-gain amplifier usually precedes either the ADC or the MUX, in order to amplify the signal level to within the range of the ADC: this is the procedure of 'scaling', which optimises the resolution of the analog–digital conversion process.

Figure 1.50 shows a variety of typical analog input configurations for both high-level and low-level input signals. Note that it is particularly difficult to multiplex low-level signals at high speeds, because of difficulties experienced with the design of satisfactory low noise low-level high-speed MUXs.

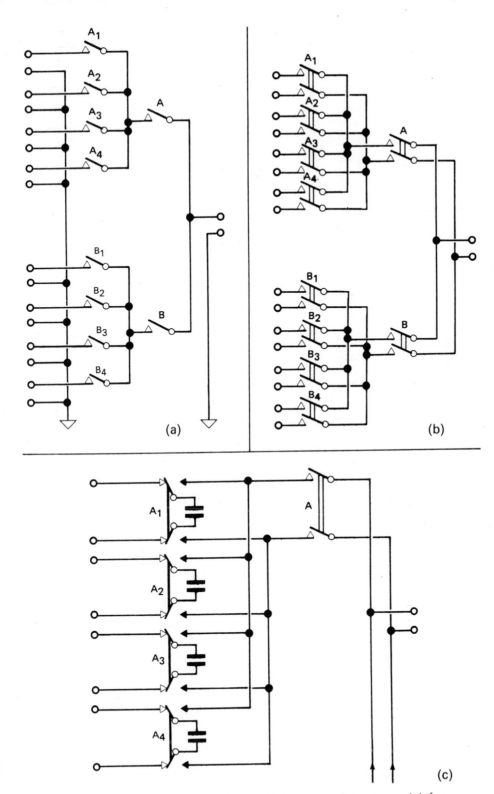

Fig. 1.49 Some alternative analog multiplexer configurations: (a) for single-ended voltage sources with group switching; (b) for double-ended voltage sources with group switching; (c) using differential flying capacitors, with group switching

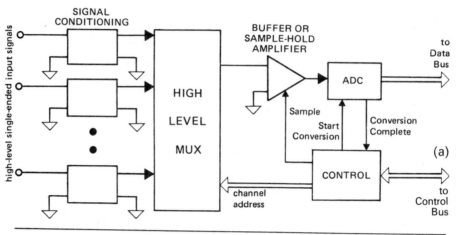

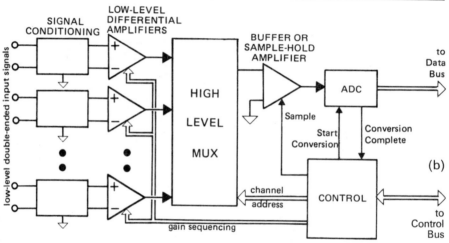

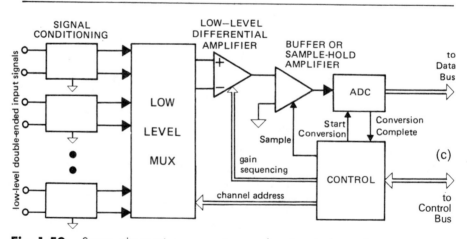

Fig. 1.50   Some alternative computer analog input channel configurations: (a) high-level analog input system; (b) low-level high-speed analog input system; (c) low-level low-speed analog input system

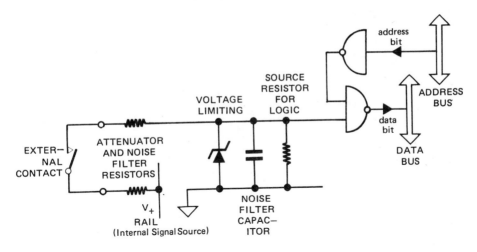

Fig. 1.51 Typical interface circuit for a single digital input channel

1.21 Analog output channels from digital processors

Analog output signals are generated by digital–analog converters, the digital data being supplied to each DAC from the processor data bus. Seldom are DACs multiplexed, so that normally one will be supplied for each analog output channel. The DAC usually generates a high-level

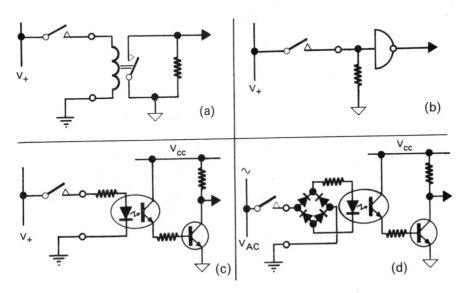

Fig. 1.52 Some altrnative techniques for providing (a) relay isolation; (b) logic gate isolation; (c) DC opto-isolation; (d) AC opto-isolation

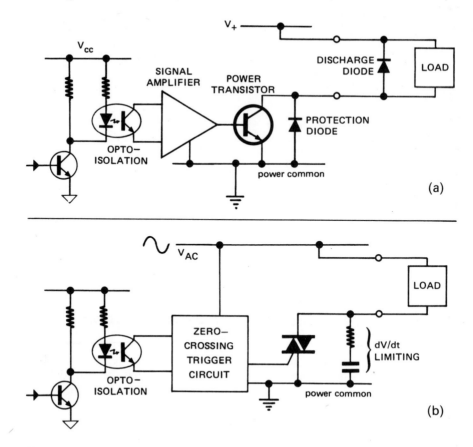

Fig. 1.53   Two alternative techniques for providing protection for single digital output channels: (a) for the case of a DC load with power transistor output drive; (b) for the case of an AC load with Triac output drive

single-ended voltage signal which, in some applications, will need to be converted to (say) a 4 to 20 mA current signal.

The output signal from the DAC would normally be used to drive either the remote set point input of an external controller, a final control element, or a recording instrument.

1.22   Single digital input channels for digital processors

Digital inputs are used to communicate the occurrence of discrete events having a binary nature. Such events include the operation of an electro-mechanical relay, the mechanical activation of a limit switch on a plant, the manual activation of a pushbutton or toggle switch on a control console, etc. The interface will require to be supplied, via the contact in question, from a suitable voltage source and will need to include a detection network and addressing logic to interrogate this network. Whenever noise and/or contact bounce are present, it will be necessary to incorporate appropriate signal conditioning circuitry: this may take the form of a low-pass filter, a Schmitt trigger IC, or a contact debouncer IC. In addition, protection circuitry similar to that in Fig. 1.48 may be required, in order to protect the interface from excessive input signals. Figure 1.51 shows a typical interface hardware arrangement for one switch contact.

Frequently, it is necessary to isolate electrically the external contact from the computer interface. This normally is a requirement whenever the signal being switched by the contact has a level which can potentially damage the interface hardware. Figure 1.52 illustrates a number of alternative arrangements for providing this isolation.

In some applications, the data required by the processor may need to represent a count of a number of events: for example, representing the number of times a particular limit switch has closed. In this situation, after conditioning, the switch signal would be used as the serial input to a digital counter the parallel output of which would represent the count size to be supplied to the processor data bus.

1.23   Single digital output channels from digital processors

Digital output signals are used to activate two-state actuators in control applications, and to activate annunciators in display applications. In some cases, the digital output may need to take the form of a pulse of variable width or, alternatively, a pulse train of variable frequency and/or duration. Like digital inputs, digital outputs can involve electromechanical (dry reed or mercury wetted relay), electronic (IC logic), or electro-optical (opto-isolator) hardware, and no further discussion on these should be necessary. For switching high-power loads, relays of sufficient rating, power transistors, power-FETs, SCRs, and Triacs may be used as alternatives for driving the load. Figure 1.53 shows two of many possible configurations.

2

Networks Sensitive to Signal Amplitude

2.1 Introduction

Occasionally, the control engineer will need to deliberately introduce into a control loop a network having a nonlinear static characteristic. Typical reasons for this include the following:

- to cancel the effect of another (unavoidable) nonlinear static characteristic which happens to be inherent in an upstream or downstream element
- to limit the demand (the reference variable) applied to the control loop, so that the steady state value of the controlled variable cannot be driven beyond preset limits
- to introduce, into the loop, a deadband which, in the steady state, will cause the passage of parasitic noise to be blocked
- to introduce a variable dynamic performance, the nature of which is arranged to be determined by the prevailing operational conditions.

Passive nonlinear shaping networks are relatively uncommon, so that consideration here will be concentrated upon active networks: that is, those incorporating one or more stages of amplification. These active networks can be divided into the following categories:

- simple active networks of resistors and signal diodes
- active resistive ladder networks
- networks using analog multipliers and dividers
- networks using logarithmic amplifiers.

The only passive networks to be considered will be those involving special types of servo potentiometer. In addition, a digital technique will be described.

2.2 Simple active networks of resistors and signal diodes

The simplest of these networks make no precise allowance for the forward volt drop across each diode used, as it conducts. Because the number of diodes used in each network is small, the resulting nonlinear static characteristic will approximate a small number of straight lines joined together: the characteristic is said to be 'piecewise-linear'. Representative examples are shown in Fig. 2.1.

All of these networks operate on the assumption that the negative input of the operational amplifier always functions as a virtual earth: this will cease to be valid if the amplifier output is driven into saturation. It will be seen that there are alternative networks available to produce a given type of nonlinearity. In every case, the actual characteristic will depart from the ideal, because of the effect of the forward characteristic of each conducting diode, and the gradient of the inclined portion of the network characteristic can be modified by changing the value of one or both of the resistors connected directly to the amplifier negative input.

Figure 2.2 shows a network in which a diode D_2 has been placed within the feedback loop with the result that, whenever D_2 is conducting, V_o will be precisely equal and opposite in value to V_i (provided that the resistor values are carefully matched): in this case, conduction always occurs whenever V_i is negative. Diode D_1 is included in order to prevent the amplifier from saturating negatively whenever V_i is positive. During the conducting half cycle, the forward volt drop across D_2 has no influence on the relationship between V_o and V_i, so that the actual static characteristic approaches the ideal. More complex precision piecewise-linear nonlinearities can be constructed using this precision half-wave rectifier network as a basis, and examples are given in Fig. 2.4.

Many other networks of this type, together with variations which can operate at frequencies much higher than those normally pertaining to control systems, are described elsewhere in the literature.

2.3 Active resistive ladder networks

If, in the last example in Fig. 2.4, resistors R_3, R_4 and R_5 had deliberately been made unequal in value, the type of piecewise-linear characteristic shown in Fig. 2.3 would have resulted: this is seen to consist of three linear segments, each having a variable gradient. Obviously, many more linear segments could be added to the characteristic, if more parallel channels of half-wave rectifiers were to be added to the network.

An alternative technique would be to create multiple parallel channels of analog comparators and solid state digitally-controlled analog switches, as shown in Fig. 2.5.

For the configuration shown, input breakpoint voltages $V_{11}, \ldots, V_{15}$ can be preset at any positive or negative value, within the range of V_i, and need not be distributed evenly. Changes in gradient may be positive or negative, depending upon whether a particular comparator, when it changes state, turns on or off the analog switch to which it is connected. The value of each gradient may be modified by changing the value of the resistor $R_1, \ldots, R_5$ concerned. The characteristic may be raised or lowered bodily, by changing the magnitude (and, if necessary, the polarity) of V_2 and/or by changing the value of resistor R_7.

NETWORKS	IDEALISED STATIC CHARACTERISTICS

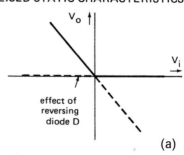

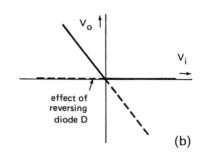

(a)

(b)

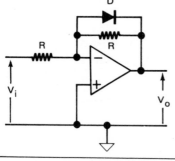

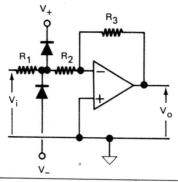

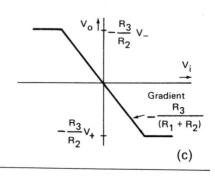

(c)

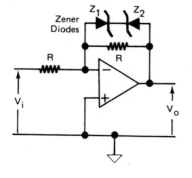

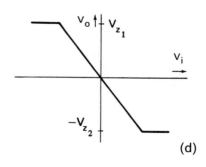

(d)

NETWORKS   IDEALISED STATIC CHARACTERISTICS

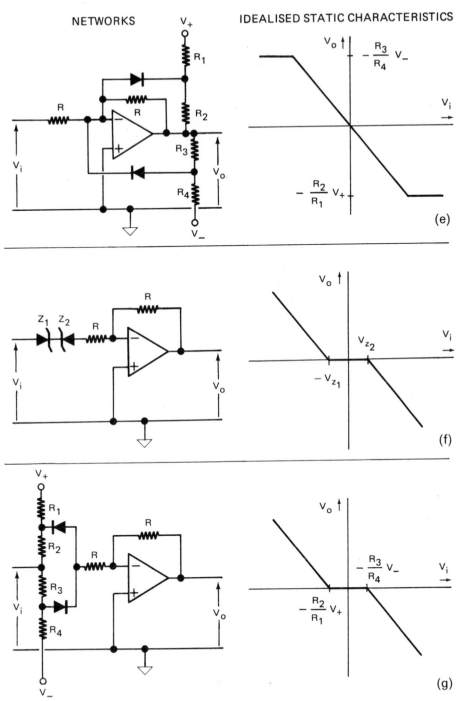

Fig. 2.1 Some representative examples of active networks of resistors and signal diodes, to achieve simple piecewise-linear nonlinear static characteristics: (a) and (b) simple half-wave rectifier; (c) (d) and (e) simple limiter; (f) and (g) simple deadband

Variations on the above configuration have been built using active ladder networks of diodes and resistors, as shown in outline in Fig. 2.6.

Blocks A and B operate for positive-going values of V_i, whilst blocks C and D operate for negative values. Whenever a diode conducts, it causes a reduction in the effective resistance between points X and Y, resulting in a negative increase in the gradient of the static characteristic.

The detail of the ladder network for block A is shown in Fig. 2.7. The number of 'rungs' shown in the ladder is not necessarily representative of a practical network: a much greater number of rungs would be more commonplace.

Diodes $D_1, \ldots, D_5$ are connected in the reverse direction for blocks B and D. The V_- supply is replaced by V_+ for blocks C and D.

Changing the values of resistors $R_{11}, \ldots, R_{15}$ will modify the breakpoint values.

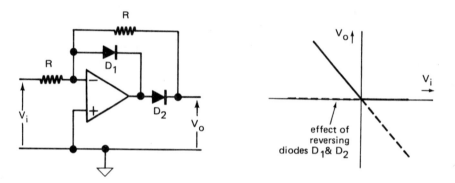

Fig. 2.2 Simple active network for a precision half-wave rectifier

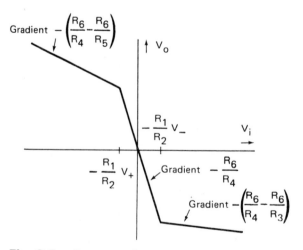

Fig. 2.3 Piecewise-linear characteristic obtained from the network of Fig. 2.4(c) for the general case of $R_3 \neq R_4 \neq R_5$

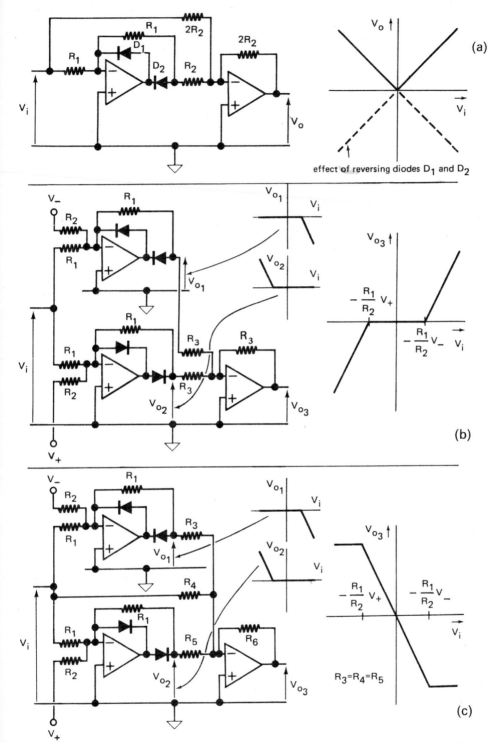

Fig. 2.4 Some representative examples of networks using precision half wave rectifier circuits, to achieve precise piecewise-linear nonlinear static characteristics: (a) precision full wave rectifier (modulus or absolute value network); (c) precision limiter (when $R_3 = R_4 = R_5$)

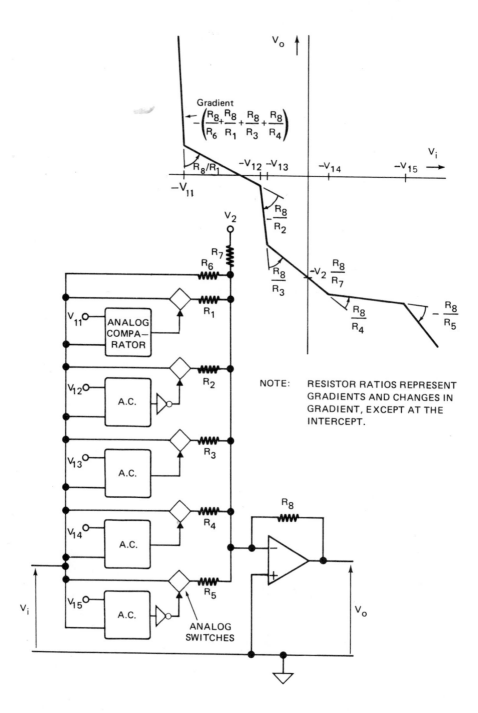

Fig. 2.5   Active network of analog comparators and switches to achieve a complex, precision, piecewise-linear nonlinear static characteristic

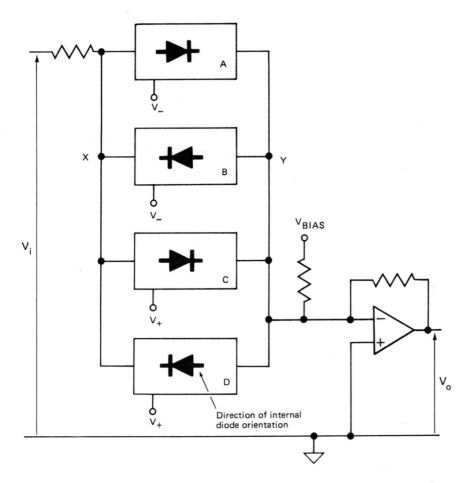

Fig. 2.6   General arrangement of a four-quadrant diode-resistor active ladder network to achieve a complex piecewise-linear nonlinear static characteristic. Blocks A, B, C and D represent ladder networks of the type shown in Fig. 2.7

Changing the values of resistors $R_{21}, \dots, R_{25}$ will modify the changes in gradient occurring at the breakpoints.

A diode will conduct, or cease to conduct, when V_i is driven through a value of potential such that the diode becomes, or ceases to become, forward biased, bearing in mind that node Y will behave as a virtual earth, if the amplifier output is unsaturated.

Ladder networks such as the one described are difficult to design with precision, because of the effect of the actual voltage dropped across each conducting diode. However, diode-resistor networks can be significantly cheaper than networks employing analog comparators and analog switches.

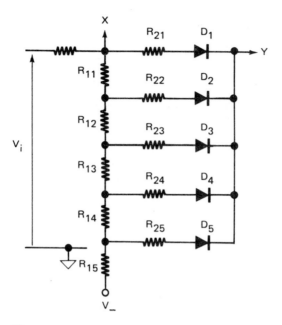

Fig. 2.7   Symbolic representation of the diode-resistor ladder network shown as blocks A, B, C and D in Fig. 2.6

2.4.   Networks using analog multipliers and dividers

Analog multipliers can be used to form the basis of nonlinear shaping networks which are not piecewise-linear. This possible advantage must be weighed against the fact that the choice of nonlinear law is rather limited. In addition to this application, multipliers may also be used in control systems, in their fundamental role of multiplying together two analog signals, in such elements as Ratio Controllers and analog adaptive controllers: in the latter case, provision may be made for enabling, for example, a DC voltage signal to vary a gain constant or time constant in the control law of a feedback controller. Analog dividers, whilst being less common than multipliers, can be used to perform similar tasks.

2.4.1   Multiplier and divider characteristics

Analog multipliers normally mechanise a relationship having the form $V_o = KV_x V_y$, where V_x and V_y are two input voltages, V_o is an output voltage, and K is a proportionality constant. Typically, $1/K$ would have a value of 10 volts (V), so that $V_o = 10\,V$ when $V_x = V_y = 10\,V$. Analog dividers normally mechanise a law of the form $V_o = KV_x/V_z$, where V_x and V_z are now the two input voltages and, in this case, K typically would have a value of 10 V, so that $V_o = 10\,V$ when $V_x = V_z = 10\,V$. A more rare type of device is the analog multiplier/divider, mentioned in Section 1.13, which has a law of the form $V_o = KV_x V_y/V_z$: in this case, the value of the proportionality constant K normally would be unity. Modern devices in these categories are usually either encapsulated circuits or monolithic integrated circuits.

The combinations of polarities of V_x, V_y, and V_z which a particular device can handle determine the alternative 'quadrants' in which the device can be operated: the choice is normally between either one, two, or four quadrant operation. Where division (by V_z) is involved, it will be common-place for V_z to be restricted to positive values, with the lower limit being that which will cause V_o just to saturate.

A number of techniques (for example, 'Quarter-Squares', variable-transconductance, Hall-effect, pulse width-pulse height, and log-antilog) have been employed for multiplying together analog signals. By far the most common method used in encapsulated and monolithic multipliers employs the variable-transconductance phenomenon applied to current signals passing through P-N junctions.

Considerations to be taken into account when selecting one of these devices, include:

- number of quadrants
- accuracy
- input and output signal ranges
- input and output null offsets
- bandwidth.

2.4.2 External circuitry required with analog multipliers

Encapsulated and monolithic multipliers usually require a significant number of external components before they can perform their specified role. Typically, these components perform such tasks as null offset adjustment, scale factor setting, input range setting and output voltage level shifting. Figure 2.8 shows typical external circuitry required by a representative monolithic multiplier.

It is most important for stable and consistent operation of such a network that the DC supply rail voltages V_+ and V_- possess long-term stability: otherwise, there may be significant drift in the offset and scale factor values. Some resistors may be required to have values which are very precise and stable and not in the preferred value range.

By making specified changes in the external circuitry, it is sometimes possible to change the function of a given encapsulated or monolithic device from multiplication to other roles such as division, squaring and square-rooting: manufacturers' data sheets will provide the necessary information.

2.4.3 Generation of power laws

The analog multiplier may be connected in a number of alternative ways, in order to generate power laws. Figure 2.9 shows some typical examples.

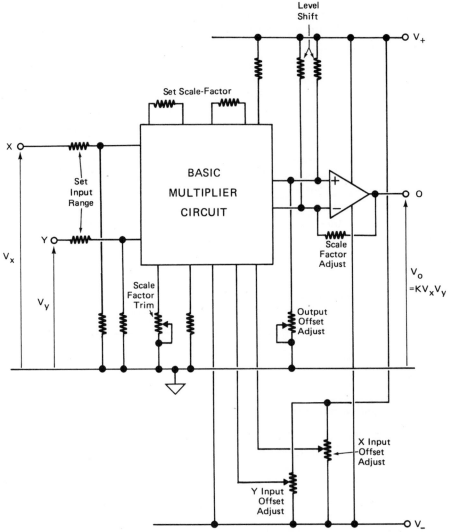

Fig. 2.8 Typical arrangement of external circuit components required to achieve correct operation of a monolithic analog multiplier

It is readily shown that:

- increasing the number of cascaded multipliers increases the index of the power law
- for certain combinations of cascaded multipliers, the output may always be unipolar, unless circuitry which will invert the output sign, and which is sensitive to the sign of the input voltage, is incorporated
- when a multiplier is placed in the feedback path of an otherwise open loop operational amplifier network, stability considerations will require that the multiplier does not sign-invert the feedback signal.

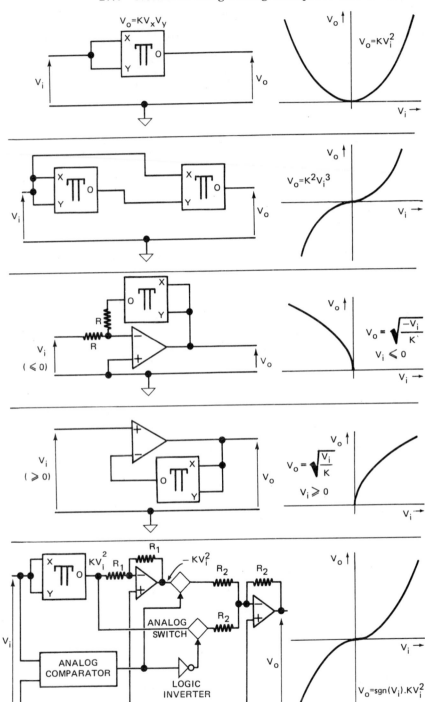

Fig. 2.9 Representative examples of the use of analog multipliers in order to generate power laws

A particular application of the square-root law network is the Square-Root Extractor. This element is used for removing the square law relating differential pressure to flowrate, which is inherent in those flowrate transducers which involve the insertion of a fixed obstruction in a pipeline. Since the extractor typically will be concerned with current signals in the 4 to 20 mA range, the actual law required for this device will be of the form

$$I_o = 4 + 16 \sqrt{\frac{(I_i - 4)}{16}} = 4[1 + \sqrt{(I_i - 4)}]\,\text{mA}$$

Figure 2.10 shows how a square-root extractor may be created, using networks discussed above and in Chapter 1.

2.4.4   Other applications of analog multipliers

Figure 2.11 shows how an analog multiplier may be used for division.

It is readly shown that, provided the operational amplifier is stable and unsaturated, $V_o = -\dfrac{1}{K}\dfrac{V_i}{V_x}$. In fact, the network will be stable only for

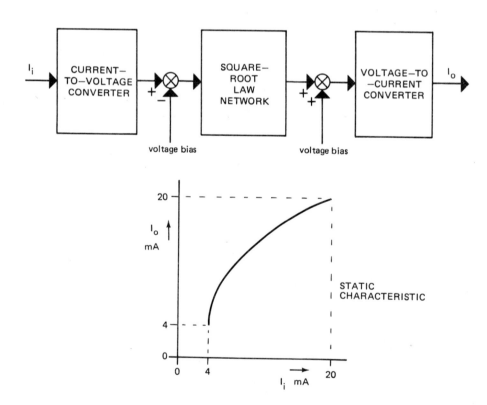

Fig. 2.10   Block diagram and static characteristic of a square-root extractor

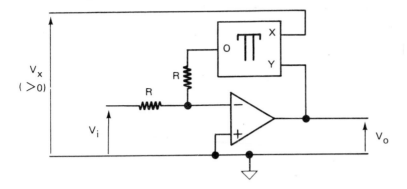

Fig. 2.11 Active analog divider network using an analog multiplier as a feedback path element

negative feedback, so that only positive values are acceptable for V_x. In addition, the amplifier will saturate if V_x is much less than V_i in value, so that the voltages V_i and V_x must be scaled carefully before the network can be applied successfully.

Another application for analog multipliers is in adaptive analog filters, which may be used, for example, to implement control laws in adaptive analog controllers. An example of an adaptive simple phase-lag network (see Section 3.3) is shown in Fig. 2.12.

For this network, it is readily shown that

$$\frac{V_{o1}}{V_i}(s) = \frac{-1/KV_x}{1 + s(CR/KV_x)} \text{ and } \frac{V_{o2}}{V_i}(s) = \frac{-1}{1 + s(CR/KV_x)}$$

provided that the network is stable and the operational amplifier is unsaturated, which requires that V_x be positive and places an upper limit on V_i/KV_x. In this event, the network behaves as a simple phase-lag with a voltage-controlled $-3\,dB$ break frequency of $KV_x/2\pi CR$ Hz.

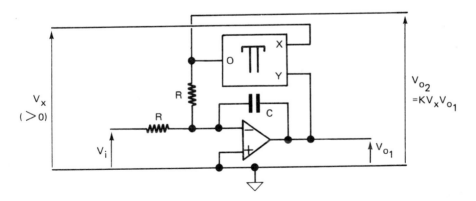

Fig. 2.12 Examples of an adaptive simple phase-log network

Another common application of analog multipliers is in modulators for use in AC-carrier control systems and this topic is covered in Section 1.4.

2.5   Networks using logarithmic amplifiers

Modern monolithic logarithmic amplifiers use the logarithmic properties which are inherent in silicon semiconductor junctions, and which afford a wide dynamic range and are amenable to temperature compensation. The circuit commonly used is based upon a transistor placed in the feedback path around an operational amplifier, as shown in Figure 2.13.

It can be shown that, for the first network, $V_o \cong \dfrac{kT}{q} \ln (I_i/I_{ES})$ and that this expression is approximately true also for the second network, provided that the transistor used has a very high value for h_{FE}. In this expression, q is unit charge ($= 1.602 \times 10^{-19}$ C), k is Boltzmann's constant ($= 1.38 \times 10^{-23}$ J/K), T is absolute temperature in Kelvin, and I_{ES} is the emitter saturation current.

Figure 2.14 shows typical characteristics for the two types of configuration. In each case, the characteristic is highly linear over at least five decades, with a gradient of 59 mV per decade. Only one polarity of input signal can be accommodated by each network, but the alternative input polarity can be handled by comparable networks using NPN instead of PNP transistors.

Figure 2.15 shows a typical temperature compensated logarithmic amplifier circuit, for which it can be shown that

$$V_o = - K \log_{10} (V_i/R_5 I_{ref}), \text{ where } K = [1 + R_1/R_2]\frac{kT}{q}\ln 10$$

The capacitors are provided for stabilisation purposes and the presence of the R_3–R_4 resistor combination permits extension of operation to cover up

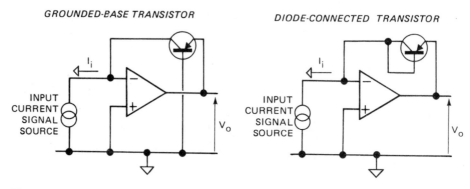

Fig. 2.13   Use of semiconductor junctions to produce logarithmic amplifiers

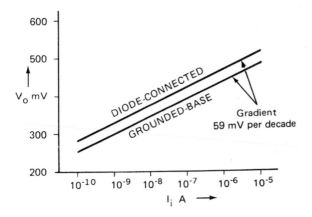

Fig. 2.14   Typical static characteristics for the logarithmic amplifier configurations of Fig. 2.13

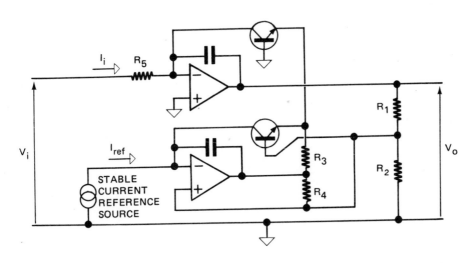

Fig. 2.15   Typical temperature compensated logarithmic amplifier circuit

to six decades of values for I_i. A typical transfer characteristic is shown in Fig. 2.16.

Figure 2.17 shows a minor circuit rearrangement which, in fact, produces an antilogarithmic amplifier with a law given by $V_o = R_5 I_{ref} 10^{-V_i/K}$. This is brought about because the potential divider R_1-R_2 and resistor R_5 have, in effect, been interchanged.

Typical applications for logarithmic amplifiers include the compression of wide-range analog signals, the linearisation of the power law static characteristics of certain transducers, and as component parts of RMS-to-

DC converters (see Section 1.13). Typical applications for antilogarithmic amplifiers include the expansion of compressed analog signals, the linearisation of the logarithmic static characteristics of certain transducers, and as component parts of RMS-to-DC converters.

2.6 Special–purpose servo potentiometers

A limited range of servo potentiometers is commercially available, having defined nonlinear laws relating brush voltage to shaft displacement. These may have a wirewound or composition track, and the nonlinear function usually is built in by graduating the cross-sectional area of the track.

Figure 2.18 shows a hypothetical nonlinear function $f(\theta)$, for a wirewound potentiometer, relating length-of-turn 1 to shaft displacement θ.

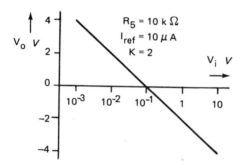

Fig. 2.16   Typical transfer characteristic of the logarithmic amplifier circuit of Fig. 2.15

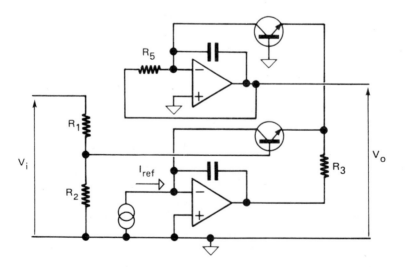

Fig. 2.17   Typical temperature compensated antilogarithmic amplifier circuit

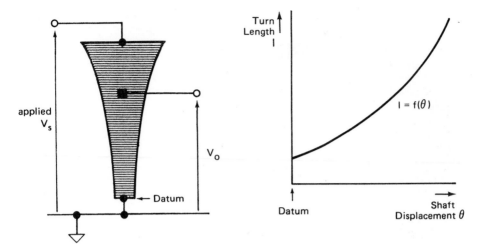

Fig. 2.18 Symbolic representation of a nonlinear wirewound servo potentiometer and its relationship of turn length v shaft displacement

Assuming the other properties (resistivity, etc.) of the winding to be constant, it follows that the resistance between the brush and the datum of the track will be given by

$$K_1 \int_0^\theta f(\theta)\,d\theta$$

where K_1 is a constant in ohm/mm (for example), and the unloaded output voltage V_o will be related to the applied voltage V_s by the law

$$V_o = V_s \int_0^\theta f(\theta)\,d\theta \Big/ \int_0^{\theta_{max}} f(\theta)\,d\theta = K_2 \int_0^\theta f(\theta)\,d\theta$$

where K_2 is a constant in V/mm (for example).

Typical laws available for potentiometers of this type would include $V_o = V_s \sin\theta$, $V_o = V_s \cos\theta$, and $V_o = V_s \log(A\theta + B)$, where A and B are constants. The sine and cosine laws can be obtained by using a continuous toroidal track with two brushes mounted on axes at right angles to one another, as shown in Fig. 2.19.

Piecewise-linear characteristics can be imposed upon nominally linear servo potentiometers, if they are provided with a suitable set of tappings along the track. Sections of the track are shunted with appropriate fixed resistance values, and the potentiometer is then said to be 'padded'. Figure 2.20 shows one possible arrangement for a potentiometer with five tappings and external padding.

The advantage with this latter technique is that the law may be modified at will by the user, but offset against this are the limitation on the maximum number of tappings which can be accommodated and the fact that only straight line segments will result, for the static characteristic.

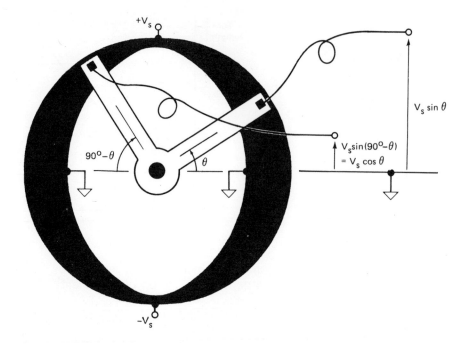

Fig. 2.19 Symbolic representation of the track and brushgear of a continuous toroidal Sine-Cosine servo potentiometer: the device will include slip rings to facilitate external connection to the brushes — these have been omitted for clarity

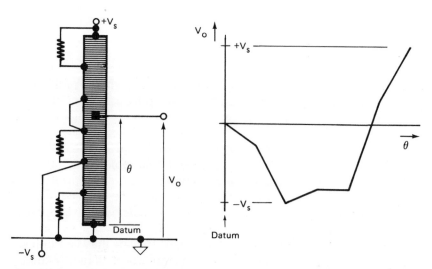

Fig. 2.20 Use of a tapped linear servo potentiometer to achieve a piecewise-linear nonlinear static characteristic

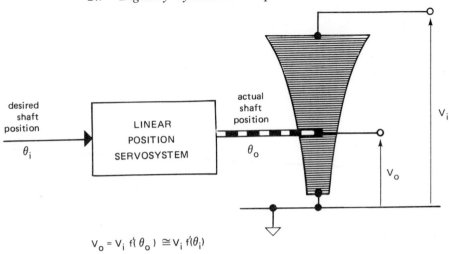

$$V_o = V_i \; f(\theta_o) \cong V_i \; f(\theta_i)$$

Fig. 2.21   Use of a linear position servosystem to position the brush of a nonlinear servo potentiometer, to achieve multiplication of one variable by a nonlinear function of a second variable

When driven by a linear servosystem but not used as the displacement feedback transducer of that system, these special potentiometers will enable one variable to be multiplied by a nonlinear function of a second variable, as shown in Fig. 2.21. Obvious shortcomings of this technique are the low bandwidth of the servosystem and the limited life and high breakaway torque of the potentiometer.

V_i need not be a DC voltage but could be, for example, an amplitude-modulated AC-carrier. The law $f'(\theta_i)$ will always become distorted if the load circuit being driven by V_o draws significant current levels from the potentiometer.

A similar effect can be achieved with a linear potentiometer, if it is configured to be driven, from the servo output shaft, through an interposed cam and cam-follower. This technique will permit special nonlinear functions to be obtained, including curved functions, but imposes certain constraints: for example, the requirement for a sudden discontinuity in the law could necessitate the follower being required to jump a radial face of the cam – an impossible undertaking.

2.7   Digitally–synthesised amplitude sensitive networks

An extremely powerful and relatively fast technique is an extension of the code-conversion application of semiconductor read-only memory (ROM) discussed in Section 1.15. Figure 2.22 indicates how the hardware would need to be organised.

The digital code representing V_i is interpreted as the address of a location in memory, whilst the (previously stored) content of that location will

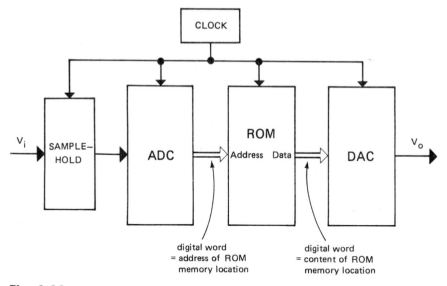

Fig. 2.22   Use of programmable ready-only memory to generate a nonlinear static characteristic

represent the corresponding value required for V_o: therefore, the relationship of V_o to V_i will have been made to represent the nonlinear function required. The size of the memory will need to be such that all possible alternative values of V_i can be accommodated, except that this size may be reduced if special digital data interpolation networks are incorporated. Memories are marketed pre-programmed with a limited number of alternative functions: for example, sine and cosine laws.

The obvious versatility of the technique means that the nonlinear function could incorporate curvature, discontinuity and four-quadrant operation. Against this must be weighed the complexity and cost of the hardware, together with the possible disadvantage of introducing sampling and quantisation processes into the control system into which the network is introduced. For maximum resolution, the input voltage V_i and output voltage V_o should each be scaled so that maximum use is made of the available voltage excursion of the ADC and DAC, respectively.

If the ROM in Fig. 2.22 is replaced by a pre-programmed microcomputer, this new arrangement can be used to generate any static characteristic capable of being defined explicitly by one or more mathematical relationships. The output from the ADC now represents input data to an appropriate pre-programmed software routine, the output data generated by which become the input to the DAC.

3

Networks Sensitive to Signal Frequency

3.1 Introduction

In many control systems, the need arises for the deliberate introduction, into the loop, of networks which exhibit gain magnitude and phase shift variations as functions of signal frequency. Such networks normally are linear, in the sense that the values of these variations are independent of input signal amplitude and rate of change except that, in practice, all active networks will exhibit saturation (if this amplitude becomes excessive) and rate limiting (if the rate of change in input signal becomes excessive).

In control systems, frequency-dependent networks are introduced normally for one or more of the following reasons:

- to establish and/or improve closed loop system stability
- to optimise closed loop dynamic performance
- to filter out (that is, to suppress) parasitic noise components in system signals and power supplies
- to correct parasitic phase shifts in the carrier components in AC-carrier control systems.

Networks which perform the first two functions listed here are often referred to as 'compensation networks' or 'compensators'.

The passive components normally used in analog filters are either resistors or capacitors. This is because these components can be manufactured to very close tolerances and usually exhibit long-term stability: for example, $\pm 1\%$ tolerance or better is readily obtained in resistor values over the range $5\,\Omega$ to $1\,M\Omega$ and in capacitor values over the range $10\,nF$ to $10\,\mu F$. For certain noise filters, electrolytic capacitors can be obtained with values up to the mF range, but the fact that they are polarised usually restricts their use to applications in which the voltages are of single polarity; however, tantalum electrolytic capacitors have occasionally been used in back-to-back pairs (that is, connected in series opposition) in bipolar applications. Inductors do not appear very frequently in these networks: only air-cored inductors can be made to close-tolerance specifications and have an inductance which is independent of signal level. Air-cored inductors having more than a small inductance value must be large physically and the inherent resistance value may be excessive for the particular application. Inductors cored with soft iron laminations, encapsulated iron dust or ferrite can achieve a high inductance/volume ratio, but then the inductance usually becomes variable as a function of signal level: for this reason, the application of such inductors is usually restricted to filters for DC power supplies.

The design of frequency-dependent networks covers a considerable field, so that the space available here permits only a brief survey of the techniques used. Most filters are designed using the transfer function as a basis, so that familiarity with the use of the Laplace operator is essential; the design of filters for DC power supplies is a possible exception, as this is often undertaken using standard graphs and algebraic formulae.

When using the transfer function as the basis for design, it is often insufficient to design for a nominal transfer function: that is, using nominal values for the parameters (gain constant, time constants, etc.) of the transfer function. Quite frequently, it is necessary to compute the sensitivities of these parameters to changes, from component nominal values, arising from such factors as manufacturing tolerances, ageing, operating conditions, etc.

As far as compensators are concerned, the advent of digital techniques has established a valuable alternative to the traditional analog networks. Digital synthesis of filters provides high precision, considerable versatility and long-term stability, but some potential disadvantages may be present and these will be discussed in Section 3.7.

3.2   Passive R–C networks

The advantage of using a passive network is the absence of the need for an amplifier. However, if the load applied to the network is significantly large, then the network transfer function for the loaded case may become markedly different from that for the unloaded case. It is the loaded version which is relevant when the network is introduced into the control system. If the load is changing during normal operation of the system, then the loaded case transfer function will be changing also, with the result that the nature of the dynamic performance of the system will be variable. This situation can be resolved by interposing a buffer amplifier between the passive network and its load, so that the relevant transfer function now becomes the (stationary) unloaded version: however, this then tends to destroy any advantage which might have arisen from using a passive network, so that an equivalent active network might be a preferable alternative to the passive network–buffer amplifier combination.

Passive R–C networks can only attenuate the input signal, so that signal amplification cannot be achieved by the network itself. This is in contrast to the situation with passive R–L–C networks, in which tuned circuits can be used to amplify voltages or currents at signal frequencies in the region of the natural frequencies of the tuned circuits.

Passive R–C circuits can assume an almost infinite variety of configurations, and the properties of these have been tabulated in many of the literature references. The unloaded transfer functions can be determined using normal mesh-nodal network analysis techniques, and often they are classified in terms of the properties of the corresponding frequency responses: typically, either the sequence of asymptote gradients on the Bode magnitude plot and/or the sequence of sign changes (that is, lag-lead, lead-lag, etc.) on the Bode phase plot.

Frequently, the simpler networks are based on an inverted-L arrangement of simple combinations of resistors and capacitors, as shown in Fig. 3.1. Examples of these R–C combinations have been tabulated in Fig. 3.2, although this list is far from exhaustive.

More complex networks may be constructed using the ladder, bridged-T, and parallel-T types of configuration shown in Fig. 3.3, using R–C combinations such as those in Fig. 3.2 to create the complex impedances.

The simple networks of Fig. 3.2 are referred to as 'two-terminal networks' or 'one-port networks' because each requires only two external connections. By similar reasoning, the networks of Fig. 3.1 and 3.3 are called 'three-terminal networks' or 'two-port networks', because, although the diagrams show four terminals, in fact the input signal source and the output load share an electrically common terminal.

The most common passive R–C networks used for compensation are the simple phase-lag, phase-lead, and lag-lead networks, shown in Figs 3.4, 3.5, and 3.6, respectively.

The phase-lag network of Fig. 3.4, when employed for compensation, is used primarily for the attenuation which it introduces: the break frequencies $1/T_1$ and $1/T_2$ rad/s are chosen such that the network will reduce the system gain crossover frequency by a greater margin than the accompanying reduction in phase crossover frequency, in order to improve the system stability margins, as discussed in Chapter 14.

The phase-lead (phase-advance) network of Fig. 3.5, when employed for compensation, is used primarily for the positive phase shift which it introduces: the break frequencies $1/T_1$ and $1/T_2$ rad/s are chosen such that the network will increase the system phase crossover frequency by a greater

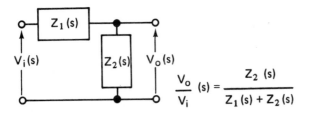

$$\frac{V_o}{V_i}(s) = \frac{Z_2(s)}{Z_1(s) + Z_2(s)}$$

Fig. 3.1   Three-terminal R–C network and its transfer function

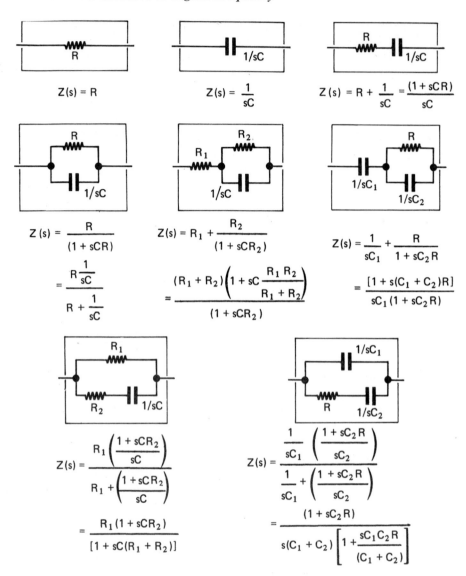

Fig. 3.2 Representative R–C combinations, with complex impedances, for use in network such as those in Figs 3.1 and 3.3

margin than the accompanying increase in gain crossover frequency. Again, as discussed in Chapter 14, this is in order to improve the system stability margins.

The lag-lead (notch) network of Fig. 3.6 is really a combination of the phase-lag network of Fig. 3.4 and the phase-lead network of Fig. 3.5. The phase response has odd-symmetry about the centre frequency $1/\sqrt{T_{11}T_{12}} = 1/\sqrt{T_{21}T_{22}}$ rad/s and the value of the maximum phase shift increases as the depth of the notch is increased: that is, by enlarging the

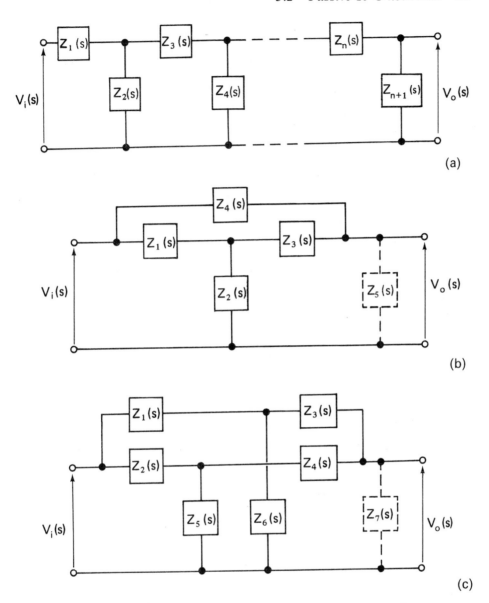

Fig. 3.3   Three alternative three-terminal passive network configurations:
(a) ladder; (b) bridged-T; (c) parallel-T

ratio $T_{21}/T_{11} = T_{12}/T_{22}$. The network, when employed for compensation, is used primarily for its attentuation in the region of the centre frequency, in combination with positive phase shift above the centre frequency: the break frequencies $1/T_{11}$, $1/T_{12}$, $1/T_{21}$, $1/T_{22}$ rad/s (which have interdependent values) are chosen such that the network will decrease the system gain crossover frequency and simultaneously increase the phase crossover

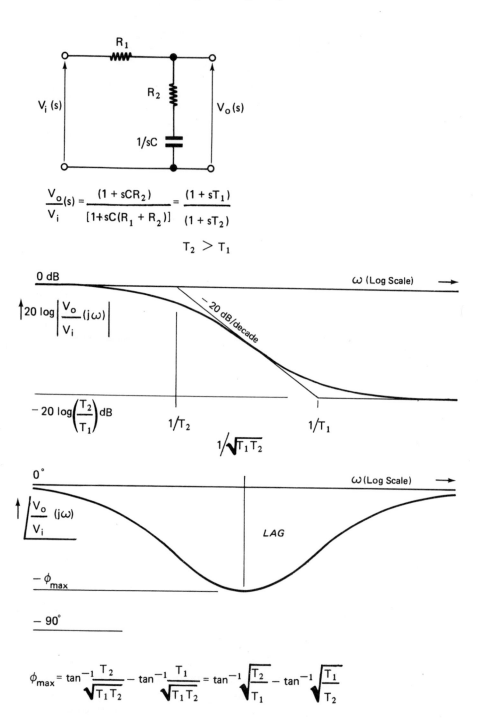

Fig. 3.4 Simple passive R–C phase-lag network, showing its transfer function and Bode diagram

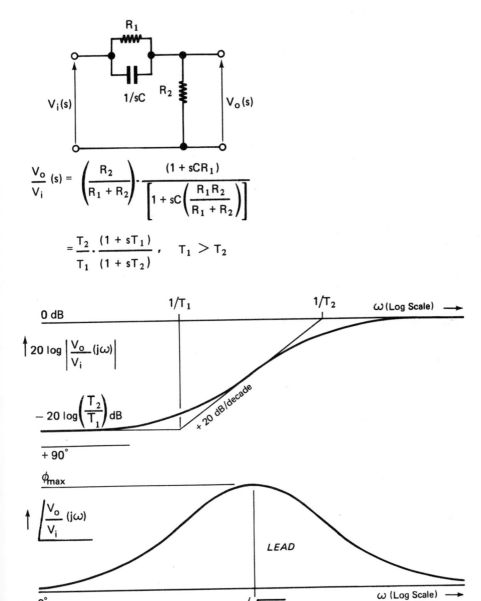

$$\frac{V_o}{V_i}(s) = \left(\frac{R_2}{R_1 + R_2}\right) \cdot \frac{(1 + sCR_1)}{\left[1 + sC\left(\frac{R_1 R_2}{R_1 + R_2}\right)\right]}$$

$$= \frac{T_2}{T_1} \cdot \frac{(1 + sT_1)}{(1 + sT_2)}, \quad T_1 > T_2$$

$$\phi_{max} = \tan^{-1}\frac{T_1}{\sqrt{T_1 T_2}} - \tan^{-1}\frac{T_2}{\sqrt{T_1 T_2}} = \tan^{-1}\sqrt{\frac{T_1}{T_2}} - \tan^{-1}\sqrt{\frac{T_2}{T_1}}$$

Fig. 3.5   Simple passive R–C phase-lead network, showing its transfer function and Bode diagram

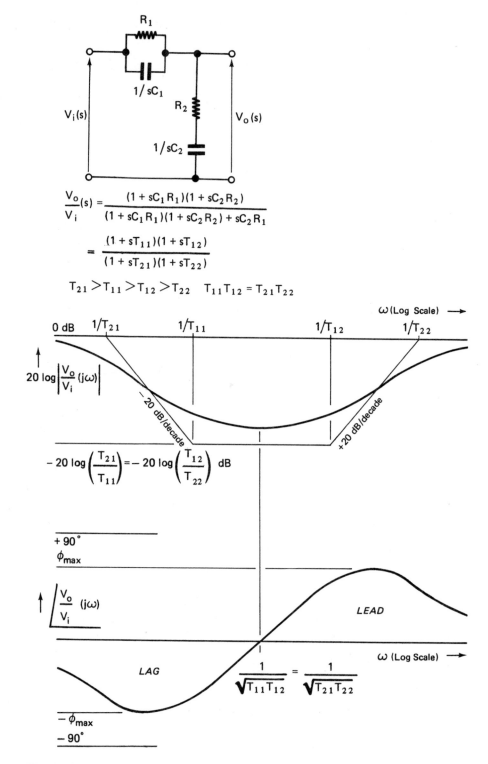

$$\frac{V_o}{V_i}(s) = \frac{(1 + sC_1 R_1)(1 + sC_2 R_2)}{(1 + sC_1 R_1)(1 + sC_2 R_2) + sC_2 R_1}$$

$$= \frac{(1 + sT_{11})(1 + sT_{12})}{(1 + sT_{21})(1 + sT_{22})}$$

$$T_{21} > T_{11} > T_{12} > T_{22} \quad T_{11}T_{12} = T_{21}T_{22}$$

Fig. 3.6 Simple passive R–C lag-lead network, showing its transfer function and Bode diagram

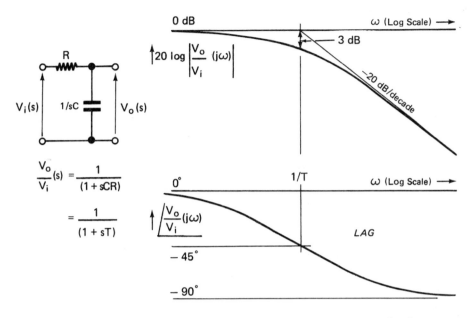

Fig. 3.7   Simple passive R–C low-pass filter, showing its transfer function and Bode diagram

frequency. Again, reference to Chapter 14 will show that the result will be an improvement in system stability margins.

The network of Fig. 3.7 is the simplest of the range of 'low-pass filters', which are used frequently as noise filters and which are passive in most instances. When employed for noise filtering, the network is used primarily for the attentuation which it introduces at the higher frequencies, which are characteristic of most noise signals: this attenuation and the accompanying negative phase shift often are deleterious to system performance (but have to be tolerated), if the filter is connected within the loop of a closed loop system. Typical applications for noise filters include:

- the attentuation of the commutator ripple component of DC tacho-generator output voltage – unfortunately, the frequency of this component is variable, in direct proportion to the shaft speed
- the attenuation of the carrier frequency component of the DC output voltage from demodulators in AC-carrier control systems (refer to Section 1.5)
- the attenuation of the AC component in the output voltage from RMS-to-DC and frequency-to-voltage converters (refer to Sections 1.13 and 1.16 respectively).

Note that compensation networks such as the phase-lead network of Fig. 3.5, which introduce at low frequencies an attenuation which is much higher in value than that at high frequencies, will tend to have an effect on parasitic noise components which is the reverse of the effect of noise filters. Thus, phase-lead compensation networks will tend to worsen signal/noise ratios and, for this reason, it is preferable to avoid using them in those systems with a high level of parasitic noise.

3.3   Active R–C networks

For all of the passive R-C networks, discussed in the previous section, there are active network counterparts. In addition, the active nature of these latter networks enables them to provide amplification as well as attenuation, and ensures that the transfer function being synthesised is independent of the loading of the network output. Active configurations of this type are normally constructed around operational amplifiers, with most single-stage configurations being sign-inverting: however, Fig. 3.8 shows general arrangements for both sign-inverting and non-inverting configurations.

Typically, the networks implied by blocks $Z_1(s)$ ... $Z_4(s)$ would be two-terminal passive R–C networks of the type indicated in Fig. 3.2 or, less commonly, $Z_1(s)$ and/or $Z_2(s)$ might be three-terminal passive R–C networks of the kind outlined in Fig. 3.3.

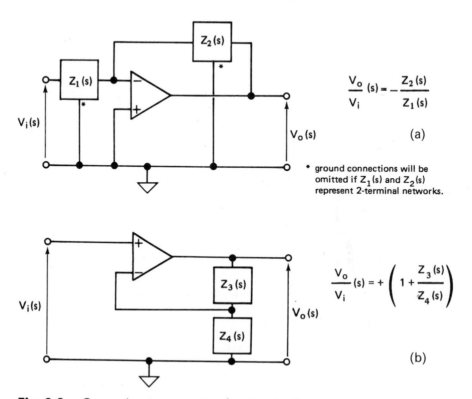

$$\frac{V_o}{V_i}(s) = -\frac{Z_2(s)}{Z_1(s)}$$

(a)

* ground connections will be omitted if $Z_1(s)$ and $Z_2(s)$ represent 2-terminal networks.

$$\frac{V_o}{V_i}(s) = +\left(1 + \frac{Z_3(s)}{Z_4(s)}\right)$$

(b)

Fig. 3.8   General arrangements of active R–C networks: (a) sign-inverting configuration; (b) non-inverting configuration

In the case of three-terminal networks in sign-inverting active configurations, complex impedances $Z_1(s)$ and $Z_2(s)$ need to be defined as 'short-circuit transfer impedances': such an impedance relates the current flowing from the network output to the applied input voltage, when the output is imagined to be short-circuited to signal common. This is because, during normal operation, the negative input terminal of the operational amplifier will be at signal common potential; in addition, the operation of the configuration will be such that there will be zero current entering this terminal.

Figure 3.9 shows four commonly used sign-inverting active counterparts of the passive networks presented in Figs 3.4, 3.5, 3.6 and 3.7, respectively. These active versions are more versatile than their passive equivalents, because:

- the gain constant K can have a wide range of values
- the lag-lead configuration may be converted to lead-lag, by a suitable choice of component values
- the numerator time constants of the lag-lead network can be adjusted independently of one another, as can the denominator time constants;
- with the lag-lead network, the high-frequency gain magnitude can have a value different from that at DC, if the product $(T_{11}T_{12})$ is made unequal to $(T_{21}T_{22})$;
- the transfer functions given for the configurations are unaffected by loading the network, provided that the output current rating of the operational amplifier is not exceeded.

Integration will always be present in the transfer function of a sign-inverting configuration, if:

(i)   there is a capacitor in the feedback network around the amplifier; and
(ii) the feedback capacitor is not shunted, directly or indirectly, by a resistor; and
(iii) there is no capacitor, in the input network of the configuration, which is not shunted, directly or indirectly, by a resistor.

Figure 3.10 shows a sign-inverting integrator for which, incidentally, no precise passive or non-inverting active counterparts exist. A common use for such a compensation network, or more complex alternatives to it, is for increasing the Type Number of a control system, as discussed in Chapter 14.

Yet another variation, sometimes used for sign-inverting configurations, involves the use of 'five-impedance' networks, as is shown in Fig. 3.11. Usually, each of the blocks denoted by $Z_1(s)\ldots Z_5(s)$ would represent either a single resistor or a single capacitor, but this does not preclude the use of other two-terminal or three-terminal networks in these positions. Techniques such as this enable a complex transfer function to be synthesised using a minimum number of operational amplifiers, although this apparent advantage has tended to become of doubtful value, because of the plummeting prices of IC operational amplifiers.

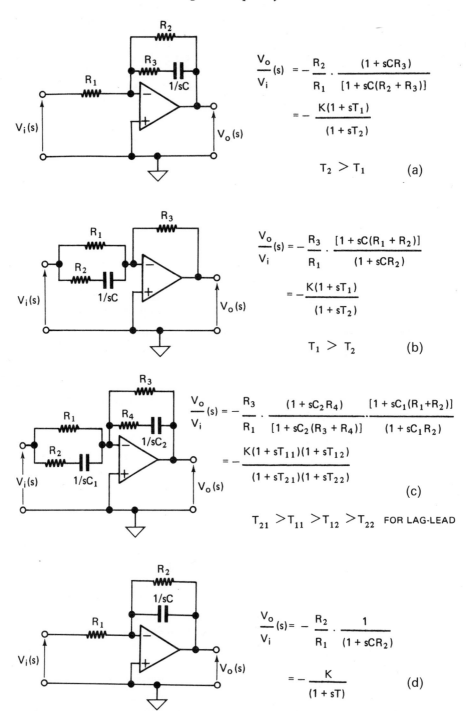

$$\frac{V_o}{V_i}(s) = -\frac{R_2}{R_1} \cdot \frac{(1 + sCR_3)}{[1 + sC(R_2 + R_3)]}$$

$$= -\frac{K(1 + sT_1)}{(1 + sT_2)}$$

$$T_2 > T_1 \qquad \text{(a)}$$

$$\frac{V_o}{V_i}(s) = -\frac{R_3}{R_1} \cdot \frac{[1 + sC(R_1 + R_2)]}{(1 + sCR_2)}$$

$$= -\frac{K(1 + sT_1)}{(1 + sT_2)}$$

$$T_1 > T_2 \qquad \text{(b)}$$

$$\frac{V_o}{V_i}(s) = -\frac{R_3}{R_1} \cdot \frac{(1 + sC_2R_4)}{[1 + sC_2(R_3 + R_4)]} \cdot \frac{[1 + sC_1(R_1 + R_2)]}{(1 + sC_1R_2)}$$

$$= -\frac{K(1 + sT_{11})(1 + sT_{12})}{(1 + sT_{21})(1 + sT_{22})}$$

(c)

$$T_{21} > T_{11} > T_{12} > T_{22} \quad \text{FOR LAG-LEAD}$$

$$\frac{V_o}{V_i}(s) = -\frac{R_2}{R_1} \cdot \frac{1}{(1 + sCR_2)}$$

$$= -\frac{K}{(1 + sT)} \qquad \text{(d)}$$

Fig. 3.9 Representative simple sign-inverting active R–C networks, with their transfer functions: (a) phase-lag; (b) phase-lead (phase-advance); (c) lag-lead (or lead-lag); (d) low-pass

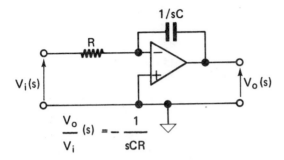

$$\frac{V_o}{V_i}(s) = -\frac{1}{sCR}$$

Fig. 3.10   Sign-inverting active integrator network, with its transfer function

The transfer functions given for all active configurations will be invalidated if the input signal level is sufficiently high for the amplifier output to be driven into saturation, either transiently or in the long term. Configurations which give increasing gain magnitude at high frequency will tend to be saturated transiently by discontinuities or other fast changes in input signal.

For all sign-inverting active configurations, it is advisable to avoid the use of any input network which places a capacitor directly between the configuration input terminal and the virtual earth point (the negative input terminal of the operational amplifier): this is because such a capacitor will present transiently, to the input signal source, the equivalent of a short-circuit to signal common, whenever the input signal changes suddenly.

With all active configurations, the precision with which a particular transfer function is synthesised depends not only upon the precision of the resistance and capacitance values but also on the performance specification of the operational amplifier. Amplifier characteristics which can be of particular concern include:

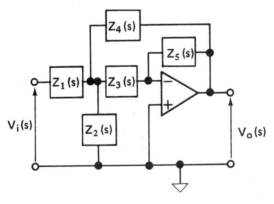

Fig. 3.11   Arrangement of a generalised 'five-impedance' active network

- input bias current
- input offset voltage
- small-signal bandwidth
- output voltage slew rate limit.

The input bias current is especially critical in those configurations which synthesise integration. In those networks not synthesising integration, the input offset voltage can particularly affect the DC behaviour, especially when the DC gain is required to be precise. Bandwidth and slew rate limit are normally significant only in high-frequency applications, which frequently do not concern the control engineer.

3.4   Filter networks for DC power supplies

Normally, DC power supplies for electronic circuits are generated from single-phase AC mains using power diodes configured for half wave, full wave or bridge operation. The output voltage from the diode network takes the form of a DC level, superimposed on which are half sinewaves, and it is the purpose of the filter to reduce the magnitude of this latter, 'ripple', component to an acceptable value. A figure will also be specified for the acceptable tolerance on the value of the DC output voltage, which will vary with changes both in load and in AC supply voltage, the effect being related to the cause by 'load regulation' and 'supply voltage regulation' figures, respectively.

Where close limits are placed upon these regulation figures, it is normal practice to employ a voltage regulator IC between the filter and the load: this regulator will also assist in the attenuation of the ripple component so that, when the regulator is used, less demand is placed upon the filter to provide the necessary attenuation of this component.

Figure 3.12 shows three alternative configurations for voltage regulated DC power supplies: the third arrangement would be the most commonly used, because it does not require a centre-tapped transformer secondary winding, and the full-wave rectified waveform is smoothed much more readily than is the corresponding half-wave version. The voltage regulator behaves as an output-voltage controlled series resistor, with the resistance automatically varied so as to keep V_o matched to the desired value, which normally is generated internally within the IC regulator: this device is thus a small closed loop (voltage) control system in its own right, with filter voltage V_f and load current I_L being wild variables.

Figure 3.13 shows how the output voltage and load current ranges of a given regulator IC may be extended, using an output potential divider for the former case and a power transistor for the latter: the two techniques may be combined, should both ranges require simultaneous extension.

Where the power supply is used for digital circuits, it is the usual practice to connect a high value (mF range) electrolytic capacitor across the power supply output: this serves as a bypass for short duration current pulses,

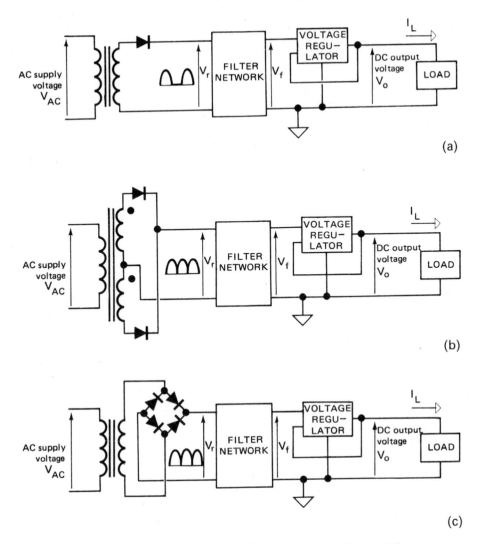

Fig. 3.12   Alternative configurations for voltage regulated DC power supplied: (a) half-wave supply; (b) full-wave supply; (c) full-wave bridge supply

which tend to circulate in the supply bus connections of digital circuits and which otherwise could cause spurious switching conditions to occur. An additional effect of this capacitor will be to cause V_o to decay only slowly after the AC supply has been switched off.

Digital circuits will frequently require the maximum safe value of V_o to be less than the peak value of V_f: one result of this is that, if the voltage regulator should fail to a condition whereby it presents a low series resistance between V_o and V_f, then V_o may consequently jump to a value which could be sufficient to destroy the digital circuits. In this situation, it is

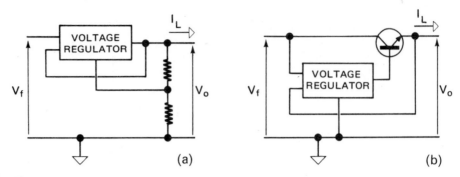

Fig. 3.13   Techniques used to extend the ranges of a given voltage regulator IC: (a) extension of output voltage range; (b) extension of output current range

imperative that protection circuits, such as 'crowbar networks', be provided in order to limit, to a safe level, the maximum value of V_o which can occur under fault conditions.

The design of the filter depends upon:

- the value of the DC component of V_r
- the nature and magnitude of the ripple component of V_r
- the value required for the DC component of V_f
- the value required for the ripple component of V_f
- the frequency of the supply
- the maximum value of load current I_L.

Most filters for this type of application would incorporate a choke, which is a ferrous or ferrite cored inductor, and one or more electrolytic capacitors. The two most commonly used types of network are shown in Figure 3.14, with the first being preferred for applications requiring close voltage regulation.

Choke input filters are typically designed using, for the particular supply frequency, a standard graph of L versus C (each plotted to a logarithmic scale) on which are superimposed contours of constant percentage ripple content in the V_f waveform, and contours of constant load resistance R_L. Capacitor input filters are typically designed in stages: the first stage, which involves C_1, would be designed using, for the particular diode configuration selected, a standard graph of (DC component of V_f)/V_{AC} versus $2\pi f C_1 R_L$ (plotted to linear versus logarithmic scales) on which are superimposed contours of constant value for the function R_L/(average source resistance presented to the filter).

Further stages of filtering may be added to either type of filter, in order to reduce further the ripple content, and typically these would resemble the choke input filter. These stages may be designed either on the basis of transfer functions or by using, for the particular supply frequency, a standard graph of percentage output ripple component versus the LC product, plotted to logarithmic scales, superimposed on which are contours of constant percentage input ripple component. However, the need for these extra stages of filtering will tend to disappear in those applications in which voltage regulators are used.

3.5   Compensation networks for AC–carrier systems

For any analog control system, the design techniques used will yield a conventional transfer function, which the compensator is required to introduce into the control loop. Sections 3.2 and 3.3 have shown how suitable networks can be created to mechanise the transfer function but, unfortunately, these network designs will be directly applicable only to those systems using DC system elements in the forward path of the loop. Whenever AC-carrier types of system element are used in the control loop between the error detection point and the final stages of amplification, the compensation hardware will need to process amplitude-modulated AC-carrier signals, and this requirement will necessitate the application of special techniques.

The problem can be demonstrated by visualising the compensators of equivalent DC and AC-carrier systems being subjected to frequency response testing as shown in Fig. 3.15.

The test signal frequency ω in the DC system becomes the modulation frequency ω_m in the AC-carrier system and the input signal to the latter system becomes a sinusoidally amplitude-modulated sinusoidal carrier: the data are represented by the envelope of the waveform (equivalent to $|A \sin \omega_m t|$) and, relative to the carrier reference supply, the phase of the carrier (representing sgn $(A \sin \omega_m t)$), as described in Section 13.9. For the output signal from the AC-carrier system to be comparable to that from

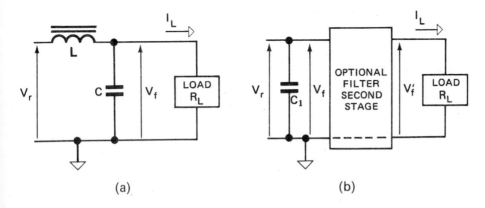

(a)   (b)

Fig. 3.14   Two commonly used power supply filter networks: (a) choke input filter; (b) capacitor input filter

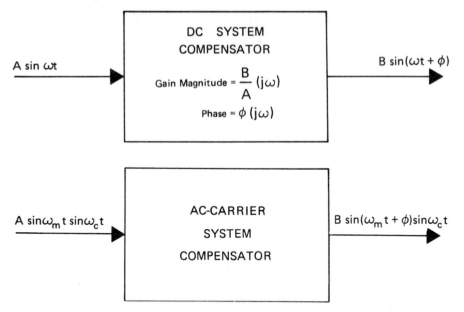

Fig. 3.15 Comparison of DC system and AC-carrier system compensators, each being subjected to frequency response test disturbances

the DC system, the signal envelope should have been modified, by the compensator, to represent $|B \sin(\omega_m t + \phi)|$ and the phase of the carrier should now represent sgn $[B \sin(\omega_m t + \phi)]$. Thus, the carrier at the output should always be either precisely inphase or precisely antiphase with respect to the carrier at the input, assuming the phasing of the input to be exactly correct, relative to carrier phasing at other points within the system.

The input signal to the AC compensator converts to $\frac{1}{2}A \cos(\omega_c - \omega_m)t - \frac{1}{2}A \cos(\omega_c + \omega_m)t$, whilst the output signal from the compensator becomes $\frac{1}{2}B \cos[(\omega_c - \omega_m)t - \phi] - \frac{1}{2}B \cos[(\omega_c + \omega_m)t + \phi]$. Thus, both signals have components at the sideband frequencies $\omega_c \pm \omega_m$.

The AC compensator is required, therefore, to have a gain magnitude B/A and phase shift $-\phi$ at the lower sideband $(\omega_c - \omega_m)$, and a gain magnitude B/A and phase shift $+\phi$ at the upper sideband $(\omega_c + \omega_m)$. If, for example, the DC compensator is required to have a phase-lead characteristic of the type indicated in Fig. 3.5, then the frequency response of the equivalent AC compensator will need to resemble that shown in Fig. 3.16.

It is immediately apparent that the gain magnitude plot must possess even symmetry about the carrier frequency ω_c, whilst the phase plot must have odd symmetry about ω_c; moreover, the symmetry is required to be *arithmetic*, not *logarithmic*.

The effect of any departure from symmetry can be demonstrated, for any particular modulation frequency ω_m, by supposing that

gain magnitude = C and phase = $-\alpha$ at $\omega = (\omega_c - \omega_m)$, and that
gain magnitude = D and phase = $+\beta$ at $\omega = (\omega_c + \omega_m)$.

The compensator output becomes

$$\tfrac{1}{2}AC\cos\left[(\omega_c - \omega_m)t - \alpha\right] - \tfrac{1}{2}AD\cos\left[(\omega_c + \omega_m)t + \beta\right]$$
$$= \tfrac{1}{2}A[\sin\omega_c t \sin\omega_m t(C\cos\alpha + D\cos\beta)$$
$$+ \sin\omega_c t \cos\omega_m t(C\sin\alpha + D\sin\beta)$$
$$+ \cos\omega_c t \sin\omega_m t(-C\sin\alpha + D\sin\beta)$$
$$+ \cos\omega_c t \cos\omega_m t(C\cos\alpha - D\cos\beta)]$$

This can be reduced to an expression of the form

$$E\sin(\omega_m t + \theta)\sin(\omega_c t + \gamma),$$

which is equivalent to the form of the ideal expression

$$B\sin(\omega_m t + \phi)\sin\omega_c t$$

only when C = D (= B/A) and $\alpha = \beta$ (= ϕ). The effects of distortion of
the symmetry of the frequency response of the compensator are therefore
threefold:

- the gain magnitude E is incorrect
- the phase shift θ applied to the modulation is incorrect

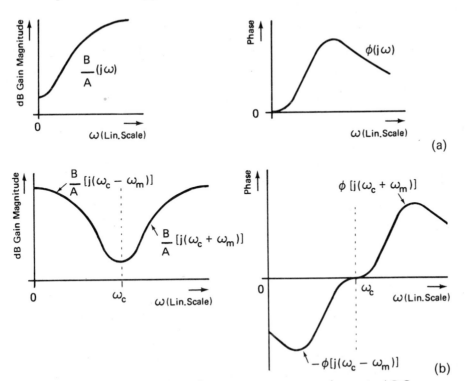

Fig. 3.16 Comparison of the frequency responses of a typical DC
compensator (a) and the equivalent AC-carrier compensator (b)

- the carrier is also phase shifted (γ).

The design of an AC compensator which can approach the desired frequency response may require considerable skill in order to minimise the distortion of the symmetry. The types of network (such as bridged-T and parallel-T) which can achieve the necessary shape of frequency response may need to be very finely tuned, particularly if the control system bandwidth is considerably less than the carrier frequency, which is often the case. Fine tuning implies that the network components must possess long-term stability; it also implies that the method used could be impracticable if the carrier reference frequency ω_c were to experience long-term drift. In any case, precise arithmetic symmetry cannot be obtained for the frequency response of an active or passive R–C network, so that this factor must always remain a source of error, which can be minimised but not eliminated.

One alternative to the use of R–C networks is an approach, sometimes quoted in the literature, which initially involves the design of the equivalent DC compensator. Subsequently, the design of the AC compensator is generated by using the substitutions shown in Fig. 3.17.

In practice, this approach may be problematical, because of the following factors:

- the AC compensator must contain inductors, with all of their attendant problems (see Section 3.1), even when the DC compensator does not contain inductors
- the resulting frequency response possesses logarithmic symmetry
- the technique requires long-term stability for the carrier frequency ω_c.

A third approach to the problem is potentially the most expensive but is likely to introduce a minimum number of difficulties: in fact, it can resolve one, as will be seen. This technique involves the conversion of the modulated AC-carrier input signal to an equivalent DC voltage, the processing of this voltage by the equivalent DC compensator, and (if necessary) the conversion of the resultant DC voltage back into an equivalent modulated AC-carrier output signal, as shown by Fig. 3.18.

This method of compensating AC-carrier systems has the following advantages:

- the operation is insensitive to drift in the carrier frequency ω_c
- the overall frequency response possesses the required arithmetic symmetry
- the compensator components are not required to have unreasonable accuracy or long-term stability specifications
- any small parasitic phase shift ε (see Section 1.5) in the carrier component of the input signal (so that, now, it becomes $A \sin \omega_m t \sin (\omega_c t + \varepsilon)$) is not propagated as a phase shift in the carrier component of the output signal (which now becomes

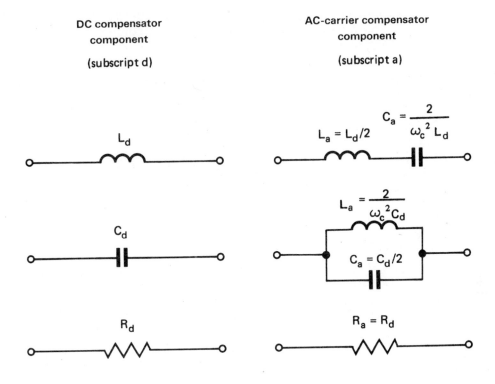

DC compensator
component

(subscript d)

AC-carrier compensator
component

(subscript a)

Fig. 3.17   Substitutions for the development of an AC-carrier compensator (subscript a) using as a basis the equivalent DC compensator (subscript d)

B′sin $(\omega_m t + \phi)$ sin $\omega_c t$, with B′ = B cos ε) but only slightly reduces the overall gain magnitude B′: thus, the ability to eliminate parasitic carrier phase shifts is a bonus arising from the compensation method

- a simple active or passive R–C network could well be adequate for the DC compensator, so that this would easily be designed and implemented

- in some applications, it might become preferable to use DC system elements downstream of the compensation, in which case the modulator would not be required.

A potential problem arises from the transfer function of the low-pass smoothing filter: the attendant time constants should either be small, in comparison to those associated with the DC compensator, or the transfer function terms to which they relate should be included in the design calculations, as part of the compensator transfer function. Moreover, it is desirable that the DC compensator should not be a phase-lead type, because the associated high-frequency amplification could well cancel out the smoothing effect of the low-pass filter, with the result that the regenerated harmonics of ω_c could saturate the DC compensator (if it is the active variety) and/or the modulator.

Because of the complexity of AC compensation design, it is quite commonplace, at least with AC-carrier servosystems, to compromise by using a rate transducer (in other words, an AC tachogenerator) to provide a damping signal in lieu of a compensation network. Problems can still arise, mainly from parasitic phase shifts in the carrier component of the tachogenerator output, with these shifts being variable as a function of shaft speed, in the case of some commercial products. The degree of damping is adjusted readily by attenuating the tachogenerator signal, but the compensation achieved may result in the system dynamic performance being far from optimal, though stable.

3.6   Phaselocked loops

In essence, a 'Phaselocked Loop', at least in its usual locked mode state, is a phase control system. In a control engineering context, these loops may be used as tracking filters but the technology is interesting because it can also be used as the basis of frequency control and synchronisation control systems for alternators and oscillators.

Figure 3.19 shows the principal components of a simple phaselocked loop. The quiescent operating condition occurs when the oscillator is running at frequency ω_o equal in value to the input frequency ω_i which, at this stage, is assumed to be constant. The output waveform from the

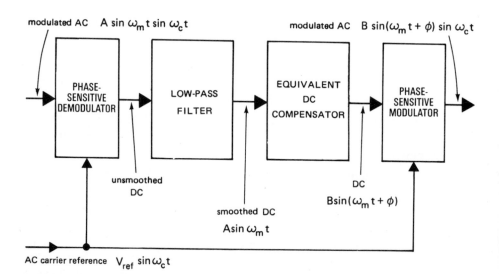

Fig. 3.18   Hardware arrangement typically involved when using a DC compensator to effect compensation of an AC-carrier system, shown being subjected to a frequency response test disturbance

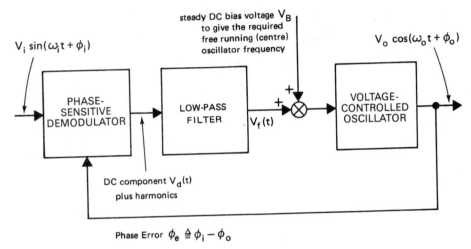

Fig. 3.19 Basic arrangement of system elements in a phaselocked loop, showing the relevant signals in the locked mode

demodulator will contain harmonics of $\omega_o\,(= \omega_i)$, together with a DC component which is proportional to $\sin \phi_e$, and therefore approximately proportional to the phase error (ϕ_e) for small values of ϕ_e. (The cosine law of the PSD has been converted to a sine law, by shifting by 90° the datum from which ϕ_e is measured: this is implicit in the definition of the VCO output as a cosinewave, rather than a sinewave.)

It is assumed that the low-pass filter attenuates to insignificance the harmonic components in the PSD output waveform, leaving a DC component V_f which is related to the DC component V_d of the PSD output by the transfer function of the filter. This signal then augments the steady DC bias voltage V_B, to yield that value of nett DC voltage necessary to make the oscillator run at frequency $\omega_o\,(= \omega_i)$.

Typically, the relevant formulae are:

PSD : $V_d(s) = K_d V_i V_o \sin \phi_e \cong K_d V_i V_o \phi_e = K'_d \phi_e(s), \; K'_d = K_d V_i V_o$

FILTER : $V_f(s) = K_f G_f(s) V_d(s)$

VCO : $\phi_e(s) = \dfrac{K_o}{s}\left[\dfrac{V_B}{s} + V_f(s)\right]$

The Ks are steady-state sensitivities (gain constants) and $K_f G_f(s)$ is the filter transfer function. The formulae can be represented by the small-signal block diagram of Fig. 3.20.

Normally, V_o will have a constant value (for the peak of the oscillator output waveform) but V_i may be a variable, in which case the loop gain will vary in proportion to input signal strength, because of the dependence of K'_d on V_i. This problem can be resolved by preconditioning the input waveform by first converting it to a squarewave having a constant peak-to-peak magnitude.

There are many variations on the basic configuration described here, and these can involve more sophisticated types of demodulator and oscillators generating waveforms which are nonsinusoidal.

Phaselocked loops can be designed to have a very narrow closed loop bandwidth, so that only relatively slow variations in $\phi_i(t)$ are propagated as variations in $\phi_o(t)$. Since frequency is the rate of change in phase, a constant component of input frequency $\omega_i(t)$ will be equivalent to a ramp in input phase $\phi_i(t)$. If $G_f(s)$ is designed to include an integration term, then $\lim_{t \to \infty} \phi_e(t) \to 0$ for a ramping $\phi_i(t)$, as is explained in Chapter 14, in connection with Type 2 systems. In this situation, the output phase $\phi_o(t)$ will also ramp and therefore the output frequency $\omega_o(t)$ will, in the steady state, have a constant component equal in value to the constant component of input frequency $\omega_i(t)$.

To summarise, the oscillator output will tend to have an undistorted waveform synchronised with and having a frequency equal to that of the fundamental component of any periodic input waveform, which may be heavily corrupted with noise. This fundamental property of the loop will be consistent for any value of input frequency ω_i occurring within a specified range: the upper practical limit will be set by the maximum frequency at which the oscillator is designed to run; the lower practical limit will be that for which the residual high-frequency components in $V_f(t)$ are just insufficiently large to drive the oscillator input voltage intermittently negative. This last limitation arises due to the fact that VCOs are designed to respond only to one polarity of input voltage, because negative frequency is meaningless in practical terms.

3.7   Digital compensators

As has been stated in the introduction to this chapter, the use of digital techniques to synthesise compensation transfer functions represents an

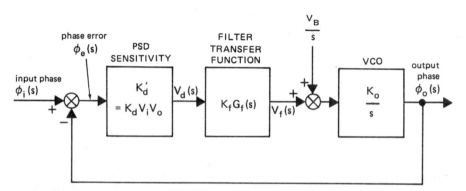

Fig. 3.20   Small-signal block diagram representation of the phaselocked loop of Fig. 3.19 in the locked mode; note that the relevant reference and feedback data are phase angles

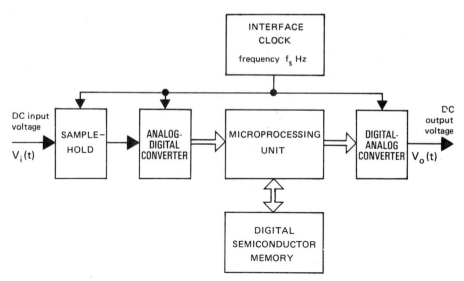

Fig. 3.21 Representation of microprocessor, memory and interface hardware configured as a digital compensator to replace an analog compensator

extremely powerful addition to the range of techniques available to the control engineer. The design of digital filters is a very large topic, so that available space can permit only a brief introduction to the subject.

Figure 3.21 shows how a microprocessor, memory and interface hardware would be configured in order to generate a digital compensator. If the clock frequency f_s Hz is chosen to be several orders of magnitude greater than the bandwidth of the system into which the compensator is to be inserted, then the variations in $V_o(t)$ will resemble a smooth waveform and the digital compensator will resemble an analog compensator. However, such an approach could be inefficient economically, because it could preclude the hardware from being time-shared between a set of control loops, thus losing one of the potential advantages of this type of hardware arrangement. If the sampling frequency is chosen to be significantly lower in value, then the variations in $V_o(t)$ will resemble a staircase waveform, with the steps occurring synchronously with the clock pulses, so that the digital compensator now tends to lose its resemblance to an analog compensator.

Figure 3.22 shows a block diagram to represent mathematically the digital compensation hardware. A new operator z has to be introduced, in order to handle the mathematical processes, and this is defined as $z \triangleq e^{sT}$, where s is the Laplace operator and sampling interval $T = 1/f_s$.

$V_i(z)$ is the Laplace transform of $V_i(t)$ after it has been subjected to the sampling process, with z replacing each e^{sT}. Note that Laplace transforms of pulse trains, or of sequences of numerical values, yield expressions which are functions of e^{sT}. $V_i(z)$ also represents the Laplace transform of the sequence of numerical values which, in coded form, is to be processed by the digital algorithm embodied in the processor program. Similarly, $X(z)$ represents the Laplace transform of the sequence of numerical values which, in coded form, is generated as a result of the application of the digital algorithm. This algorithm may therefore be modelled mathematically by a 'pulse transfer function' $D(z) \triangleq X(z)/V_i(z)$.

The first stage of the design procedure would be to choose a value for the sampling frequency, f_s, bearing in mind:

- the probable bandwidth of the final control system design
- the requirement (if any) for time sharing, and thus multiplexing, of equipment
- the conversion time of the ADC and settling time of the DAC.

One approach for the second design stage would be to devise, using conventional design techniques, an analog compensator $G_c(s) \left(= \dfrac{V_o}{V_i}(s) \right)$ for the control loop, and then to use appropriate transformations to arrive at an expression for an equivalent pulse transfer function $D(z)$: this approach only yields an approximately equivalent function, and the approximation becomes worse as the sampling frequency is lowered towards the bandwidth of the ultimate closed loop system. Note that, for any given $G_c(s)$, there is no physically realisable $D(z)$ which will exhibit a frequency response which is identical to that of $G_c(s)$ at all frequencies: the choice of $D(z)$ therefore must be subject to compromise. Many different transformations have been proposed, some of which are tabulated below in Table 3.1, which assumes that $G_c(s)$ has been expressed as the ratio of two polynomials in the variable $(1/s)$. The variable Δ can be considered to be defined as $\Delta \triangleq z^{-1}$.

In any situation, the choice of the 'best' transformation will depend upon the likely nature of the variations in $V_i(t)$ to be expected. A transformation which is effective for one type of variation (for example, a ramp in $V_i(t)$) may prove to be poor for a second type of variation (for example, a sinusoidal variation in $V_i(t)$).

The alternative approach for the second stage in the design process would be to design the control loop as a complete entity using the z

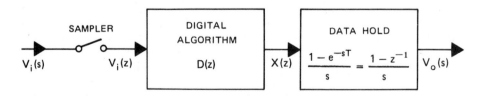

Fig. 3.22 Block diagram providing mathematical representation of the digital compensator of Fig. 3.21

Table 3.1 Alternative transformations from $G_c(s)$ to $D(z)$

Method	Substitution for:			
	$1/s$	$1/s^2$	$1/s^3$	$1/s^4$
First difference	$\dfrac{T}{1-\Delta}$	$\left(\dfrac{T}{1-\Delta}\right)^2$	$\left(\dfrac{T}{1-\Delta}\right)^3$	$\left(\dfrac{T}{1-\Delta}\right)^4$
z Transform	$\dfrac{T}{1-\Delta}$	$T^2\dfrac{\Delta}{(1-\Delta)^2}$	$\dfrac{T^3\,\Delta(1+\Delta)}{2\,(1-\Delta)^3}$	$\dfrac{T^4\,\Delta(1+4\Delta+\Delta^2)}{6\,(1-\Delta)^4}$
Tustin	$\dfrac{T}{2}\dfrac{1+\Delta}{1-\Delta}$	$\left(\dfrac{T}{2}\dfrac{1+\Delta}{1-\Delta}\right)^2$	$\left(\dfrac{T}{2}\dfrac{1+\Delta}{1-\Delta}\right)^3$	$\left(\dfrac{T}{2}\dfrac{1+\Delta}{1-\Delta}\right)^4$
Boxer-Thaler	$\dfrac{T}{2}\dfrac{1+\Delta}{1-\Delta}$	$\dfrac{T^2}{12}\dfrac{1+10\Delta+\Delta^2}{(1-\Delta)^2}$	$\dfrac{T^3}{2}\dfrac{\Delta(1+\Delta)}{(1-\Delta)^3}$	$\dfrac{T^4}{6}\dfrac{\Delta(1+4\Delta+\Delta^2)}{(1-\Delta)^4}-\dfrac{T^4}{720}$
Madwed-Truxal	$\dfrac{T}{2}\dfrac{1+\Delta}{1-\Delta}$	$\dfrac{T^2}{6}\dfrac{1+4\Delta+\Delta^2}{(1-\Delta)^2}$	$\dfrac{T^3}{24}\dfrac{1+11\Delta+11\Delta^2+\Delta^3}{(1-\Delta)^3}$	$\dfrac{T^4}{120}\dfrac{1+26\Delta+66\Delta^2+26\Delta^3+\Delta^4}{(1-\Delta)^4}$

operator as a basis, to optimise that design, and finally to ascertain the expression for $D(z)$ needed to implement that optimisation. This is a much more precise approach to the problem, but it requires the engineer to have a sound working knowledge of sampled data system design techniques.

The third stage of the design process is to convert the pulse transfer function $D(z)$, however obtained, into an equivalent 'finite difference equation', which is a formula relating the sequence of numerical values generated by the digital algorithm to the sequence of numerical values being supplied to the algorithm. This can best be demonstrated by means of an example.

Suppose we require $D(z) \triangleq \dfrac{X(z)}{V_i(z)} = \dfrac{20(z - 0.2)}{(z + 0 \cdot 7)} = \dfrac{20 - 4z^{-1}}{1 + 0.7z^{-1}}$

Cross-multiplying yields $X(z) + 0.7z^{-1}X(z) = 20V_i(z) - 4z^{-1}V_i(z)$

so that $X(z) = 20V_i(z) - 4z^{-1}V_i(z) - 0.7z^{-1}X(z)$

In the time domain, this transforms to a finite difference equation:

$$x_n \quad = 20\ v_{i_n} - 4v_{i_{n-1}} - 0.7x_{n-1}$$

where x_n = current value, of the algorithm output, to be computed

x_{n-1} = previous value, of the algorithm output; previously computed and stored, for use at the current time

v_{i_n} = current value of the input to the algorithm

$v_{i_{n-1}}$ = previous value of the input to the algorithm, stored for use at the current time

The difference equation would then be rewritten as a set of program statements (using machine language, assembly language, or a high-level language such as FORTH, C, or PASCAL) which would be stored in the program area of the semiconductor memory.

The advantages of using digital compensators include the following:

- the wide range of types of algorithm which can be implemented; (there are even some realisable digital algorithms which do not have a physically realisable analog compensator counterpart)
- the ability to incorporate amplitude sensitive algorithms as well, along lines comparable to those described in Section 2.7
- the ability to make the algorithms adaptive, by using program statements to cause the values of the difference equation coefficients to be varied.

The potential disadvantages would be:

- the complexity and cost of the hardware, although, in some installations, this can largely be offset by timesharing the hardware between many control loops

- the deleterious effects arising from the introduction of sampling and quantisation processes into the control system; the effects of sampling can be minimised by maximising the sampling frequency, whilst resolution can be maximised by scaling the input and output DC voltages to use fully the available voltage excursions of the ADC and DAC.

4

Microprocessor Families

4.1 Introduction

The microprocessor, since its early development in the early 1970s, has made rapid progress. The first microprocessor developed was the 4-bit 4004 produced by the Intel Corporation. In twenty short years the development of 32-bit microprocessors is challenging the domain of mini and mainframe computers. The growth in complexity to the current generation of micro-processors has been made possible by advances in Very Large Scale Integration (VLSI), allowing up to 1 million transistors to be packed on to a piece of silicon 25 mm square. Paralleling the development of the microprocessor is the development of those devices which go hand-in-hand to build the complex computers of today: for example, memory, peripheral devices, and storage technology.

4.2 8-Bit microprocessor families

There are two main microprocessor families that have evolved in the 8-bit category: the 6800 family from Motorola, and the 8080 family developed by Intel. The architecture of the two families differs significantly.

4.2.1 6800/6502 family

The Motorola 6800 was first introduced in 1974. It is manufactured using NMOS technology and was the first microprocessor to utilise a single 5 volt power supply. This device is capable of driving up to ten 6800 family support chips (see Section 7.7.3) without buffering. It also has TTL-compatible pins. The 6800 has 72 basic instructions and five addressing modes. Note that not all addressing modes are applicable to each instruction. The result is a total of 197 different operations that may be performed with this device.

The 6800 μP is representative of a family of devices such as the 6801, 6802 and 6805. The difference between these devices lies mainly in the on-chip read-only memory (ROM), serial ports and peripheral interface

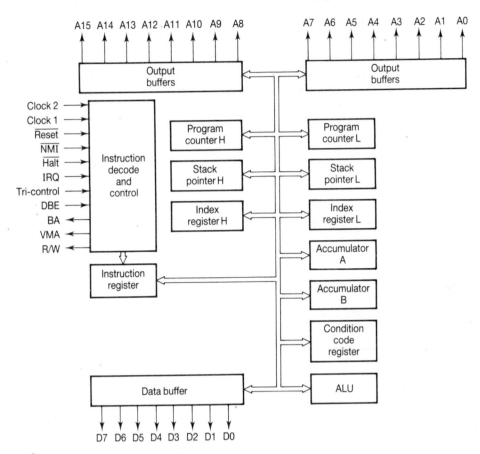

Fig. 4.1   Internal block diagram of the 6800 microprocessor

facilities provided. The instructions and internal architecture are all very similar; thus, in this section only the 6800 μP will be discussed.

The internal architecture of the 6800 μP is best discussed with the aid of Fig. 4.1.

The internal architecture of the 6800 is very similar to that of minicomputers such as the DEC PDP-11. The interface to the outside world consists of three buses: an 8-bit non-multiplexed data bus, a 16-bit address bus providing access to 65 535 address locations, and a control bus.

Internally, there is a 16-bit Program Counter used for holding the next address of the instruction to be executed. A 16-bit Stack Pointer is used to hold the address of the next available location of a last-in first-out stack.

The Index Register is also 16 bits wide and is used in the index addressing mode, where the register is used to store a base address. Operations may then be performed relative to the base address. Index addressing is a very convenient way in which to access tables of information. For example, the address of the start of the table can be stored in the Index Register; each table location is then a known offset from the start.

Two 8-bit Accumulators, known as A and B, are used to store the results of arithmetic instructions such as ADD and SUBTRACT.

The Arithmetic Logic Unit (ALU) is the device that performs all maths functions before they are stored in the accumulators.

Depending on the result of an arithmetic operation, the bits of the 8-bit Condition Code Register (CC) will be set or reset as appropriate. Only six of the eight bits of the CC are used and are defined as:

```
7 6 5 4 3 2 1 0   Bit Numbers
1 1 H I N Z V C
```

- Carry Flag (C) – The carry flag is set to 1 if there is a carry-out of the most-significant bit of an operation; otherwise it is 0.
- Overflow Flag (V) – This flag is set to a 1 if a result exceeds $+ 127$ or $- 128$; otherwise it is 0.
- Zero Flag (Z) – The zero flag is set to a 1 when the result of an operation is zero; otherwise it is 0.
- Negative Flag (N) – If the most significant bit of an operation is a 1 then $N = 1$; otherwise it is 0.
- Interrupt Flag (I) – The setting of this bit enables $(I = 0)$ or disables $(I = 1)$ interrupts.
- Half-Carry Flag (H) – The half-carry flag is set to a 1 if there is a carry from bit 3 to bit 4 of an operation; otherwise is is 0.

The final major components within the μP are the Instruction Decoder and Bus Controller. The Instruction Decoder, as the name suggests, is used to decode the instruction and take the appropriate action internally. The Bus Controller is used to control the following:

- Valid Memory Address (VMA) – When true, a valid memory address is on the address bus.
- Read/Write (R/$\overline{W}$) – Read(1) or Write(0) operation is taking place.
- Interrupt Request ($\overline{IRQ}$) – When low, an external device is requesting to interrupt the current operation. If the I bit is set in the CC, an interrupt will occur; otherwise, no action takes place.
- Reset ($\overline{RESET}$) – When low, the processor will reset and start operation from the address pointed to at address locations $FFFE and $FFFF (specified in hexadecimal format).
- Phase 1/Phase 2 – These are the clock signals used by the 6800 μP. It has a two-phase clock with each phase in quadrature.
- Bus Available (BA) – When high, the bus is available for use by other devices.
- Tri-State Control (TSC) – This input is used to control the tri-state internal data and address buffers.

- Data Bus Enable (DBE) – This input signals to the μP that valid data exists on the data bus.
- Non-Maskable Interrupt ($\overline{\text{NMI}}$) – When low, this input causes the μP to interrupt at the end of the currently executing instruction, regardless of the setting of the I bit in the CC register.

From a programmer's point of view the internals of this μP are represented as shown in Fig. 4.2.

As mentioned previously, there are five addressing modes available with 6800 μP. Register Addressing is used when performing operations on the internal registers of the microprocessor; for example, SBA. This instruction is interpreted as subtracting the content of Accumulator B from the content of Accumulator and storing the result in A. Another example is ADD A 5, where 5 is added to the content of Accumulator A and the result stored in A.

Direct Addressing allows operations to be performed on any of the valid 65535 addresses; for example, ADD B $FF00. This instruction adds the value that is stored in address $FF00 to the content of Accumulator B and stores the result in B. Thus, if the content of B were 02 and the content of $FF00 were 07, the above instruction would leave 09 in the B Accumulator.

Direct Page Addressing is a modified form of Direct Addressing, with the only modification being that the memory addresses accessed can only lie within the first 255 bytes. Since the first 255 adddresses can be represented by an 8-bit hexadecimal number, one byte can be saved in each instruction using this mode. An example is ADD A $10: that is, add the content of address $10 to the content of Accumulator A and store the result in A.

Indexed Addressing makes use of the Index Register X. An example is ADD A 5, X: this instruction is interpreted as 'add the value stored in the memory location, pointed to by adding 5 to the value stored in the X register, to the content of Accumulator A and store the result in A'. If A holds $02 and X holds $C000, the content of A will be added to the content of $C005, say $10: thus, at the end of this instruction A will hold $12.

Immediate Addressing is used to handle constant operands: for example LDX 1934. This instruction loads the X register with 1934 hexadecimal.

The final addressing mode is known as Relative Addressing and is used primarily by branching instructions: that is, those instructions which modify the next instruction to be executed. These instructions are the only ones which modify the Program Counter. An example of this mode is BEQ 12; interpreted, this instruction checks the zero bit of the Condition Code register and, if it is set, causes the program to start executing 12 bytes from the location specified by the current contents of the Program Counter. Thus, this class of instruction is dependent on the setting of the appropriate bits in the CC register by the previous instruction.

Subroutines are sections of programs that can be used over-and-over by different parts of the program, without having to re-enter the same code. An example of the use of a subroutine is a floating-point subroutine where two numbers are passed to the routine and it returns with the answer. Obviously, to re-enter the same microprocessor instructions every time one

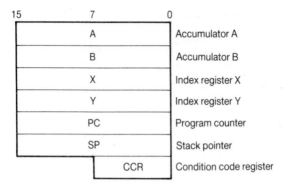

Fig. 4.2   Programming model for the 6800 microprocessor

needed to add together two floating-point numbers would make a program very long indeed. The 6800 μP handles subroutines with the Stack Pointer. When a subroutine is called (the JSR instruction), the position where the program was executing is stored in the two consecutive addresses pointed to by the Stack Pointer, and then the Stack Pointer is decremented by two to point to the next free space on the stack. Note that the stack grows downward in memory. When the subroutine has finished executing (the RTS instruction), the Program Counter is loaded with the address pointed to by the Stack Pointer plus two: that is, the memory address where the program was executing before the subroutine was called. For example,

> JSR $1000; Jump to the subroutine starting at ... address $1000
> LDA 1234
> etc.

at address $1000:
> ADD A B
> etc.
> RTS

The final point to note about the 6800 μP is that it has no separate instructions for accessing peripheral devices: all devices are mapped within the 65 535 valid address locations. Thus, there is no distinction made between memory and peripheral devices such as serial ports, A/D converters, and so on.

The 6502 μP from MOS Technology was introduced after the 6800 and is similar in many respects. The 6502 has 56 basic instructions and 13 different addressing modes to give a total of 151 different instructions. The main difference between the two processors is in the architecture, where the 6502 has only one accumulator but has two index registers.

The 6502 offers the same addressing modes as the 6800, plus extras. The 6502 offers 16-bit offsets from an Index Register, as opposed to the 8-bit

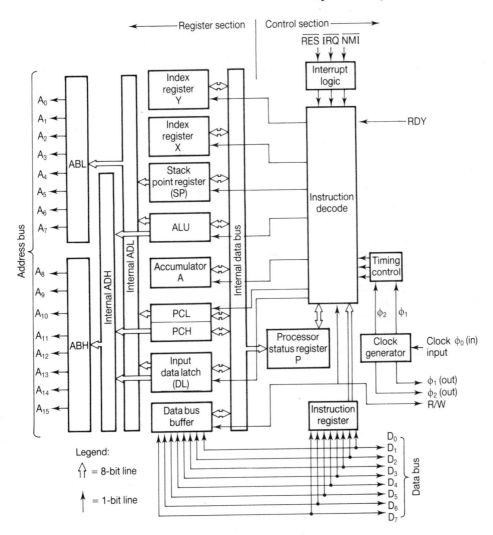

Fig. 4.3 Architecture of the 6502 miocroprocessor

offered by the 6800. The 6502 also offers other forms of indexed addressing which will be discussed shortly. As opposed to the 6800 stack which could be located anywhere within the 64 kb boundary, the 6502 must have its stack located in the first 255 bytes of memory: that is, Page Zero.

In 6502 terminology, the processor uses a Processor Status Register (PSR) to indicate the same information as the Condition Code Register of the 6800. The PSR is organised as follows:

```
7  6  5  4  3  2  1  0   Bit Numbers
N  V     B  D  I  Z  C
```

- Carry Flag (C) – The carry flag is set to 1 if there is a carry-out of the most-significant bit of an operation; otherwise it is 0.

- Zero Flag (Z) – This flag is set to a 1 if an instruction produces a zero result.
- Interrupt Flag (I) – The setting of this bit disables (I = 1) or enables (I = 0) interrupts.
- Decimal Flag (D) – The setting of the decimal flag determines whether the ALU will perform operations using binary numbers (D = 0) or decimal numbers (D = 1). This differs from the 6800, which has special instructions for decimal maths.

The 6502 has a 16-bit address bus, giving it access to 65 535 separate address locations. The address bus lines are TTL-compatible. The data bus is 8 bits wide; these lines are tri-state.

Shown in Fig. 4.4 are the pin-outs for the 6502. The clocking arrangements of the 6502 are similar to those of the 6800 in that a two-phase clock $\emptyset1$ and $\emptyset2$ is used.

The $\overline{\text{RES}}$ line (RESET) is normally at logic 1. When taken low, the processor will perform a start-up procedure consisting of fetching the contents of locations $FFFC and $FFFD, to determine the starting address of the first instruction.

The SYNC (SYNCHRONISE) output is used to signal when it is fetching op-codes from memory.

The SO (SET OVERFLOW) input is used to set or clear the V flag of the PSR.

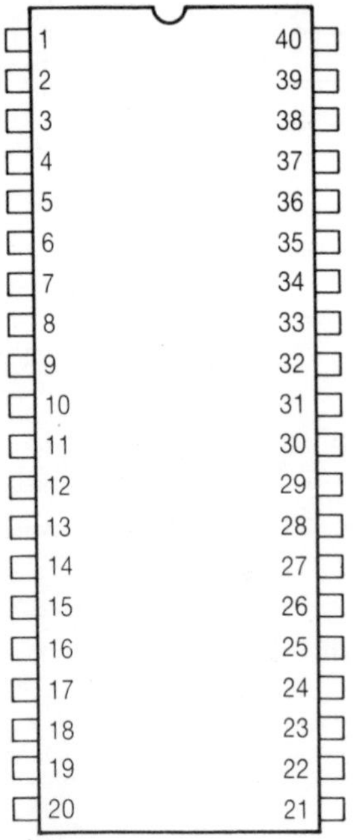

Fig. 4.4   Pin-outs of the 6502 microprocessor

The NMI (NON-MASKABLE INTERRUPT) input will interrupt the processor at the end of the current instruction and save the content of the PSR and program counter on the stack. The contents of $FFFA and $FFFB will then be fetched, to determine the address of an interrupt service routine. The setting of the I flag has no effect on the NMI status. Note however that, after setting the NMI line low, the I flag is set to a logic 1 to disable interrupts from the IRQ line. An RTI (RETURN FROM INTERRUPT) restores the status of the processor to the state it was in before the interrupt was received.

The IRQ (INTERRUPT REQUEST) line will perform an action similar to that produced by an NMI, except that recognition of an interrupt is dependent on the setting of the I flag. The address at which the interrupt service routine is located is held in locations $FFFE and $FFFF.

The R/W line is the same as for the 6800.

The programmer's model of the 6502 is shown in Fig. 4.5.

The addressing modes available in the 6502 are:

- Absolute or Direct
- Zero Page
- Immediate
- Indexed Absolute
- Indexed Zero Page
- Implied
- Indirect
- Indirect Indexed
- Indexed Indirect
- Relative
- Accumulator.

Only those modes that differ from those of the 6800 will be discussed.

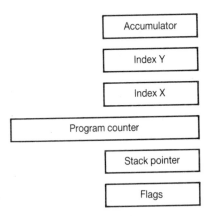

Fig. 4.5   Programmer's model of the 6502 microprocessor

In Indirect addressing, the address following the op-code is used to calculate the adddress from which the data is finally fetched (the effective address): for example, LDA [$1234]. The accumulator is loaded with the byte stored in the address location pointed to by the content of addresses $1234 and $1235.

The Indexed Indirect mode differs from the above only by the fact that the final address will always be in Page Zero. This mode can only use the X index register.

The Indirect Indexed mode can only use the Y index register. In this mode, the indirect address is used to determine a base address in Page Zero, which, when added to the content of the Y register, yields the final address.

4.2.2   8080 device

The 8080 was Intel's entry device into the 8-bit microprocessor market. It was the predecessor to the highly successful 16-bit 8088/8086 devices, which have become the most commonly installed microprocessors in use today. This is probably due to the success of IBM PC and compatible personal computers. These two devices are described in detail in Section 4.3.2.

4.2.3   Z80 family

The Zilog Z80 was introduced to compete directly with microprocessors like the 6800 and the 6502. It was a further development of the 8080 architecture pioneered by Intel, and thus is predominantly an 8-bit device. Internally this microprocessor appears as shown in Fig. 4.6.

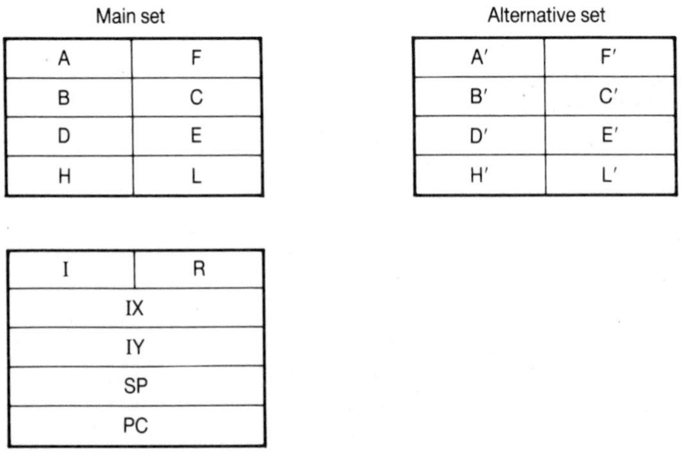

Main set	
A	F
B	C
D	E
H	L

Alternative set	
A′	F′
B′	C′
D′	E′
H′	L′

I	R
IX	
IY	
SP	
PC	

Fig. 4.6   Internal organisation of the Z80 microprocessor

The A register is the traditional accumulator register and is used for the majority of operations with this device. The F register holds the 'flags'. There are six bits to this flag register, defined as:

C   Z   P/V   S   N   H

where

C   = Carry flag
Z   = Zero flag
P/V = Parity or Overflow flag. Logical operations affect the Parity
      flag, whilst arithmetic operations affect the Overflow flag
S   = Sign flag
N   = Add/Subtract flag
H   = Half-Carry flag

The B, C, D, E, F, H and L registers form a bank of seven general-purpose registers. Note that there also exists an alternative set of registers to those described thus far. Only one set can be active at any one time.

The final group of registers is:

- R: The R register is known as the Memory-Refresh register and is used in conjunction with dynamic refresh memory (see Section 4.6.2).
- I: The I register is known as the Interrupt Page Address register and is used to provide the high-order address of an indirect call to memory, after an interrupt. This register is only used in special cases.
- IX, IY: These are two sets of index registers, used to support indexed addressing.
- SP: Stack Pointer
- PC: Program Counter.

The Z80 supports the following addressing modes:

- Implicit addressing, also known as Implied or Register addressing.
- Immediate addressing
- Absolute addressing
- Direct addressing
- Relative addressing
- Indexed addressing
- Indirect addressing.

4.3   16-Bit microprocessor families

The 16-bit microprocessors were a natural evolution from the 8-bit microprocessors, which served to demonstrate the power and versatility of these devices. As manufacturing techniques improved, designers were able to offer more power on a single chip. This section deals with the next stage in their evolution.

4.3.1   68000 device

The 68000 μP from Motorola is an NMOS microprocessor that has a 16-bit external data path, but internally the data path is predominantly thirty-two bits wide. The 68000 has sixty-one basic instructions and, when coupled with the fourteen addressing modes, provides the programmer with over one thousand different instructions. The major difference between this processor and earlier 8-bit offerings from Motorola, apart from the data path, is that no longer is there a separate accumulator. Instead, a variety of general-purpose registers, any of which may be used as an accumulator, is provided. In addition, the instruction set and addressing modes are more orthogonal: that is, almost all of the available instructions are applicable to all registers.

The internal architecture of the 68000 is shown in Fig. 4.7.

The 68000 has a 16-bit non-multiplexed data bus, and a 23-bit address bus providing access to 16 megabytes of memory. As with the 6800, the 68000 treats all input/output as if it were part of memory: thus no special I/O instructions are required.

As mentioned previously, the major difference from earlier 8-bit microprocessors is that no longer are there dedicated accumulators or index registers. Instead, eight general-purpose registers, each of which is 32 bits wide, are provided. The 68000 handles four basic types of data:

- 4-bit BCD nibbles, packed 8 per register
- 8-bit bytes, packed 4 per register
- 16-bit words, packed 2 per register
- 32-bit double-word, one per register.

As well as the versatile group of data registers, the processor also provides a further eight address registers. Register 7 of this set is, in fact, two sets of 32-bit registers implemented as stack pointers, although only one can ever be in operation at any one time. A 16-bit status register is provided, with each bit defined as shown in Fig. 4.8.

Of these flags only, T, S and I are new. The definitions of these bits are as follows:

- I – The 68000 provides eight prioritised interrupt levels. The settings of these bits determine which levels will be acted upon.
- S – The setting of this bit indicates that the processor is operating in Supervisor mode. This mode is typically used by multi-tasking/multi-user operating systems, to prevent one program from overwriting another's memory space.
- T – This bit determines if the processor is operating in Trace mode: that is, executing a single operation at a time. This is useful in debugging both application hardware and software.

External control/input to the microprocessor consists of six buses, apart from the data and address buses. The Processor Status bus consists of three

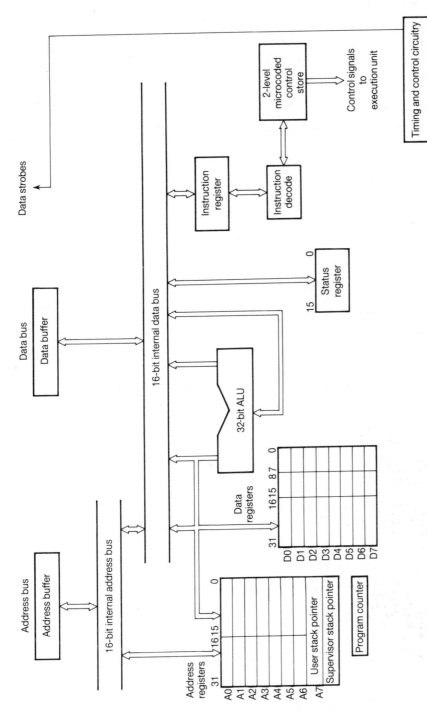

Fig. 4.7   Internal architecture of the 68000 microprocessor

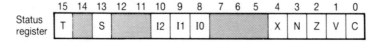

Fig. 4.8   68000 status register format

output lines: FC0, FC1, and FC2. The Peripheral Control bus consists of two output lines – E (Enable) and VMA (Valid Memory Address), and one input line – VPA (Valid Peripheral Address), to control external device access to the address and data buses.

The System Control bus consists of three input lines: BERR (Bus Error Input), RESET and HALT. As mentioned previously, the 68000 provides eight interrupt levels: thus, there are three interrupt lines.

The Bus Arbitration bus is used to control contention on the Address bus. It consists of two input lines – BR (Bus Request), BGACK (Bus Grant Acknowledge), and one output line – BG (Bus Grant).

The Final Control bus is used for placement of data on the Data bus. It consists of four output lines – AS (Address Strobe), R/W (Read/Write), UDS and LDS (Data Strobe Output). The one input line consists of DTACK (Data Transfer Acknowledge).

It is beyond the scope of this work to detail the intricate operation of the above control signals, and the interested reader is referred to the appropriate Motorola Data Book.

Fourteen addressing modes are provided in this microprocessor:

- Data Register Direct
- Address Register Direct
- Address Register Indirect
- Indirect Postincremented
- Indirect Predecremented
- Indirect with Displacement
- Indirect with Index
- Absolute Short Address
- Absolute Long Address
- Program Counter with Displacement
- Program Counter with Index
- Immediate
- Condition Code
- Implied Reference.

4.3.2   8088 and 8086 devices

The 8088 and 8086 were Intel's entry devices into the 16-bit microprocessor market. The only difference between the 8086 and the 8088 microprocessors is that the latter has an 8-bit external databus, whereas the 8086 has a 16-bit external databus.

The standard features of this microprocessor are:

- 1 048 576 bytes (1 Mb) of addressable memory
- 16-bit internal registers
- segmented architecture.

The internal registers as seen by the programmer of this microprocessor are as shown in Fig. 4.9.

There is a total of fourteen 16-bit registers, which can be logically divided into four distinct functions: the Data registers, Pointer and Index registers, Segment registers and the Instruction Pointer and Flags register.

There are four data registers – AX, BX, CX, and DX – each of which can be accessed in terms of either bytes or words, thus giving rise to the nomenclature of AH for the upper byte and AL for the lower byte. The data registers behave almost identically for each instruction; however, there are some differences. The AX register is generally used for the accumulator, although all of the other registers can serve equally well. The BX register has special application in addressing memory – hence its name of Base register. The CX register is often used to hold the value of a loop variable –

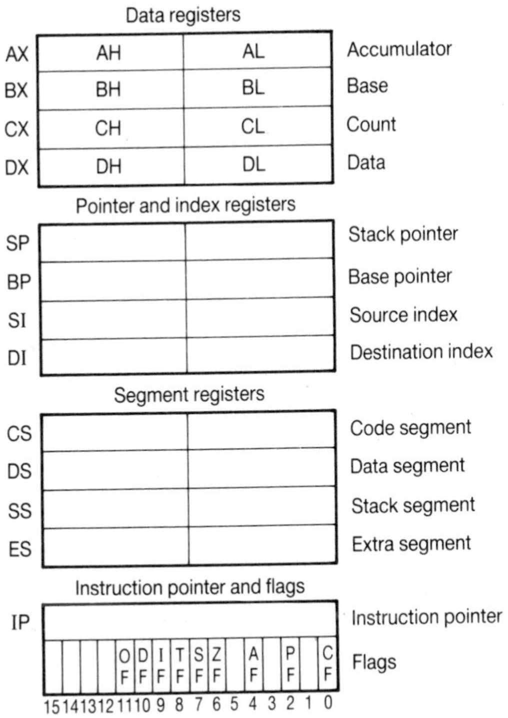

Data registers

AX	AH	AL	Accumulator
BX	BH	BL	Base
CX	CH	CL	Count
DX	DH	DL	Data

Pointer and index registers

SP		Stack pointer
BP		Base pointer
SI		Source index
DI		Destination index

Segment registers

CS		Code segment
DS		Data segment
SS		Stack segment
ES		Extra segment

Instruction pointer and flags

IP	Instruction pointer

| O F | D F | I F | T F | S F | Z F | | A F | | P F | | C F | Flags |

15 14 13 12 11 10 9 8 7 6 5 4 3 2 1 0

Fig. 4.9   8088/8086 register set

hence the name Count register. The DX register's special instructions refer to I/O operations.

The Pointer and Index register set also contains four 16-bit registers; however, these registers cannot be used for byte components. This register set is used to hold the Stack Pointer, the Base Pointer, Source Index, and Destination Index. The last two are especially useful in moving large amounts of information quickly. The first two are used to implement the various addressing schemes previously discussed.

The Segment register set also consists of four 16-bit registers and is used to implement the segmented architecture referred to earlier; this architecture is a methodology adopted by Intel to increase the amount of addressable memory. Most other manufactures implement a 'flat' addressing scheme: that is, the Instruction Pointer is as long as is required to address all memory. Note that all registers within the 8088/8086 are 16-bit: thus, the theoretical maximum amount of memory that can be addressed using a flat scheme is only 64 Kb or 2^{16}. The 8088/8086, however, is capable of addressing 1 Mb of memory. To achieve this result, the Segment Registers and Instruction Pointer are used in the manner shown in Figure 4.10 to provide 1 Mb addressing.

Thus, the Segment registers are used to map four distinct 64 kb areas of memory within the 1 Mb of addressable memory. The Code or Program Segment register is used to point to that area of memory containing the executable program. The area within memory is found by combining the CS register with the IP register, as indicated above, to determine the actual address; for example:

CS = $C800   IP = $0100
memory location = $C8000 + $0100 = $C8100

The Data Segment is used to point, as the name suggests, to the data area of a program. This area may or may not be the same as the Code Segment. Similarly, the SS segment is used to locate the stack anywhere within the

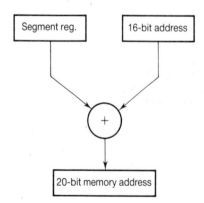

Fig. 4.10   Segmentation principles

1 Mb address space. The Extra Segment register, ES, is used primarily for certain 'string primitive' operations.

The use of segment registers allows the programmer to break the 64 Kb barrier, at the expense of increased complexity in the memory addressing scheme. In most cases, however, the Segment registers can be ignored because the program size will not exceed 64 Kb.

The final two 16-bit registers are the familiar Instruction Pointer (IP) and Flags registers.

There are twenty-four different addressing modes available to the programmer of this microprocessor. Thankfully, most of these modes are simply variations of modes already discussed.

The microprocessor has forty pins for interfacing to the outside world. Of these, twenty are address pins, A0 through to A19. Note, however, that A0 to A7 have a dual function: firstly as the address lines and secondly as the data lines. The data and address information is multiplexed together. Similarly, A16 to A19 operate as the high-nibble of the address bus; they are also used as the Status lines. Control of memory and the I/O ports is achieved via the control bus, S0 to S2. These three lines interface directly with a peripheral chip known as the 8288 Bus Controller. The basic configuration of an 8088-based computer is shown in Fig. 4.11.

4.3.3   TMS9900 family

The TMS9900 was one of the first 16-bit μP devices delivered to the marketplace. Manufactured by Texas Instruments, it is fabricated using NMOS technology. There are 64 basic instructions provided with this μP. The one striking feature of this device is that its working registers are placed off the processor chip itself in normal read–write memory; thus, its main use has been in signal processing applications, where its flexible architecture has been used to great effect. However, the use of this processor is decreasing, as specialised and far more powerful μP and signal processing chips are delivered.

A diagram of the internal architecture of this device is shown in Fig. 4.12.

Only five internal registers are provided with this processor. The Workspace register (WSP) points to sixteen contiguous 16-bit memory locations. These memory locations then function as general-purpose registers which can be used to provide accumulators or index registers as required. It is usual to reserve the last three words of this group for storage of the Program Counter (PC), WSP register, and Status register (STATUS) during subroutine operations.

A 16-bit program counter is provided, giving access to 32 Kb words of program memory. Note that a word is 16 bits long. This small amount of memory (by today's standards) is one reason why this μP has not achieved great popularity.

The Status register (STATUS) is also 16 bits long, providing the usual array of Condition and Interrupt Control flags.

The Communications Register Unit (CRU) is used to provide an on-chip serial communications capability.

4.3.4   NS16000 family

This family consists of three microprocessors, basically differing only in the size of the data bus. The three members are NS16008, NS16016 and NS16032.

This family of microprocessors is manufactured by National Semi-conductor, and is meant to compete for the segment of the market ranging from low-end embedded controllers (that is, against the 6800) up to the high-end workstation market (for example, against the 68000).

The instruction set provides over one hundred basic instructions and nine addressing modes.

Since each member of this family is basically the same, only the NS16032 will be covered from this point on. The internal view of this microprocessor, as far as the programmer is concerned, is as shown in Fig. 4.13.

As with other modern microprocessors, there is no dedicated accumulator. Instead, a group of eight general-purpose 32-bit registers is used for

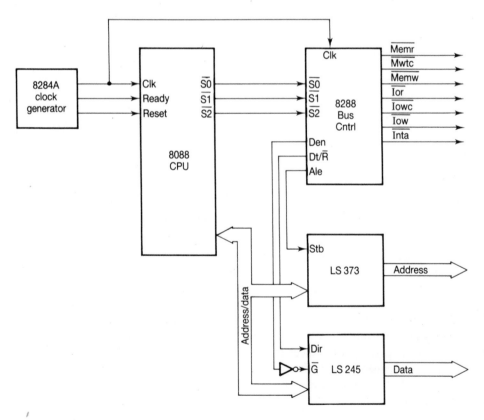

Fig. 4.11   8088-based computer

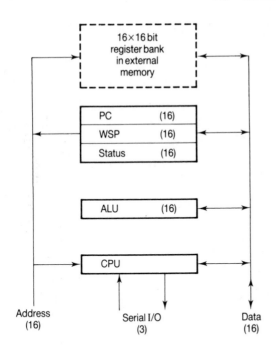

Fig. 4.12   Internal architecture of the TMS 9900 microprocessor

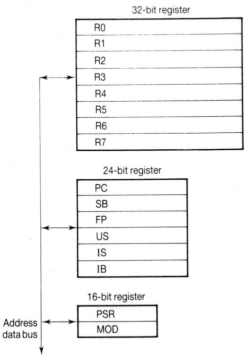

Fig. 4.13   Internal register structure of the NS16032 microprocessor

data handling. As well as this group of data registers, there are also seven 24-bit address registers used as follows:

- PC – Program Counter
- SB – Static Base register, used to point at an area in data memory
- FP – Frame Pointer register, which points to a stack containing the parameters of the currently-executing subroutine
- US – User Stack register, pointing to the user stack
- IS – Interrupt Stack register, used when the processor is operating in Supervisor mode. Either the IS or the US register will be active at any one time, but not both
- IB – Interrupt Base register, which points to a table of interrupt vectors.

Finally there are two 16-bit registers which are used for:

- PSR – Program Status register, also known as the Condition Code register on other microprocessors
- MOD – Module register, pointing to the currently executing segment of program.

4.3.5   Z8000 device

The Z8000 from Zilog is fabricated in NMOS and is typical of the 16-bit microprocessors produced during its day, in that its internal architecture closely mimics that found on minicomputers. This processor provides 110 basic instruction types which, when combined with the various addressing modes, provide over 400 different instructions.

Like other μPs discussed in this section, it has registers dedicated to act as accumulators: it uses a bank of sixteen general-purpose 16-bit registers, any of which may be used as an accumulator. A programmer's view of the internal registers is shown in Fig. 4.14.

The first eight registers, designated R0 through R7, may be used as upper and lower halves, providing access to 8-bit data types. Similarly, R0/R1, R2/R3, R4/R5, R6/R7 may be combined to provide access to 32-bit data types. In addition, R0/R1/R2/R3 and R4/R5/R6/R7 provide access to 64-bit data types. Registers R14 and R15 are used to provide access to a segmented stack structure. Note that Fig. 4.14 indicates two sets of R14 and R15: a User Stack Pointer and a System Stack Pointer. Only one set can be active at any one time, depending on the mode in which the processor is currently operating.

The status and control of operations in the processor is determined by a 16-bit status register known as the Flag Control register. A Program Counter is provided by way of a segment address as well as the applicable offset address. Note that two versions of the Z8000 μP are available. Firstly there is a segmented version providing access to 16 Mb of memory, and secondly there is the non-segmented version providing access to 64 Kb of

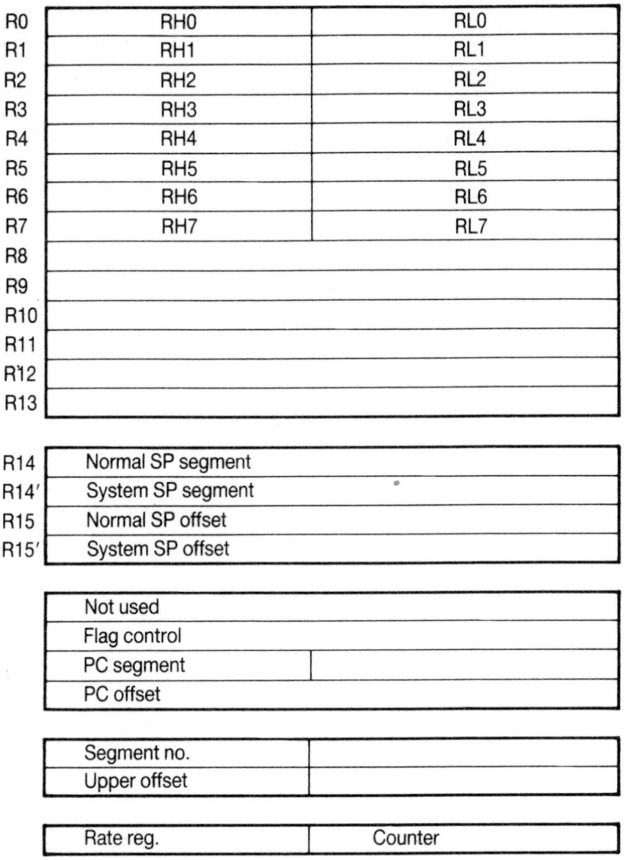

R0	RH0	RL0
R1	RH1	RL1
R2	RH2	RL2
R3	RH3	RL3
R4	RH4	RL4
R5	RH5	RL5
R6	RH6	RL6
R7	RH7	RL7
R8		
R9		
R10		
R11		
R12		
R13		

R14	Normal SP segment
R14'	System SP segment
R15	Normal SP offset
R15'	System SP offset

Not used
Flag control
PC segment
PC offset

Segment no.
Upper offset

Rate reg.	Counter

Fig. 4.14   Z8000 register architecture

memory. Clearly, the latter version is primarily for embedded control applications.

Of the remaining three registers, the two titled Segment No. and Upper Offset are used to point to an area of memory which contains vectors for interrupt processing. The processor supports three types of interrupts – non-maskable, vectored or non-vectored – depending on the appropriate input line status.

The remaining register is used to control the refresh rate of dynamic random-access memory (DRAM).

The addressing modes provided by this μP are as follows:

- Register
- Indirect Register
- Direct
- Immediate
- Indexed

- Relative
- Base Address
- Base Indexed.

4.4   32-Bit microprocessor families

As with the change from eight bits to sixteen bits, the 32-bit microprocessor represents a quantum leap in processing power, so much so that modern 32-bit microprocessors offer as much power as some minicomputer configurations.

4.4.1   68020 device

Like the 80386, the 68020 is the first of the 32-bit microprocessors developed by Motorola, offering full 32-bit data and address paths. It is object code compatible with Motorola's earlier 68000 family. The programming model for the 68020 is shown in Fig. 4.15.

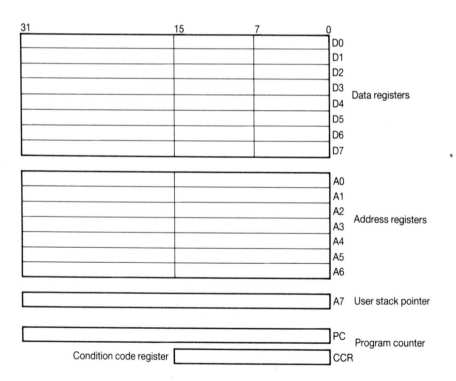

Fig. 4.15   68020 programmer's model

The eight data registers (D0 to D7) are used for general computational work on the following seven basic data types:

- bits
- bit fields
- BCD digits
- byte integer (8 bits)
- word integer (16 bits)
- long word integer (32 bits)
- quad word integer (64 bits).

The eight address registers (A7 is the User Stack Pointer) are used as pointers, base and index registers. The Program Counter and Condition Code Register are conventional in their operation.

As with other microprocessors in the 68000 family, the 68020 operates in either a User or a Supervisor mode. The User mode is for use by application programs, whereas Supervisor mode is for use by operating systems.

The Condition Code Register, while described as 'conventional' in nature, actually consists of two bytes known as the 'system byte' and the 'user byte'. The user byte corresponds to the content of the conventional Condition Code Register, giving indication of status such as negative, zero, overflow, etc. The system byte is only accessible when operating in the Supervisor mode.

The addressing modes offered by this microprocessor are extensive, with too many to provide details here. However, the modes fall into eight broad categories:

- Register Direct
- Register Indirect
- Register Indirect with Index
- Memory Indirect
- Program Counter Indirect with Displacement
- Program Counter Indirect with Index
- Absolute
- Immediate.

4.4.2   68030 Device

The 68030 is an enhancement of the previous Motorola 32-bit offering, the 68020. The 68030 is directly code-compatible with the 68020 and the earlier MC68000 microprocessor. Major improvements have occurred in this chip, one of which is the use of the internal non-multiplexed address and data buses. The use of 256-byte instruction and data caches (which may be

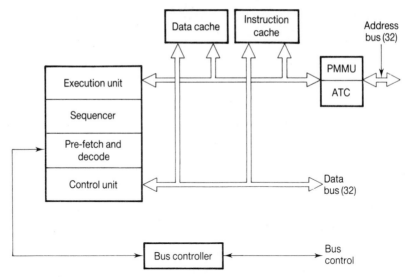

Fig. 4.16   68030 Block Diagram

accessed simultaneously) has boosted performance, by improving the flow into the execution unit. A block diagram of the 68030 is shown in Fig. 4.16.

The major features of the 68030, according to Motorola, are summarised as follows:

- 32-bit non-multiplexed address and data buses
- 16 × 32-bit general-purpose registers
- 2 × 32-bit supervisor stack pointers
- 10 special-purpose control registers
- paged memory-management unit (PMMU)
- enhanced bus controller capable of operating in a variety of modes
- dynamic bus sizing to support 8/16/32-bit memories and peripherals
- 4 Gbyte of addressable memory.

Clearly, the 68030 has been designed to operate as the CPU of multitasking systems. This is further evidenced by the programming model for this microprocessor. As with other advanced devices, the programming model consists of two parts: the supervisor's model and the user's model. It is beyond the scope of this volume to provide a full explanation of the supervisor's model; however, for information purposes the model is depicted graphically in Fig. 4.17.

The user's programming model for the 68030 is identical to that for the 68020, as expected. For completeness, the addressing modes are provided as follows:

- Register Direct

- Register Indirect
- Register Indirect with Index
- Memory Indirect
- Program Counter Indirect with Displacement
- Program Counter Indirect with Index
- Program Counter Memory Indirect
- Absolute
- Immediate.

4.4.3   68040 device

The 68040 is Motorola's latest offering in the 680xx family. This device contains over 1.2 million transistors and is one hundred percent code compatible with other members of the family. The 68040 is claimed to be, on average, three times faster than its predecessor, the 68030. Like the 68030, the 68040 has separate cache for data and instructions, and each cache is 4 kilobytes in size. Bus snooping – required in systems with direct memory access or multiple processors – in order to provide cache coherency is implemented on-chip. This is a major improvement over the 68030, in which the data cache has to be disabled in systems which share data.

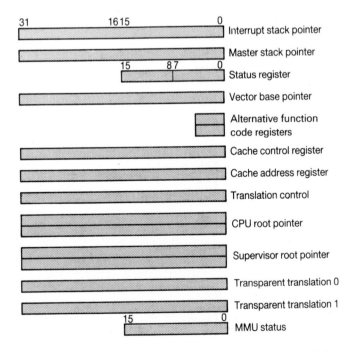

Fig. 4.17   68030 Supervisor's Programming Model

Also implemented on-chip is the floating-point unit (FPU), which is compatible with the 68882 FPU with one exception: the trigonometric functions are not supported directly and must be implemented in software.

4.4.4 80386 device

The 80386 microprocessor is the third in the 80×86 series from Intel, offering full 32-bit wide address and data paths and the capability of running the earlier 8086 and 80286 programs. The introduction of this device has seen the distinction between microprocessors and minicomputers slowly vanishing.

The programmer's model of the 80386 is shown in Fig. 4.18.

This processor offers eight general-purpose registers (EAX through to EDX, ESI, EDI, EBP, and ESP) for general computational work as well as a Stack Index (ESI), Data Index (EDI), Base Pointer (EBP) and Stack Pointer (ESP). The six segment registers allow access to 64 terabytes (trillion bytes) of physical memory. It is possible to have six separate logical address spaces operating at any one time. The Flag and IP registers are conventional in their operation.

There are four inherent data types supported by this device:

- ordinal

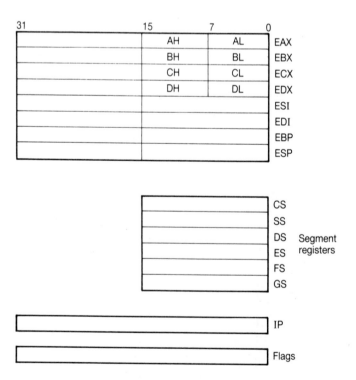

Fig. 4.18   80386 programmer's model

- integer
- string
- bit string.

As well as the above, the Intel maths co-processors, the 80287 and 80387, are also fully supported to provide full access to a wide range of floating-point mathematical functions.

The 80386 has been designed for use in high-end personal computers, network file servers and engineering workstations. As such, the operating systems required to support these functions must be sophisticated. To facilitate faster and simpler operating systems for these environments, the designers of this microprocessor have incorporated many features which greatly ease and speed up development of these applications. Examples of some of these features are:

- hardware-enforced segment/page/task isolation
- fast hardware task-switching
- multiple execution and addressing environment.

The 80386 is a very complex device, and further discussion into its capabilities is beyond the scope of this work. Suffice it to say, however, that the features offered by this chip, and others offering 32-bit processingh capability, represent the direction in which high-end microprocessor development is moving.

4.4.5   80486 device

The 80486 is Intel's latest offering in the 80×86 series. The device contains 1.18 million transistors and is implemented using 1 micron CMOS technology. It is designed to be one hundred percent code compatible with the 80×86 series.

Contained on the chip are:

- integer unit
- floating-point unit (compatible with the 80387 co-processor)
- data and instruction cache (8 kilobytes total)
- virtual memory management unit.

Performance of the 80486 is claimed to be two to four times that of the 80386. Note that first releases of the chip will have a clock speed of 25 MHz: future releases are stated to be 33 MHz and 40 MHz.

There are no significant architecture changes between the 80386 and 80486, with the exception of additional instructions to provide support for the cache and multiprocessor functions.

4.4.6   NS32000 family

The NS32000 family is National Semiconductor's response to the market-place requirement for 32-bit microprocessors. It is typical of the 32-bit

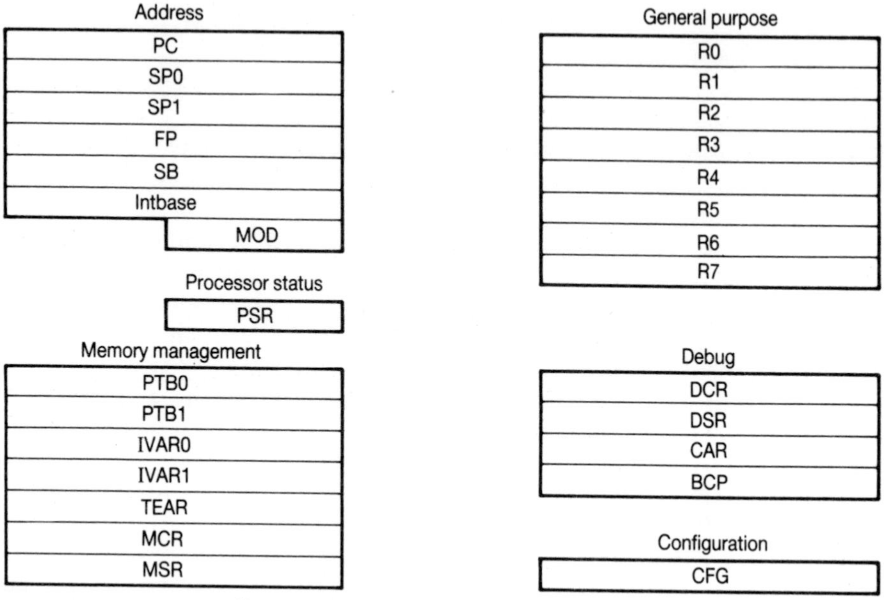

Address	
PC	
SP0	
SP1	
FP	
SB	
Intbase	
	MOD

General purpose
R0
R1
R2
R3
R4
R5
R6
R7

Processor status
PSR

Memory management
PTB0
PTB1
IVAR0
IVAR1
TEAR
MCR
MSR

Debug
DCR
DSR
CAR
BCP

Configuration
CFG

Fig. 4.19   Programmer's model of the NS32532

processors available from other suppliers in that its main application lies in high-end workstations where large amounts of computing power are required.

Features of the NS32000 family include:

- 32-bit architecture
- demand paged virtual memory
- fast floating-point capability
- high-level language support
- symmetrical architecture.

A representative member of this family is the NS32532, available in 20, 25 and 30 MHz versions. This device features more than 370 000 transistors using 1.25 μm double metal CMOS fabrication technology. According to National Semiconductor, this microprocessor is suitable for applications ranging from realtime controllers to high-end computer workstations.

The programmer's view of this microprocessor is as shown in Fig. 4.19.

The NS32532 has 28 internal registers grouped as follows:

- 8 general-purpose registers
- 7 address registers
- 1 processor status registers
- 1 configuration register
- 7 memory management registers
- 4 debug register.

All registers are 32 bits wide except for the Module and Processor Status registers which are 16 bits wide.

The general-purpose registers are used for high-speed temporary storage and are free for any use by the programmer. Even though these registers are 32 bits wide, they may be addressed as either 32, 16 or 8-bit components.

The Address registers serve to implement:

- Program Counter – Pointing to the first byte of the instruction currently being executed
- Stack Pointers – SPO is normally used by operating systems and is used primarily to hold return addresses for interrupt service routines. The SP1 register is used to point to the last item on the 'User Stack'. The User Stack is used by user programs as opposed to operating system programs.
- Frame Pointer – Used by procedures to access parameters and variables on the stack.
- Static Base – Points to the global variables of a software module. This register is used to implement relocatable global variables.
- Interrupt Base – This register holds the address of the Interupt table.
- Module Register – Holds the address of the module descriptor of the currently executing software module.

The Processor Status Register is 16 bits wide and can be divided in two. The lower 8 bits are accessible to all programs and hold the 'normal' status information. The upper 8 bits are accessible only to those programs operating with Supervisor privileges and contain 'system' information.

The Configuration Register is 32 bits wide; however, only nine of the thirty-two bits are implemented. The settings of these bits determine the operating mode of the CPU.

The Memory Management Registers provide much of the power provided by 32-bit processors. It is beyond the scope of this work to provide details on the functions of these registers; suffice to say that these registers implement the complex memory management required to support multi-user, multi-tasking operating systems.

4.5 Co-processors

'Co-processor' is a term for describing what is essentially a specialised microprocessor. Its prime function is to relieve a 'host' microprocessor from having to execute a particular task. Co-processors are designed such that execution of the particular task is achieved far more effectively than with a general-purpose microprocessor. Typical applications of co-processors are for the implementation of:

- maths functions
- digital signal processing
- disc sub-systems
- graphics systems.

The co-processors that have firstly found wide acceptance in the personal computer marketplace, where they are used in conjunction with the Intel 8088/8086, 80286, and 80386 microprocessors, are the 8087, 80287 and 80387, respectively. Essentially, all three co-processors perform the same function, the major difference being that each co-processor is optimised for its corresponding microprocessor.

From a programmer's point of view, the 8087 looks as shown in Fig. 4.20.

There are eight Data registers, each of which is eighty bits long. Associated with each Data register is a Tag field, which is used to store the status of each Data register. The Control register is used to determine the operating mode of the co-processor. The Status register is similar to the Flag register found in the associated microprocessor, and the Instruction Pointer and Data Pointer are used to point to the current instruction and associated data within the microprocessor's addressable memory. The 8087 is capable of working with seven different possible data types:

- word integer, 15 bits and 1 sign bit
- short integer, 31 bits and 1 sign bit
- long integer, 63 bits and 1 sign bit
- short real, 24-bit significand, 8-bit exponent, 1 sign bit

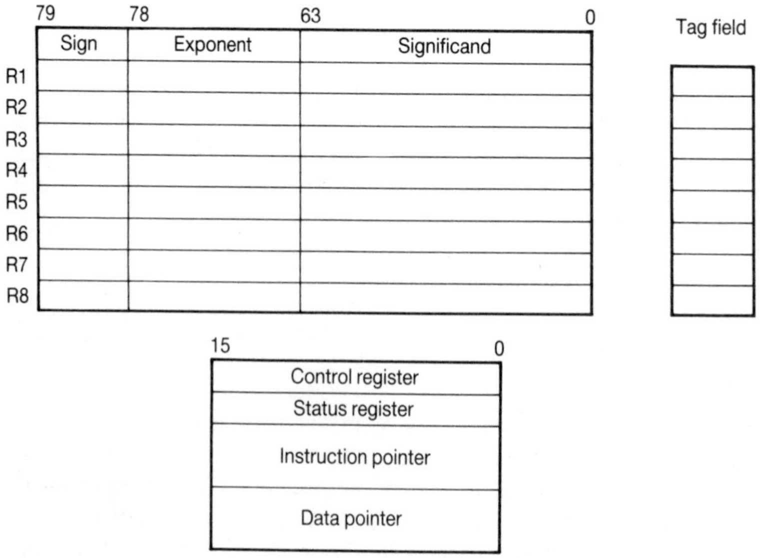

Fig. 4.20   8087 register set

- long real, 53-bit significand, 11-bit exponent, 1 sign bit
- temporary real, 64-bit significand, 16-bit exponent, 1 sign bit.

The functions supported by this co-processor are:

- addition
- subtraction
- multiplication
- division
- square-root
- absolute
- transcendentals.

Typically, performance improvements of about one hundred times can be expected with this co-processor.

4.6   Memory devices

In the previous sections, where the various microprocessors were described, a common parameter was the size of the addressable memory. The following sections describe the types of memory storage available.

4.6.1   Read-only memory

Read-only memory (ROM) describes a family of memory storage devices that generally are programmed only once with information: this information is then available to a computer on a read-only basis, and is not lost when the power sources are removed. Typical applications of ROMs include storage of microprocessor operating systems, code conversion, and character generation.

There are three broad family categories available in a ROM. First is the ROM which is mask programmed during manufacture and can never be changed. The term 'mask programmed' refers to the manner of etching the silicon during manufacture.

The second family has a generic name of 'PROM' – programmable read-only memory. These devices perform the same function as ROMs: that is, permanent storage of information. However, because of the method of fabrication, the information can be erased, usually by exposure to ultra-violet light, and then re-programmed.

The third and final family is known as 'EEPROM' or 'E^{2}PROM' – electrically-erasable programmable read-only memory. These devices are very similar to the PROM, except that erasure is by electrical means. The consequence here, though, is that an EEPROM can be re-programmed under normal operating conditions, thus allowing it to keep information which may be changed periodically. A typical application would be where configuration information for a microprocessor is kept in an EEPROM.

4.6.2   Random-access memory

Random-access memory (RAM) is the major component of all memory fitted to microprocessor systems. RAM is read/write memory and may be logically divided into two distinct types: static RAM and dynamic RAM.

Static RAM may be defined as a memory device which will retain its content whilst the appropriate power sources remain available. In contrast, dynamic RAM requires that its content be refreshed occasionally. The content of both is lost when the power sources are removed.

A static RAM chip typically is larger and more expensive than a similar capacity dynamic RAM chip, due to the method by which static RAMs are manufactured. A typical static RAM is made using flip-flops.

Dynamic RAMs, on the other hand, are implemented using what are essentially small capacitors: hence the smaller chip area and the need for 'refreshing'. Because the charge on each capacitor (memory cell) tends to leak away, each cell needs to be read and then re-written to boost the charge held. Many microprocessors provide support for the refreshing operation, by directly providing a control signal to indicate that a refresh operation is in progress.

4.6.3   Disc storage

RAM and ROM provide a microprocessor with both volatile and non-volatile storage of information; however, because of the cost, it is uneconomical to provide large amounts of this type of storage technology for off-line information. In most cases of computer applications, the programs and associated data exceed the available memory. Thus, storage techniques have been developed whereby these programs can be stored off-line: that is, not held in memory all the time. Clearly, small computer applications do not fall into this category. The following two categories describe this technology.

Soft discs
'Soft disc' is a term used to describe a device, first developed by IBM, whereby a magnetic medium is coated onto a thin plastic disc. The disc is then enclosed by a semi-rigid plastic jacket, such that the disc is capable of revolving inside the jacket. Because of the thinness of the plastic disc, the terms 'soft', 'floppy' and 'flexible' disc have evolved. Recent trends with this technology have seen the development of rigid jackets so that the term 'floppy disc' no longer applies.

Floppy discs generally store medium amounts of information, typically less than 1.5 megabytes of data. The method by which the information is written to and read from the discs is very similar to audio cassette technology, as shown in Fig. 4.21.

Soft disc technology is typified by the relatively large (compared to hard discs) air gap between the read/write head and the magnetic surface and it is for this reason that soft discs have much lower storage capacity than hard

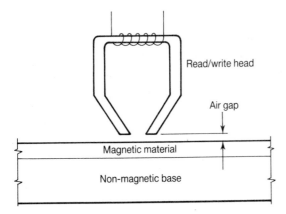

Fig. 4.21   Typical read/write head configuration

discs. Information if stored as a serial bit stream on a set of concentric tracks. There are many techniques available whereby the data are encoded to both improve the storage of data as well as to provide synchronisation as the disc revolves.

Hard discs

Hard discs are very similar to soft discs, with the major differences being:

- the magnetic medium is coated onto a hard non-magnetic base
- the read/write heads are far closer to the surface of the disc (the separation is less than the size of a smoke particle)
- a much faster disc rotational speed.

Hard discs vary in size from 20 megabytes up to 1 gigabyte. A 1 gigabyte hard disc consists of a number of hard discs stacked on top of each other to form a 'platter'. Each disc has its own read/write head.

Hard discs have far greater storage capacity and data transfer speeds, mainly due to the high rotational speeds and the smaller distance between the heads and disc surface, but because of this performance improvement, hard discs are much more susceptible to damage.

5

Specialist Microprocessors

5.1 Introduction

This chapter deals with those specialist microprocessors which are currently available, including microcontrollers, digital signal processors, transputers and graphics controllers. These devices are dedicated microprocessors which have been refined to suit the particular application intended.

5.2 Microcontrollers

Microcontrollers are dedicated microprocessors tailored to function in the control environment. Typical applications of such devices include microwave ovens, washing machines, automotive applications, and so on. Such devices are typified by on-board integration of at least one if not all of the following functions:

- RAM
- ROM
- A/D converter
- D/A converter
- serial port
- parallel port
- programmable timer.

It is beyond the scope of this work to provide details of the different types of microcontroller available. Instead, typical controllers available from Motorola and Intel are discussed.

5.2.1 Motorola microcontrollers

Three types of Motorola microcontroller are described in this Section: the 6801, 6803 and 6805.

The MC6801 is an 8-bit single-chip microcontroller which, according to Motorola, offers the following features:

- instruction-set compatible with the MC6800 family
- 8-bit × 8-bit multiply instruction
- serial communications interface
- 16-bit three-function programmable timer
- bus-compatibility with the MC6800 family
- 2 kilobytes of ROM
- 128 bytes of RAM
- 29 parallel I/O lines
- 2 handshake control lines
- 5 volt power operation
- TTL compatiblity.

The internal architecture of the MC6801 is shown in Fig. 5.1.

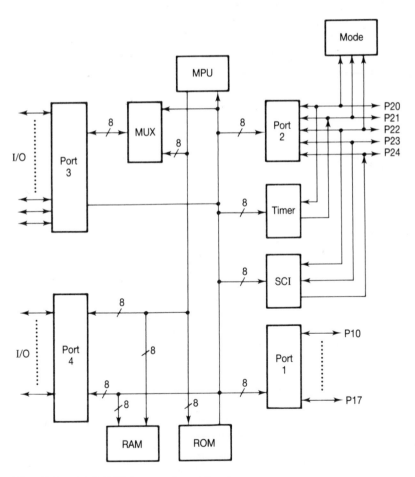

Fig. 5.1   MC6801 internal architecture

The MC6801 is capable of operating in eight modes, each of which determines:

● the available on-chip resources
● the memory map
● the location of interrupt vectors
● the type of external bus.

The eight modes can be divided into three fundamental groups:

● single chip
● expanded, non-multiplexed bus
● expanded, multiplexed bus.

The programming model of the MC6801 is shown in Fig. 5.2.
    As can be seen, the model is similar to that of the MC6800.
    The MC6803 is very similar to the MC6801, the only difference between the two devices being that the MC6803 operates in Modes 2 and 3 only. These modes specify expanded memory space (64 kilobytes) via a multiplexed bus with *no* internal ROM capability.
    The MC6805 is another microcontroller offered by Motorola. The features of this device are:

● internal 8-bit timer with 7-bit pre-scaler
● on-chip oscillator
● memory mapped I/O
● versatile interrupt handling
● vectored interrupts

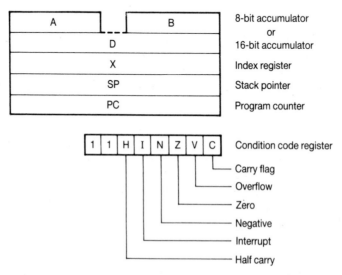

Fig. 5.2   Programming model of the MC6801

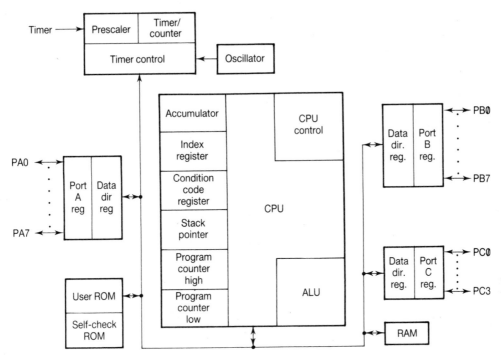

Fig. 5.3 Internal architecture of the MC6805

- internal ROM
- internal RAM
- 24 bi-directional I/O ports
- A/D converter.

The amount of internal RAM and ROM varies between revision levels. For example:

        MC6805 R2 – 2048 bytes of ROM, 64 bytes of RAM
        MC6805 R3 – 3776 bytes of ROM, 112 bytes of RAM.

The internal architecture of the MC6805 is shown in Fig. 5.3.

A variation, known as the MC6805 S2, of the MC6805 family is similar to other members of the family, with the exception that it offers a serial interface and the internal timer is seven bits wide, with a 15-bit pre-scaler.

A further variation of the MC6805 family is the MC6805 P2, which again is similar to other family members with the exception that it has no A/D converter or internal oscillator.

5.2.2 Intel microcontrollers

The Intel family of microcontrollers includes the 8031, 8032, 8051, 8052, 8751 and 8752 devices, which are specialised microprocessors implemented in HMOS. As with the Motorola family, it is possible to have a variety of different features on the chip. According to Intel, the features offered by the MCS51 family (as it is known) include the following:

- 8-bit microprocessor optimised for control
- 64 kilobytes of program space
- 64 kilobytes of data space
- 4 kilobyte program memory on-chip
- 128 byte data memory on-chip
- 32 I/O lines
- two 16-bit counter/timers
- full-duplex UART for serial communications.

The MCS51 family is very popular with designers, as is evidenced by the marketing of large amounts of third-party support products, such as in-circuit emulators, debuggers, assemblers and high-level languages.

The instruction set of this microprocessor contains a range of instructions typically found on 8-bit microprocessors. However, as with other microcontrollers, it includes instructions in the area of bit manipulation and movement.

The addressing schemes offered by these devices are:

- direct addressing
- indirect addressing
- register addressing
- immediate addressing
- indexed addressing.

The internal architecture of these devices supports five interrupt types – two internal, two external and one from the serial port. The architecture is shown in Fig. 5.4.

5.3   Digital signal processors

One of the fastest developing fields in the computer industry is that of *Digital Signal Processors* (*DSP*s), which find wide application in the following areas:

- FIR filters
- IIR filters
- FFT engines
- echo cancellers
- data encryptors
- split-band modems
- image processors.

DSPs are available from a wide range of suppliers such as AT & T, Fujitsu Microelectronics, Motorola, Texas Instruments and NEC.

DSPs have developed an architecture which allows fast, efficient manipulation of mathematical operations (typically addition and multiplication) and explains their popularity for the types of application listed above.

One of the first uses for the DSP was in the implementation of *Finite-Impulse Response* (*FIR*) and *Infinite-Duration-Impulse Response* (*IIR*) filters. The definitions of the one-dimensional form of the equations for these filters are:

$$\text{FIR:} \quad Y(n) = \sum_{k=0}^{N-1} \{h(k).X(n-k)\}$$

$$\text{IIR:} \quad Y(n) = \sum_{k=1}^{M} \{a_k Y(n-k)\} + \sum_{k=0}^{N} \{Sb_k X(n-k)\}$$

Clearly, these applications require the significant computational capabilities which the DSP can provide. The rest of this section looks in more detail at two specific DSPs.

5.3.1   Motorola DSP56001

This is an 88-pin CMOS device with a 20 MHz clock rate, capable of 10 million instructions per second executing on 24-bit data. The main subsections of this device are:

- program controller
- data ALU
- *x* and *y* memories.

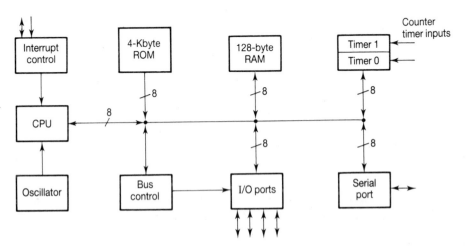

Fig. 5.4   Internal architecture of the MCS51 family

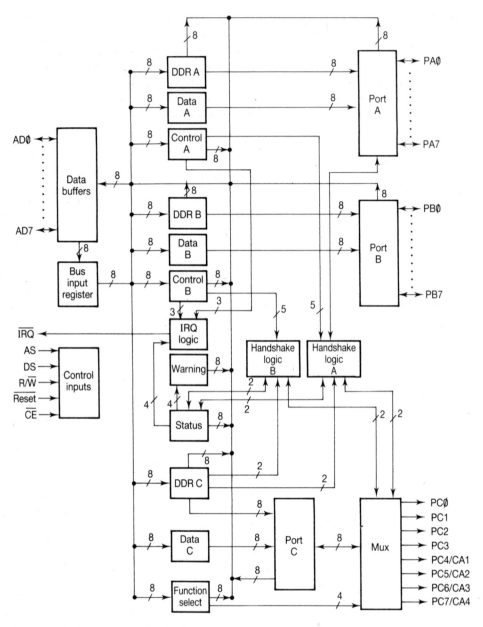

Fig. 5.5   DSP56001 block diagram

The x and y memories significantly ease graphics tasks (involving x and y pixel co-ordinates) and Fourier transformations (involving real and imaginary numbers).

The internals of this device are shown in Fig. 5.5.

Capabilities of this chip as quoted by the manufacturer are:

- 1024-point complex FFT in 3.5 milliseconds
- graphic equalisation of 10 bands of stereo audio data in realtime.

5.3.2   Texas instruments TMS320 family

Texas Instruments has been heavily involved in DSPs since the creation of the concept, to the extent that there is a large variety of such TI devices, offering a wide range of features from which to choose. Of more importance, however, is the considerable range of development support and third-party technical support available for this family of devices.

At the current time, there are three main groups within the TMS320 family.

The TMS320C1x, features:

- 16-bit CPU
- 160 microsecond instruction cycle
- 256 word data RAM
- 4 kiloword ROM/EPROM
- 32-bit multiply (16-bit × 16-bit)
- 2 serial ports
- companding hardware
- co-processor interface.

The TMS320C2x, featuring:

- 16-bit CPU
- 100 microsecond instruction cycle
- 544 word data RAM
- 4 kiloword ROM
- 32-bit multiply (16-bit × 16-bit)
- serial port
- timer
- multiprocessor interface.

The most powerful in the family, the TMS320C3x, featuring:

- 32-bit CPU
- 60 microsecond instruction cycle
- 2 kiloword RAM
- 4 kiloword ROM
- 64 word instruction cache
- 16 megaword total memory
- 40-bit multiply (32-bit × 32-bit)

- 2 serial ports
- 2 timers
- DMA support.

5.4   The Transputer family

The Transputer was developed by INMOS Ltd to be a powerful single-chip computer which could be directly connected to others of the same type. Thus, a network of transputers can form a parallel computer capable of working on a single problem. There is no limit to the size of the array of transputers. The surprisingly simple but very innovative architecture of the chip gives rise to a processor of enormous power, and with very fast task switching capability. When the T414 was released in 1986 it was well ahead of its time.

The connection between transputers is via an *INMOS link*, which is an autonomous handshaken serial port which can operate at 5, 10 or 20 megabits per second.

The language OCCAM was developed to facilitate the programming of the transputer, and this poses something of a dilemma for the system developer. In order to fully exploit the parallel capabilities of the chip it is necessary to program in OCCAM, but this requires a considerable commitment in terms of expense, time and other resources. Becoming proficient in using the development software takes some time. Many developers prefer to use a 'C' environment in order to obtain results more quickly, but unfortunately this compromises the on-chip parallelism which can be obtained.

The Transputer family is comprised of three processors: the T800, the T414 and the T222. All other processors are variants on these three, although the first processor, the T414, has now been superseded by the T425.

Most processors are available in Military Standard and Space Proven versions.

The T800 Processor features:
- 32-bit internal and external architecture
- 4 INMOS link interfaces
- 32-bit multiplexed memory bus
- 4 Kbytes of fast on-chip RAM
- 30 MIPS peak instruction rate
- 4.3 Megaflops peak processing rate
- internal timers for realtime operation
- configurable memory interface timing
- IEEE standard 64-bit floating-point co-processor on-chip.

The internal architecture of the T800 is shown in Fig. 5.6.

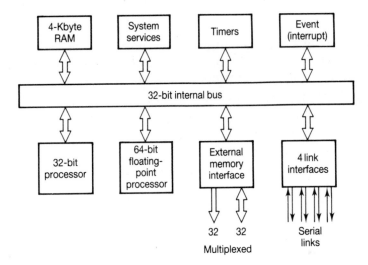

Fig. 5.6   Internal architecture of the T800 transputer

The T414/425 Processor features:
- 32-bit internal and external architecture
- 4 INMOS link interfaces
- 32-bit multiplexed memory bus
- 4 Kbytes of fast on-chip RAM
- 30 MIPS peak instruction rate
- internal timers for realtime operation
- configurable memory interface timing.

The T222 processor features:
- 16-bit internal and external architecture
- 4 INMOS link interfaces
- 16-bit data bus and 16-bit address bus
- 4 Kbytes of fast on-chip RAM
- 20 MIPS peak instruction rate
- internal timers for realtime operation.

5.5   Graphical co-processors

Recent times have seen an increase in the number of specialised co-processors; an example of this trend is the graphical co-processor. This type of device is a highly-specialised microprocessor, which has been optimised to relieve the main processor from having to perform the following types of task:

- perform the updating of a high-resolution screen (in the case of screens with resolutions such as 1024×768 pixels with 256 colours, updating the screen can occupy a significant proportion of CPU time)
- implement basic drawing primitives, such as line, circle, and area-fill (again, this feature enables the main CPU to be relieved from having to execute these primitives in software).

Examples of such graphical co-processors which are gaining popularity are those in the Texas Instruments TMS320xx family of devices.

6

Backplane Bus Structures for Control Microcomputers

6.1 Introduction

For the purposes of the design and manufacture of digital control systems, it is often necessary to use a microcomputer which is built for the purpose. Such a computer is usually 'dedicated': that is, committed to one particular function throughout its working life. Embedded microcomputers are microprocessor systems which are built into other items of equipment.

Most equipment manufacturers find it more convenient and economical to build embedded microcomputers from component parts, which are purchased from other manufacturers, rather than manufacture all the parts themselves. Standard backplane buses provide a means by which components from a variety of sources may be assembled to form a specialised dedicated microcomputer.

There is, naturally, a range of requirements for backplane bus systems which has given rise to a multiplicity of standards. From time to time a particular bus may lose support and become obsolete and, as technology progresses, new bus standards are created.

6.2 General specification for a bus standard

6.2.1 Address and data lines

Since the principal information conveyed by the backplane bus is address and data information, the first item to be specified is usually the number of address and data lines. The sizes of the address and data bus sections of the backplane may influence the choice of microprocessor to be used with the bus, but this choice is not necessarily restricted to a microprocessor having matching numbers of data and address lines.

There have, in the past, been backplane buses with unorthodox numbers of data lines but in modern buses the number is usually a multiple of eight (one byte).

The most common are: 8 lines or bits, 16 lines or bits, and 32 lines or bits. The term 'word' is often used to denote an ordered set of 16 data lines and a set of 32 bits is often called a 'long word'. However, the term 'word' more correctly means the width, in bits, of the internal data bus of a computer, wherein word lengths of 12 bits or 36 bits are common.

The more straightforward of the backplane buses are *simplex*, which means that the data and the address lines are independent. Some of the standards, however, use a *multiplex* structure, whereby the data and addresses are transmitted on the same lines but at different times. Examples of multiplexed buses are: Q-BUS and FUTUREBUS.

It is usual to divide microprocessor memory into bytes, and accordingly, the more popular memory chips tend to be byte-wide. As a result, one memory address is accorded to one byte, and a 32-bit word occupies four addresses.

There is an unresolved difference of opinion regarding the mapping of bytes onto data line numbers. Different manufacturers and different bus standards adopt different methods. The *little-endian* mapping has the least-significant byte mapped to the lowest memory address section of a word. The Intel 8086 is an example of a little-endian microprocessor. The *big-endian* mapping has the most-significant byte mapped to the lowest memory address section of a word. The Motorola 68000 is a big-endian microprocessor.

Some microprocessors always have the least-significant byte occurring at a particular address (the 68000 has it on the even-numbered addresses) and others (such as the 8088) can have the least-significant byte at any address.

In a similar way, some buses have the least-significant byte mapped to the lowest eight numbered data lines. These buses are *justified*. *Non-justified* buses can have the least-significant byte on any byte-wide lane.

6.2.2   Power supplies

Most backplane buses are supplied with the standard GND (0 volts), + 5 volts DC and auxiliary voltages of + 12 and − 12 volts which are used for RS 232 biassing and sometimes for analog circuitry.

Other voltages which are common are − 5 volts, + 15 and − 15 volts.

Many buses also have the option of battery standby supplies for + 5 volts and + 12 and − 12 volts.

The S100 bus is unusual in having unregulated supplies of + 8, + 16 and − 16 volts.

6.2.3   Control, command and status signals

Control signals vary enormously from one bus to another. The range includes the following:

Resets, occasionally at more than one level.
Read and write (or read/write) lines for memory, ports and peripherals.
Memory and port address expansion .
Bus and other error condition signals.

Microprocessor status signals.
Processor halt.
Power supply loss.
Other signals.

6.2.4   Timing

Once again, there is a variety of timing regimes for the different buses. Timing signals include:

Microprocessor system clock.
Slower peripheral clocks.
Memory refresh cycle timing.
Strobe lines for address and data lines.
Read and write cycle synchronisation.
Microprocessor handshaking (request and acknowledge).
Wait state flags or data transfer acknowledge.
Strobes for interrupts.

The read and write signals included in Section 6.2.3 can sometimes be regarded as timing since they are often used for strobes, especially with multiplexed buses.

6.2.5   Interrupts

Provision for interrupts on buses varies from none (Multibus 2, which uses message-passing instead) to two interrupt lines and a strobe (STD bus) to highly organised systems of prioritised interrupts, interrupt handshaking and provision for daisy-chain interrupts.

A typical method of organising a daisy-chain interrupt is shown in Fig. 6.1. The two lines required for the daisy-chain (*chain in* and *chain out*) are not connected right through on the backplane (6.1c). The through connection is made via the board so that any empty slot must be fitted with a dummy board which connects the daisy-chain (6.1b). If two interrupt requests occur simultaneously, the board with the highest priority is granted the interrupt. The higher priority boards are positioned earlier in the daisy-chain and are shown at the left. The order of priority decreases towards the right.

6.2.6   Bus arbitration

If a bus has the capability of accommodating more than one microprocessor it is necessary to decide, on a continuous basis, which processor is permitted to use the bus at any time. The protocol to take care of this is called *bus arbitration* and several lines are usually allocated for the purpose.

The most common arbitration scheme is to have one processor as the bus master, which allocates the bus to another processor after a bus request. After use, the control of the bus is returned to the bus master. This scheme

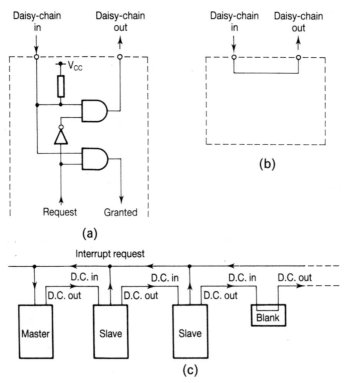

Fig. 6.1   Conventional method of implementing a prioritised system of interrupts or bus arbitration, using a daisy-chain (a) conventional daisy-chain circuit for a regular board; (b) daisy-chain connection for a blank or dummy board; (c) backplane connections for daisy-chained prioritised interrupts

is referred to as centralised arbitration. The centralised arbitration scheme on some buses may use a daisy-chain for allocating priority. Some bus specifications permit an unused interrupt daisy-chain to be re-assigned for bus arbitration. The buses tend to vary in the way priority is allocated to each processor, with a fixed priority hierarchy being common. With this scheme, some processors are considered to be more important than others, and thus, have a greater need for rapid access to the bus. Also common is a round-robin 'fair share' method of allocating the use of the bus, whereby each processor has equal priority. Some buses allow the network master to request a premature return of the bus from a low-priority user.

Another bus arbitration scheme uses *token passing*. With token passing there is no bus master but, instead, a token (which is just a simple message) is circulated between all the microprocessors on the bus. Upon receipt of the token, any processor is allowed to decide whether it needs to use the bus before passing the token on. One processor must assume the responsibility for checking and correcting an accidental loss of the token or duplication of the token.

6.3   Descriptions of common backplane buses

There are very many backplane buses in use and many of them are used by particular equipment manufacturers. The buses described in the following sections are the more common ones and the list is not intended to be fully comprehensive.

The buses are placed approximately in order of cost and complexity, (although the two attributes are not necessarily synonymous).

6.3.1   STD bus (IEEE P961)

The STD is the most popular low-cost backplane bus on the market. The bus was originally designed by the Pro-Log Corporation and the name STD is not an abbreviation for 'standard' but merely a collection of three letters. The bus is now supported by about 300 manufacturers around the world and there are several thousand different cards available. The bus is more popular with Intel and Zilog processors but ranges of other processor cards, including some with Motorola processors, are made.

Variations in implementation can cause some difficulty with cards from different manufacturers. Typical differences are:

- use of the − 5 volts line for + 5 volts battery-backed supply
- use of I/O expansion lines for extra memory addresses
- multiplexing of data lines for extra memory address lines
- multiplexing of address lines for 16-bit data bus use
- use of a daisy-chain for arbitration rather than for interrupts

Most of the differences stem from the desire to push the performance beyond its original requirements, and potential users should check card specifications carefully.

The STD bus is mainly used for lower-cost general-purpose industrial control and instrumentation applications.

Specification
Board sizes: 115mm wide by 165mm long.

Connector is a double sided (28 pins each side) printed circuit connector with 0.18 inch pitch between centres of adjacent conductors.

There are eight data lines and sixteen address lines.

Power supplies are + 5 volts, − 5 volts, GND, Aux + 12 volts, Aux − 12 volts, Aux GND, + 5 volts battery backup (pin 5).

Control and command signals are:

memory request, I/O request, memory expansion, I/O expansion, status, read, write, wait request, system reset, pushbutton reset, DC power loss (pin 6).

The four timing signals are:

processor clock, machine sync. (address latch enable for demultiplexing), control (auxiliary or peripheral timing), refresh.

Interrupts are:

interrupt request, non-maskable interrupt request, interrupt acknowledge (vector read strobe), two daisy-chain connections.

Arbitration uses:

bus request, bus grant. Daisy-chain can be used.

Connector pinouts for the STD bus

Signal	Pin			Pin	Signal
+ 5 volts	1	●	●	2	+ 5 volts
GND	3	●	●	4	GND
− 5 V or V battery	5	●	●	6	− 5v or DC pwr OK*
D3	7	●	●	8	D7
D2	9	●	●	10	D6
D1	11	●	●	12	D5
D0	13	●	●	14	D4
A7	15	●	●	16	A15
A6	17	●	●	18	A14
A5	19	●	●	20	A13
A4	21	●	●	22	A12
A3	23	●	●	24	A11
A2	25	●	●	26	A10
A1	27	●	●	28	A9
A0	29	●	●	30	A8
Write*	31	●	●	32	Read*
I/ORq1s*	33	●	●	34	Mem Rq*
I/O exp	35	●	●	36	Mem exp
Refresh	37	●	●	38	McSynch
Status 1	39	●	●	40	Status 0
Bus Ack	41	●	●	42	Bus Requ
Int Ack	43	●	●	44	Int Requ
Wait Requ	45	●	●	46	N M Irq
Sys Reset	47	●	●	48	P/b reset
Clock	49	●	●	50	Control
Daisy-chain PCo	51	●	●	52	Daisy-chain PCi
Aux Gnd	53	●	●	54	Aux Gnd
Aux + 12 volts	55	●	●	56	Aux − 12 volts

5.3.2   STE bus (IEEE P1000)

Many early users of the STD bus felt that the Eurocard format was tougher and better engineered than the STD rack. The Eurocard edge connector is more reliable than the printed edge connector and costs about the same. The objectives of the designers of this bus (which is sponsored by the IEEE Computer Society) were similar to those of the STD bus developers, with the additions of four extra address lines, a vectored interrupt system, encoded status lines and data strobes. The bus is modern and streamlined and it is gaining in popularity.

The interrupt system is similar to that for the Motorola 68000 processor and is quite complicated for a simple bus. The bus will handle multiprocessor operation for up to three master processors via a complex independent decentralised arbitration system.

Specification

Board size: single-height Eurocard – 100mm wide by 160mm long

Connector is a DIN 41612-C64, 2-row, 64-pin connector with military specification versions available.

The bus has eight data and twenty address lines.

Power supplies are: 0 volts, + 5 volts, + 12 volts, − 12 volts and + 5 volts standby power.

The three command lines are clocked by the data strobe and can encode eight command signals. Of these, five are used. They are: vector fetch (011), I/O write (100), I/O read (101), memory write (110), memory read (111).

Control is exercised through data handshaking of which the data acknowledge line (DTACK) is the most important. DTACK informs the processor that correct data are available. Error information is transferred via the system error line.

Timing is via two strobe lines (address strobe and data strobe) which control the read and write cycles. A system clock is also provided, and a system reset.

There are eight prioritised interrupt lines of which 0 is the highest priority and 7 is the lowest. These are also known as 'attention request'.

The bus can support three processors by a decentralised arbitration system. There are two bus request and two bus acknowledge lines.

Connector pinouts for the STE bus

Signal	Pin		Pin	Signal
0 volts	a1	● ●	c1	0 volts
+ 5 volts	a2	● ●	c2	+ 5 volts
D0	a3	● ●	c3	D1
D2	a4	● ●	c4	D3
D4	a5	● ●	c5	D5

D6	a6	● ●	c6	D7
A0	a7	● ●	c7	0 volts
A2	a8	● ●	c8	A1
A4	a9	● ●	c9	A3
A6	a10	● ●	c10	A5
A8	a11	● ●	c11	A7
A10	a12	● ●	c12	A9
A12	a13	● ●	c13	A11
A14	a14	● ●	c14	A13
A16	a15	● ●	c15	A15
A18	a16	● ●	c16	A17
command 0	a17	● ●	c17	A19
command 2	a18	● ●	c18	command 1
address strobe	a19	● ●	c19	0 volts
data acknowledge	a20	● ●	c20	data strobe
system error	a21	● ●	c21	0 volts
interrupt requ 0	a22	● ●	c22	system reset
interrupt requ 2	a23	● ●	c23	interrupt requ 1
interrupt requ 4	a24	● ●	c24	interrupt requ 3
interrupt requ 6	a25	● ●	c25	interrupt requ 5
0 volts	a26	● ●	c26	interrupt requ 7
bus requ 0	a27	● ●	c27	bus requ 1
bus grant 0	a28	● ●	c28	bus grant 1
system clock	a29	● ●	c29	+ 5 volts standby
− 12 volts	a30	● ●	c30	+ 12 volts
+ 5 volts	a31	● ●	c31	+ 5 volts
0 volts	a32	● ●	c32	0 volts

6.3.3   RM 65 bus

The RM bus is an 8-bit bus sponsored by Rockwell International Corporation. The bus was developed to provide support for the Rockwell integrated circuit chips, particularly the 6502 microprocessor family and the 6522 Versatile Interface Adaptor. It was available from about 1982 onwards.

Two versions of the bus were created, one for the American market, and one for the European market. In both cases the card is a single-height Eurocard, but the American version uses a 72-pin edge connector and the European version has a 64-pin Eurocard connector.

The bus is supported by a few manufacturers, but far less so than for the STD and STE buses.

Specification
Board size: single height Eurocard, 100mm wide by 160mm long.

Connector (American version): 72-pin edge connector on 0.1 inch centres – EBTBH4DD36-13 (Burndy). Key slot between pins 5 and 6.

Connector (European version): 64 pin connector – DIN 41612-C64 (rows a and c only).

There are eight data lines and seventeen address lines. The seventeenth address line is a bank switching line (BADR).

Power supplies are 0 volts and + 5 volts only.

Control and command signals consist of a system reset (BRES), bus active (BACT), a read/not write line (BR/W*), and an inverted read/not write line (BR*/W). The bus active line indicates that a card has been addressed and buffer chips can be enabled in either direction. The read/not write is standard. The set overflow (BSO) signal is an input to the processor and sets the overflow flag. The ready signal (BRDY) halts the processor during the time it is held low.

The timing signals are: phase 2 clock (Bϕ2), an inverted phase 2 clock (Bϕ2*) for peripheral chip timing, a phase 1 clock (Bϕ1) and a phase 0 clock (Bϕ0). There is a timing synchronism (BSYNC) signal which goes high at the start of an op-code fetch and remains high over the whole cycle.

There are two interrupts: interrupt request (BIRQ) and non-maskable interrupt (BNMI).

The bus arbitration is via four Direct Memory Access handshake signals: DMA request 1 (BDMA1), DMA request 2 (BDMA2), DMA terminate (BDMT) and bus float (BFLT).

Connector pin-outs for the RM-65 bus

Signal	Pin		Pin	Signal
0 volts	a1	● ●	c1	+ 5 volts
bank address switch	a2	● ●	c2	A15
0 volts	a3	● ●	c3	A12
A13	a4	● ●	c4	A12
A11	a5	● ●	c5	0 volts
A10	a6	● ●	c6	A9
A8	a7	● ●	c7	A7
0 volts	a8	● ●	c8	A6
A5	a9	● ●	c9	A4
A3	a10	● ●	c10	0 volts
A2	a11	● ●	c11	A1
A0	a12	● ●	c12	phase 1 clock
0 volts	a13	● ●	c13	sync
set overflow	a14	● ●	c14	DMA request 1
Ready	a15	● ●	c15	0 volts
user spare 1	a16	● ●	c16	− 12 volts
+ 12 volts	a17	● ●	c17	user spare 2
0 volts	a18	● ●	c18	bus float
DMA terminate	a19	● ●	c19	phase 0 clock

		• •		
user spare 3	a20	• •	c20	0 volts
not read/write	a21	• •	c21	system spare
DMA request 2	a22	• •	c22	read/not write
0 volts	a23	• •	c23	bus active
IRQ	a24	• •	c24	NMI
not phase 2 clock	a25	• •	c25	0 volts
phase 2 clock	a26	• •	c26	reset
D7	a27	• •	c27	D6
0 volts	a28	• •	c28	D5
D4	a29	• •	c29	D3
D2	a30	• •	c30	0 volts
D1	a31	• •	c31	D0
+ 5 volts	a32	• •	c32	0 volts

6.4   Descriptions of common 16-bit and 32-bit buses

6.4.1   G-64 bus

The G-64 bus is another bus based on the Eurocard hardware. The design originated in Switzerland and is sponsored by Gespac, 3 Chemin des Aulx, Geneva. The bus is big-endian and the structure is most suitable for the Motorola MC6809 chip and the MC68000 family. This is also a modern streamlined bus and is popular in Europe, where it is well supported.

The bus is unusual in that it has sixteen data lines and only sixteen address lines.

The data format is non-justified, which means that a byte-wide peripheral may be installed in either the upper or lower half of the word but can only use half of the available addresses. Because of this, data strobes are available for timing of upper and lower bytes.

Specification
Board size: single-height Eurocard – 100mm wide by 160mm long.

Connector: 96 pin DIN-41612-C96, rows a and c only are used.

There are sixteen data lines and sixteen address lines. A seventeenth address line is the page select line.

Power supplies are: + 5 volts, + 12 volts, − 12 volts, 0 volts and a battery-backed + 5 volts supply.

Control and command signals are: reset, halt processor, bus error, read/not write. The bus timing can be either synchronous (similar to the MC6809) or handshaken (like the MC68000).

The timing signals are: master clock (enable), memory clock or system clock, valid memory address, valid peripheral address, data transfer acknowledge. The data strobe 0 is for the even-numbered byte and data strobe 1 is for the odd-numbered byte.

Interrupts are: non-maskable interrupt and three other interrupt lines. An interrupt acknowledge line is used to indicate a vector-fetch cycle. A daisy-chain is available for interrupts, but this may alternatively be used for bus arbitration.

Bus arbitration uses three lines: bus request, bus granted, bus grant acknowledge (bus busy).

Connector pin-outs for the G-64 bus

Signal	Pin			Pin	Signal
0 volts	a1	●	●	c1	0 volts
A0	a2	●	●	c2	A8
A1	a3	●	●	c3	A9
A2	a4	●	●	c4	A10
A3	a5	●	●	c5	A11
A4	a6	●	●	c6	A12
A5	a7	●	●	c7	A13
A6	a8	●	●	c8	A14
A7	a9	●	●	c9	A15
bus granted	a10	●	●	c10	bus request
data strobe 0	a11	●	●	c11	data strobe 1
halt	a12	●	●	c12	BGA (bus busy)
sys clock	a13	●	●	c13	enable
valid periph address	a14	●	●	c14	reset
DTack (ready)	a15	●	●	c15	NMI
valid memory address	a16	●	●	c16	IRQ1
r/w*	a17	●	●	c17	IRQ2
IRQ3	a18	●	●	c18	I ack
D8	a19	●	●	c19	D12
D9	a20	●	●	c20	D13
D10	a21	●	●	c21	D14
D11	a22	●	●	c22	D15
D0	a23	●	●	c23	D4
D1	a24	●	●	c24	D5
D2	a25	●	●	c25	D6
D3	a26	●	●	c26	D7
page	a27	●	●	c27	bus error
chain out	a28	●	●	c28	chain in
power fail	a29	●	●	c29	+ 5 volts battery
+ 12 volts	a30	●	●	c30	− 12 volts
+ 5 volts	a31	●	●	c31	+ 5 volts
0 volts	a32	●	●	c32	0 volts

6.4.2   The G96 bus

The G-96 bus also uses the Eurocard hardware and is closely related to the G-64 bus, although it has a different structure. The design is also sponsored by Gespac. It has eight more address lines than the G-64 and a prioritised self-selection arbitration which allows many more processors to be run in parallel.

Specification
Board size: single-height Eurocard – 100mm wide by 160mm long.

Connector: 96 pin DIN-41612-C96, all three rows are used.

There are sixteen data lines and twenty four address lines. A twenty-fifth address line is the page select line.

Power supplies are: + 5 volts, + 12 volts, − 12 volts, 0 volts and a battery-backed + 5 volts supply.

Control and command signals are: reset, halt processor, read/not write, bus error, valid event data (VED).

   The timing signals are similar to those of the G-64 bus and are: enable, valid peripheral address, valid memory address, ready, memory clock, data transfer acknowledge, data strobe 0, data strobe 1.
   There are two more interrupts than with the G-64 bus: IRQ1, IRQ2, IRQ3, IRQ4, IRQ5, NMI. In addition, there is a special interrupt line – system failure (SYSFAIL).
   The bus arbitration is quite different from that with the G-64 bus. This bus will support up to thirty two independent masters. Bus arbitration is clocked by a timing signal – the arbiter clock for self selection (ARBCLK). Automatic selection is carried out by way of six priority encoding lines, P0 to P5.
   The c row of pins also has eight spare or reserved lines.

Connector pin-outs for the G-96 bus

Component side						Solder side	
Signal	Pin		Pin	Signal		Pin	Signal
0 v	a1	● ●	b1	0 v	●	c1	0 v
A0	a2	● ●	b2	A8	●	c2	A16
A1	a3	● ●	b3	A9	●	c3	A17
A2	a4	● ●	b4	A10	●	c4	A18
A3	a5	● ●	b5	A11	●	c5	A19
A4	a6	● ●	b6	A12	●	c6	A20
A5	a7	● ●	b7	A13	●	c7	A21
A6	a8	● ●	b8	A14	●	c8	A22
A7	a9	● ●	b9	A15	●	c9	A23

BGRT	a10	• •	b10	B req	•	c10	
DSO	a11	• •	b11	DS1	•	c11	
halt	a12	• •	b12	BGA	•	c12	
sys clk	a13	• •	b13	enable	•	c13	0v
VPA	a14	• •	b14	reset	•	c14	
DTack (rdy)	a15	• •	b15	NMI	•	c15	
VMA	a16	• •	b16	IRQ1	•	c16	IRQ3
r/w*	a17	• •	b17	IRQ2	•	c17	IRQ5
IRQ4	a18	• •	b18	I ack	•	c18	VED
D8	a19	• •	b19	D12	•	c19	0v
D9	a20	• •	b20	D13	•	c20	P5
D10	a21	• •	b21	D14	•	c21	P4
D11	a22	• •	b22	D15	•	c22	P3
D0	a23	• •	b23	D4	•	c23	P2
D1	a24	• •	b24	D5	•	c24	P1
D2	a25	• •	b25	D6	•	c25	P0
D3	a26	• •	b26	D7	•	c26	
page	a27	• •	b27	BERR.	•	c27	SYSFAIL
chain out	a28	• •	b28	chain in	•	c28	ARBCLK
power fail	a29	• •	b29	+ 5v batt	•	c29	
+ 12v	a30	• •	b30	− 12v	•	c30	
+ 5v	a31	• •	b31	+ 5v	•	c31	+ 5v
0v	a23	• •	b32	0v	•	c32	0v

Note: 0v, + 5v, + 12v, − 12v, + 5v batt are all voltages.

6.4.3 Q-bus

The Q-bus was developed by the Digital Equipment Corporation for the LSI-11 series of minicomputers, which have been widely used as both standalone computers and as embedded systems. As a result, the Q-bus is to be found in much industrial equipment, such as CNC machine tools and robots.

The original version had sixteen address lines, but this was later extended to eighteen and then to twenty two.

The bus has the following unusual features:
There are two edge connectors, not one.
The data lines are all multiplexed with the address lines.

Specification
Board size: there are two sizes – 132mm by 214mm, and a double size of 265mm by 214mm.

There are two similar connectors which are 36-contact printed circuit edge connectors.

The bus has sixteen data lines and twenty two address lines. The sixteen data lines and address lines 0 to 15 are multiplexed. Address lines 16 and 17 are multiplexed with two bus error lines. Address lines 18 to 21 are non-multiplexed.

Power supplies are: + 5 volts, + 12 volts, − 12 volts, GND and also + 5 volts and + 12 volts battery-backed supplies.

Control and command signals are: initialise (reset), halt processor, DC power fail, AC power fail, write transfer and byte transfer, input/output page selection, memory refresh.

The backplane timing uses separate read and write lines.

Data transfer has a read-modify-write cycle as well as read and write cycles. Timing signals are: address latch (sync) for demultiplexing the address and data, data write strobe, data read strobe, bus acknowledge.

Interrupts consist of a non-maskable interrupt (event), four prioritised interrupt levels (IRQ4, IRQ5, IRQ6, IRQ7), and two daisy-chain interrupt lines for operation within one interrupt level.

Arbitration is by a round-robin token-passing direct memory access scheme. The token passing is daisy-chained. There are four arbitration lines: token in (DMA grant in), token out (DMA grant out), DMA request, DMA granted (or acknowledge).

Connector pin-outs for the Q-bus

Component side				Solder side
Signal	Pin		Pin	Signal
	A1		A2	
IRQ5	a	● ●	a	+ 5 volts
IRQ6	b	● ●	b	− 12 volts
D/A16	c	● ●	c	0 volts
D/A17	d	● ●	d	+ 12 volts
spare 1	e	● ●	e	data output h/s
spare 2 or run	f	● ●	f	data reply h/s
spare 3	h	● ●	h	data in
0 volts	j	● ●	j	sync
maint spare	k	● ●	k	write byte
maint spare	l	● ●	l	IRQ 4
0 volts	m	● ●	m	int ack input
DMA request	n	● ●	n	int ack output
halt	p	● ●	p	bank select 7
mem refresh	r	● ●	r	DMA grant input
+ 12 volts batt	s	● ●	s	DMA grant o/p
0 volts	t	● ●	t	init (reset)
not assigned 1	u	● ●	u	D/AO
+ 5 volts batt	v	● ●	v	D/A1

	B1			B2	
DC power OK	a	●	●	a	+ 5 volts
power OK	b	●	●	b	− 12 volts
spare 4 or D/A18	c	●	●	c	0 volts
spare 5 or D/A19	d	●	●	d	+ 12 volts
spare 6 or D/A20	e	●	●	e	D/A2
spare 7 or D/A21	f	●	●	f	D/A3
spare 8	h	●	●	h	D/A4
0 volts	j	●	●	j	D/A5
maint spare	k	●	●	k	D/A6
maint spare	l	●	●	l	D/A7
0 volts	m	●	●	m	D/A8
DMA bus ack	n	●	●	n	D/A9
IRQ7	p	●	●	p	D/A10
external event IRQ	r	●	●	r	D/A11
not assigned 4	s	●	●	s	D/A12
0 volts	t	●	●	t	D/A13
not assigned 2	u	●	●	u	D/A14
+ 5 volts	v	●	●	v	D/A15

6.4.4   VME bus

The VME bus was developed in the early 1980s, from the Versabus, by a consortium of companies, many with European affiliations. Since Versabus was supported by Motorola the VME bus also is best suited to the Motorola microprocessors, especially the MC 68000. The bus is designed around the popular Eurocard hardware. However, there are several options with the bus which compromise the standardisation and which can cause difficulties for the unwary purchaser of boards.

The two main options are a single-connector, single-height card, 16-bit data bus, 24-bit address bus option and a double-connector, double-height, 32-bit data bus, 32-bit address bus option.

The bus carries extra data highways. There is a serial bus consisting of two lines, one for the clock and one for the data. The second connector can be fitted with extension buses (two possibilities): the VMX or the MVMX.

Specification
The board size for the single-height version is a standard Eurocard – 100mm by 160mm. The double size is 233mm by 160mm.

The connectors are standard DIN 41612–C96; ninety six pins are used. The single-height card uses one connector and the double-height uses two.

Numbers of data and address lines vary according to the option. There are two main options:

32 address lines and 32 data lines (double connector).
24 address lines or 16 address lines and 16 data lines.

In addition, there are six address modifier lines (addr mod 0 – addr mod 5) for controlling access to various address spaces for data, program, peripherals etc. Everything is memory mapped.

Power supplies are 0 volts, + 12 volts, – 12 volts, + 5 volts battery-backed.

Control and command signals are straightforward. They are reset, AC power failure, system failure, long word indication, write line (0 volts to assert), bus error, and handshake lines for timing.

Timing is identical to that for the MC 68000 processor: non-synchronous, handshaken, with data strobes. A valid address is indicated by the processor by AS (address strobe) and the slave acknowledges valid data with DTack (data transfer acknowledge).

The bus has seven prioritised interrupt request lines. Further interrupts are organised by daisy-chain with the two lines interrupt acknowledge input and interrupt acknowledge output serving as the daisy-chain lines. The bus master generates an interrupt acknowledge signal.

Arbitration is also by prioritised daisy-chain with four independent daisy-chain pairs (bus grant in 0–3 and bus grant out 0–3) being available. There are four straight bus request lines for use by bidding masters, and two organisational lines – bus clear and bus busy. The bus specification allows for several options, including: fixed allocated priority, round robin bus allocation, single-level priority only, bus release on request, bus release when finished. The multiplicity of options has unfortunately compromised an otherwise popular bus standard.

Connector pin-outs for the VME bus

Mandatory connector J1/P1

Component side				Component side				Solder side	
Signal	Pin			Pin	Signal		Pin	Signal
D 0	a1	●	●	b1	bus busy	●	c1	D 8
D 1	a2	●	●	b2	bus clear	●	c2	D 9
D 2	a3	●	●	b3	AC pwr fail	●	c3	D10
D 3	a4	●	●	b4	BG 0 in	●	c4	D11
D 4	a5	●	●	b5	BG 0 out	●	c5	D12
D 5	a6	●	●	b6	BG 1 in	●	c6	D13
D 6	a7	●	●	b7	BG 1 out	●	c7	D14
D 7	a8	●	●	b8	BG 2 in	●	c8	D15
0 volts	a9	●	●	b9	BG 2 out	●	c9	0 volts
sysclock	a10	●	●	b10	BG 3 in	●	c10	sysfail
0 volts	a11	●	●	b11	BG 3 out	●	c11	buserr

data strobe1	a12	● ●	b12	bus req 0	●	c12	reset
data strobe2	a13	● ●	b13	bus req 1	●	c13	long wrd
write	a14	● ●	b14	bus req 2	●	c14	add mod 5
0 volts	a15	● ●	b15	bus req 3	●	c15	A23
DTack	a16	● ●	b16	addr mod 0	●	c16	A22
0 volts	a17	● ●	b17	addr mod 1	●	c17	A21
addr strobe	a18	● ●	b18	addr mod 2	●	c18	A20
0 volts	a19	● ●	b19	addr mod 3	●	c19	A19
int ack	a20	● ●	b20	0 volts	●	c20	A18
int ack in	a21	● ●	b21	serial clk	●	c21	A17
int ack out	a22	● ●	b22	serial data	●	c22	A16
addr mod 4	a23	● ●	b23	0 volts	●	c23	A15
A 7	a24	● ●	b24	IRQ 7	●	c24	A14
A 6	a25	● ●	b25	IRQ 6	●	c25	A13
A 5	a26	● ●	b26	IRQ 5	●	c26	A12
A 4	a27	● ●	b27	IRQ 4	●	c27	A11
A 3	a28	● ●	b28	IRQ 3	●	c28	A10
A 2	a29	● ●	b29	IRQ 2	●	c29	A 9
A 1	a30	● ●	b30	IRQ 1	●	c30	A 8
− 12 volts	a31	● ●	b31	+ 5 v batt	●	c31	+ 12 volts
+ 5 volts	a32	● ●	b32	+ 5 volts	●	c32	+ 5 volts

The optional expanded VME bus connector J2/P2 also uses a DIN 41612 96 pin connector.

Optional connector J2/P2

Component side							Solder side
Signal	Pin		Pin	Signal		Pin	Signal
user I/O	a1	● ●	b1	+ 5 volts	●	c1	user I/O
user I/O	a2	● ●	b2	0 volts	●	c2	user I/O
user I/O	a3	● ●	b3	reserved	●	c3	user I/O
user I/O	a4	● ●	b4	A 24	●	c4	user I/O
user I/O	a5	● ●	b5	A 25	●	c5	user I/O
user I/O	a6	● ●	b6	A 26	●	c6	user I/O
user I/O	a7	● ●	b7	A 27	●	c7	user I/O
user I/O	a8	● ●	b8	A 28	●	c8	user I/O
user I/O	a9	● ●	b9	A 29	●	c9	user I/O
user I/O	a10	● ●	b10	A 30	●	c10	user I/O
user I/O	a11	● ●	b11	A 31	●	c11	user I/O
user I/O	a12	● ●	b12	0 volts	●	c12	user I/O
user I/O	a13	● ●	b13	+ 5 volts	●	c13	user I/O
user I/O	a14	● ●	b14	D 16	●	c14	user I/O
user I/O	a15	● ●	b15	D 17	●	c15	user I/O
user I/O	a16	● ●	b16	D 18	●	c16	user I/O

user I/O	a17	● ●	b17	D 19	●	c17	user I/O
user I/O	a18	● ●	b18	D 20	●	c18	user I/O
user I/O	a19	● ●	b19	D 21	●	c19	user I/O
user I/O	a20	● ●	b20	D 22	●	c20	user I/O
user I/O	a21	● ●	b21	D 23	●	c21	user I/O
user I/O	a22	● ●	b22	0 volts	●	c22	user I/O
user I/O	a23	● ●	b23	D 24	●	c23	user I/O
user I/O	a24	● ●	b24	D 25	●	c24	user I/O
user I/O	a25	● ●	b25	D 26	●	c25	user I/O
user I/O	a26	● ●	b26	D 27	●	c26	user I/O
user I/O	a27	● ●	b27	D 28	●	c27	user I/O
user I/O	a28	● ●	b28	D 29	●	c28	user I/O
user I/O	a29	● ●	b29	D 30	●	c29	user I/O
user I/O	a30	● ●	b30	D 31	●	c30	user I/O
user I/O	a31	● ●	b31	0 volts	●	c31	user I/O
user I/O	a32	● ●	b32	+ 5 volts	●	c32	user I/O

6.4.5 Multibus

Multibus has been by far the most popular backplane bus in industry, although with the advent of 32-bit microprocessors and the wider use of personal computers its popularity is waning. The bus was devised by Intel in 1976 and is also supported in Europe by Siemens in a Eurocard version. There are several variations of the bus from the published standard (IEEE 796) including the Eurocard connector version and the use of a second connector (P2) to extend the addressing from twenty to twenty four lines and to add services such as error reporting.

The bus is justified, which means that an 8-bit peripheral can be accessed by any address and it appears on the lower byte by default. There is a high byte enable line to select a byte on the data lines D8 to D15.

Specification
The board size is 305mm by 170mm (233 × 160mm for the double-height Eurocard version).

The Intel version uses an 86-pin card edge connector (DIN 41612-C96 for the Siemens version).

There are several variations in the number of data and address lines. As is usual for Intel architecture, there are different data and address line standards for memory and for input/output. The most common standard provides for twenty address lines and sixteen data lines, but there can be sixteen, twenty or twenty four address lines with eight lines or sixteen lines reserved for I/O. Data can be 8-bit, or little-endian 16-bit justified.

Power supplies are 0 volts, + 5 volts, + 12 volts, − 12 volts. The − 5 volts lines on connector P1 have now been reassigned as reserved. The non-standard connector P2 can carry + 5 volts battery-backed, − 5 volts battery-backed, + 12 volts battery-backed, − 12 volts battery-backed, + 15 volts, − 15 volts.

Control and command consist of: reset (init), byte high enable (BHen), a bus locking line for use with multimaster multiprocessing (lock), two prioritised inhibit lines for slave substitution (inh1, inh2).

Multibus has several timing lines for different purposes. There is a 10 MHz clock (constant clock) for unspecified functions, a 10 MHz clock pulse for the arbitration, memory read strobe, memory write strobe, input/output read strobe, input/output write strobe, slave positive acknowledge, and the interrupt strobe mentioned below.

There are eight prioritised interrupt lines (Int0 to Int7) and one interrupt acknowledge (Int ack) which serves as a strobe for the interrupt vector fetch cycle on the data bus.

The bus allows various arbitration schemes. The two most common are a single-master centralised arbitration scheme with a bus request and a bus grant line, or alternatively a daisy-chain system with daisy-chain in and out lines. The bus grant and the daisy-chain in share the same line (Bgnt/DaisyCin). There are also options for synchronised transfer using the arbitration clock, pipelined data transfer using the bus busy line, and default bus master indication line for multiple master requests (MBreq). The bus is well known for its multiprocessor support and there are several systems on the market. The variations in the standard can lead to some difficulties for unwary users.

Connector pin-outs for multibus

Component side			Solder side
Signal	Pin	Pin	Signal
CONNECTOR P1			
0 volts	1	2	0 volts
+ 5 volts	3	4	+ 5 volts
+ 5 volts	5	6	+ 5 volts
+ 12 volts	7	8	+ 12 volts
reserved	9	10	reserved
0 volts	11	12	0 volts
arbitration clock	13	14	reset (init)
Bus gnt/DaisyCin	15	16	DaisyCout
bus busy	17	18	bus request
memory read	19	20	memory write
input/output read	21	22	input/output write
slave pos ack	23	24	inhibit (inh1)
lock	25	26	inhibit(inh2)
high byte enable	27	28	A16

bus request	29	●	●	30	A17
constant clock	31	●	●	32	A18
interrupt ack	33	●	●	34	A19
interrupt 6	35	●	●	36	interrupt 7
interrupt 4	37	●	●	38	interrupt 5
interrupt 2	39	●	●	40	interrupt 3
interrupt 0	41	●	●	42	interrupt 1
A 14	43	●	●	44	A 15
A 12	45	●	●	46	A 13
A 10	47	●	●	48	A 11
A 8	49	●	●	50	A 9
A 6	51	●	●	52	A 7
A 4	53	●	●	54	A 5
A 2	55	●	●	56	A 3
A 0	57	●	●	58	A 1
D 14	59	●	●	60	D 15
D 12	61	●	●	62	D 13
D 10	63	●	●	64	D 11
D 8	65	●	●	66	D 9
D 6	67	●	●	68	D 7
D 4	69	●	●	70	D 5
D 2	71	●	●	72	D 3
D 0	73	●	●	74	D 1
0 volts	75	●	●	76	0 volts
reserved	77	●	●	78	reserved
− 12 volts	79	●	●	80	− 12 volts
+ 5 volts	81	●	●	82	+ 5 volts
+ 5 volts	83	●	●	84	+ 5 volts
0 volts	85	●	●	86	0 volts

CONNECTOR P2

0 volts	1	●	●	2	0 volts
+ 5 volts batt	3	●	●	4	+ 5 volts batt
reserved	5	●	●	6	EEPROM power
− 5 volts batt	7	●	●	8	− 5 volts batt
reserved	9	●	●	10	reserved
+ 12 volts batt	11	●	●	12	+ 12 volts batt
power fail SR	13	●	●	14	reserved
− 12 volts batt	15	●	●	16	− 12 volts batt
power fail SN	17	●	●	18	AC low
power fail IN	19	●	●	20	MPRO
0 volts	21	●	●	22	0 volts
+ 15 volts	23	●	●	24	+ 15 volts
parallel 1	25	●	●	26	inhibit(inh2)
parallel 2	27	●	●	28	halt
bus request	29	●	●	30	wait
PLC	31	●	●	32	ALE

reserved	33	● ●	34	reserved
reserved	35	● ●	36	reserved
reserved	37	● ●	38	reserved
reserved	39	● ●	40	reserved
reserved	41	● ●	42	reserved
reserved	43	● ●	44	reserved
reserved	45	● ●	46	reserved
reserved	47	● ●	48	reserved
reserved	49	● ●	50	reserved
reserved	51	● ●	52	reserved
reserved	53	● ●	54	reserved
A 22	55	● ●	56	A 23
A 20	57	● ●	58	A 21
reserved	59	● ●	60	reserved

6.4.6   Futurebus

Futurebus has the highest performance of the backplane buses described here. It was developed by the IEEE Computer Society as a modern streamlined bus standard which is independent of any manufacturers. It is a true standard since there are no options or permitted variations. Futurebus is expensive to implement, so it is used in high-quality applications such as military, aerospace and some industrial situations.

The bus is supported by a few companies, mainly amongst the subcontractors to Government and military establishments.

Specification
The board is a triple-height extended Eurocard, 280mm long by 366mm high.

One of the three connectors has all the pins assigned. The other two are unassigned. The connector is a standard DIN 41612-C96.

The bus has thirty two data and the thirty two address lines which are multiplexed, sharing the same numbering. Data lines are unjustified and the memory map is unspecified, although input/output is memory mapped. The large size of board means that the backplane is not generally used for routine memory access, but, instead, it is designed for multiprocessor systems and for interprocessor communication. Because of this many of the normal services, such as routine interrupts, are absent.

Power supplies are + 5 volts and 0 volts regulated supply. There is a separate logic ground.

Most of the control and command signals are arranged to support the multiprocessor philosophy, for which the arbitration is quite complex. There are five address lines (command 0 to 4) which are time multiplexed. Such functions as read, broadcast, single destination, block transfer, extended command, lock and byte lane selection are covered by these.

There is also an extra command line for the command parity. There are three slave status reply lines which encode slave status response signals (S status 0, 1, 2), such as illegal, parity error, end of data, access error, busy, valid. There is an error detection valid line to confirm that there is a genuine error (ED valid). The system has one reset line.

There is one extra bus, a serial bus, with two lines only. These are serial clock and serial data.

Futurebus does not carry a clock, but since the address and data lines are multiplexed, it has an address strobe and a data strobe. There are two acknowledge lines for single-destination operation (address positive acknowledge and data positive acknowledge) and also two acknowledge lines for broadcast operation (address inverse acknowledge and data inverse acknowledge).

A multiprocessor philosophy is somewhat incompatible with interruption of processors, since there must be orderly message passing between processes on different processors. The equivalent of interrupts is generally handled by process scheduling on individual processors. The bus, therefore, does not have any interrupt lines.

Bus arbitration is a prioritised self-selection mechanism which uses an arbitration condition (or busy) line. There are three handshake lines (h/shake P, Q, R) and seven priority or arbitration number lines (arbitration number 0 to 6). On the connector there are also five pins which are individually hard wired to define the absolute address of the connector, for address tagging of each processor (GS address 0 to 4).

Connector pin-outs for futurebus

Component side						Solder side
Signal	Pin		Pin	Signal	Pin	Signal
0 volts	a1	● ●	b1	0 volts	● c1	0 volts
+ 5 volts	a2	● ●	b2	+ 5 volts	● c2	+ 5 volts
A/D 0	a3	● ●	b3	A/D 1	● c3	A/D 2
A/D 3	a4	● ●	b4	GS addr 0	● c4	A/D 4
A/D 5	a5	● ●	b5	A/D 6	● c5	A/D 7
ground	a6	● ●	b6	err det 0	● c6	A/D 8
A/D 9	a7	● ●	b7	A/D 10	● c7	ground
A/D 11	a8	● ●	b8	A/D 12	● c8	A/D 13
A/D 14	a9	● ●	b9	GS addr 1	● c9	A/D 15
err dedt 1	a10	● ●	b10	A/D 16	● c10	A/D 17
ground	a11	● ●	b11	A/D 18	● c11	A/D 19
A/D 20	a12	● ●	b12	A/D 21	● c12	ground
A/D 22	a13	● ●	b13	A/D 23	● c13	errdet 2
A/D 24	a14	● ●	b14	GS addr 2	● c14	A/D 25

A/D 26	a15	••	b15	A/D 27	•	c15	A/D 28
ground	a16	••	b16	A/D 29	•	c16	A/D 30
A/D 31	a17	••	b17	err det 3	•	c17	ground
command 0	a18	••	b18	command 1	•	c18	command 2
command 3	a19	••	b19	GS addr 3	•	c19	command 4
comnd parity	a20	••	b20	ED valid	•	c20	S stat 0
ground	a21	••	b21	S status 1	•	c21	S stat 2
addr strobe	a22	••	b22	addr ack	•	c22	ground
addr inv ack	a23	••	b23	data strobe	•	c23	data ack
data inv ack	a24	••	b24	GS addr 4	•	c24	h/shakeP
h/shake Q	a25	••	b25	h/shake R	•	c25	arb cond
ground	a26	••	b26	arbit no 0	•	c26	arb no 1
arbit no 2	a27	••	b27	arbit no 3	•	c27	ground
arbit no 4	a28	••	b28	arbit no 5	•	c28	arb no 6
serial clock	a29	••	b29	reset	•	c29	ser data
reserved 0	a30	••	b30	reserved 1	•	c30	resvd 2
+ 5 volts	a31	••	b31	+ 5 volts	•	c31	+ 5 volts
0 volts	a32	••	b32	0 volts	•	c32	0 volts

6.5   Buses not discussed in detail

There are many other backplane buses which have not been discussed in this chapter. Some of these are briefly discussed here in the interests of background information.

6.5.1   The S100 bus

This bus was developed for the Altair microcomputer and was appropriate for the 8080 processor. There were two quite different versions – the original S100 and the IEEE 696. The bus was characterised by a 100-pin edge connector, twenty four address lines, eight data read lines, eight data write lines and + 8 volts, + 16 volts and − 16 volts unregulated power supplies. It has now been superseded by Multibus.

6.5.2   The PC bus

The PC bus is the backplane bus used in all IBM and compatible personal computers, and so is one of the commonest backplane buses on the market. There is a very wide range of boards available from many manufacturers worldwide. It is not discussed in detail here since there is considerable literature available elsewhere and the bus was never designed to be used for industrial work, control systems or for modular computer construction.

The PC bus was developed around the Intel processor chips and is optimised for them. It uses a 62-pin edge connector and has eight data lines and twenty address lines plus clocks, interrupts, control, command and direct memory request lines.

6.5.3   The AT bus

The AT bus is a logical development of the PC bus, with the addition of a second (32-bit) edge connector. It has eight more data lines, four extra address lines and a few extra interrupts. Some pins on the 62-pin connector are reassigned, mainly to allow the use of a wider range of memory chips, especially zero-wait-state memory.

6.5.4   Nubus

Nubus is a synchronous bus with thirty two address lines multiplexed with thirty two data lines designed for multiprocessor support. It evolved from a non-synchronous bus designed at the Massachusetts Institute of Technology, and it is now used by Western Digital and Texas Instruments and is the bus used in the Apple Macintosh 2 computer. The board size and connector are similar to that of Futurebus but, although Nubus is simpler, it has some difficulty with byte alignment.

6.5.5   Multibus 2

Multibus 2 is not a development from Multibus, but rather was developed by Intel as a competitor to Futurebus and Nubus. It is a message-passing multiprocessor bus which is optimised for Intel architecture processor chips and which runs synchronously on a 10 MHz clock. The synchronous operation renders the bus simple to use, but limits its performance in several ways.

7

Data Communications for Control

7.1 Introduction

Data communications are becoming increasingly important in all fields of engineering, as the dependence upon computers grows. For control systems, an understanding of data communications is essential, since they are involved in many applications; for example, 'smart' transducers must be able to transmit the measurement information and distributed controllers must have a fast, reliable, method for passing information to other controllers.

This chapter describes the various International Standards, protocols and communication devices. The section on Local Area Networks is perhaps the most important, since it is this field which is experiencing the most growth and offers the greatest prospects.

7.2 Transmission media

There are three basic types of transmission medium available for data communications:

- copper-based media
- glass-based media
- ether-based media.

7.2.1 Copper-based media

Copper-based media include both twisted-pair cable and co-axial cable: these types of medium have been available for many years. Twisted-pair cabling is traditionally used for low-speed long-distance applications. Co-axial cables are used where relatively high speeds and short distances are involved. Fig. 7.1 shows cross-sections through the two different types.

Copper-based systems form the major proportion of all currently installed transmission systems. The major advantages with copper-based

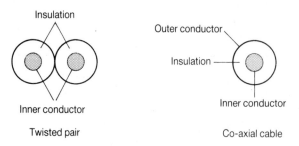

Fig. 7.1   Cross-sections through copper-based transmission media

systems have been the maturity of the technology as well as the cost benefit. The greatest disadvantage is the inability to support the high-speed communication demanded by modern applications, but another disadvantage which is becoming more important is the lack of security of information offered by these systems.

7.2.2   Glass-based media

The use of optical fibres is becoming more prevalent in industry. Glass-based media offer the advantages of high signal bandwidth, high signal-carrying capacity, small cable diameter, absence of electromagnetic radiation and extremely high information security. The principal disadvantages are the high cost of splicing, that is, joining the fibres end-to-end, and their relative fragility; however, these disadvantages are becoming less significant as the technology progresses.

A laser or similar light-emitting transmitting device converts computer-generated signals into a beam of light, which is carried along the fibre. A light-sensitive receiving device senses the beam and converts it back into electrical data for subsequent processing.

Figure 7.2 shows the two types of fibre-optic cable: *single-mode* and *multi-mode*. A single-mode system is one which only propagates light

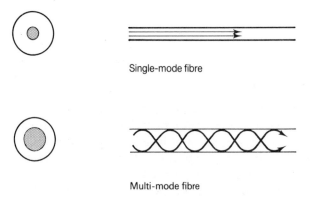

Fig. 7.2   Glass-based transmission media

having a single coherent wavelength. Single-mode systems are character-ised by very high frequency bandwidth, relatively long cable runs (approximately 50 km maximum), and expensive transmitting and receiving equipment.

Multi-mode systems permit a number of different wavelengths to be propagated simultaneously. Multi-mode systems are characterised by much lower frequency bandwidth, medium-length cable runs (approximately 10 km maximum), and modestly priced transmitting and receiving equipment.

Single-mode systems are typically used by PTT (Postal, Telephone and Telegraph) authorities, whereas multi-mode systems find wider application in the industrial arena.

7.2.3   Ether-based media

'Ether-based media' is a generic term to describe any transmission system which uses the atmosphere as the transmission medium. Examples of such systems include:

- HF, VHF and UHF radio
- microwave
- infra-red light.

Such systems operate on the basis of *wave theory*; they find limited use in automatic control applications. Their main use tends to be in situations in which cable-based systems are impractical: for example, with mobile machinery. The advantages with this type of system are:

- high frequency bandwidth
- long transmission distance
- mature technology.

The main disadvantages are relatively high cost and susceptibility to multi-path propagation.

7.3   Modems

Modem is a word which has come into common usage, and stands for *Modulator/demodulator*. Modems are used for converting digital information from one form to another.

Before proceeding any further, it is important to establish one of the most crucial relationships of Communication Engineering – both analog and digital: *the capacity of a communication 'channel' determines the maximum rate of information transfer*. From the early work of Hartley and Shannon, the capacity of a communication channel can be expressed as:

$$C = W \log_2(1 + S/N)$$

where   C = channel capacity, in bits/second

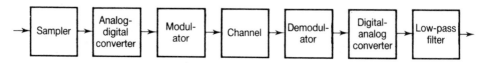

Fig. 7.3   A typical information channel

> W = channel bandwidth, in Hz
> S = signal power
> N = noise power

This equation applies specifically to an analog channel that is linear, non-dispersive, and with additive Gaussian noise; it is known as *Shannon's Theorem*. In general, the capacity of a channel increases with an increase in bandwidth.

The other concept requiring a short explanation at this point is that of the *Nyquist Criterion*, which in layperson's terms may be stated thus:

> *Information must be encoded (in a digital format) at a rate greater than – or equal to – twice the bandwidth of the information source, in order to be able to accurately reproduce the information after transmission.*

If the sampling rate should be less than the Nyquist rate, the effect would resemble the example shown in Fig. 7.4.

Thus, for a typical voice channel having a bandwidth of 3.4 kHz, the sampling rate chosen is 8 kHz.

Typically, modems are used for the transmission of digital information over analog circuits, using techniques such as *frequency shift keying (FSK)*, *phase shift keying (PSK)*, *quadrature amplitude modulation (QAM)* and *trellis coding*. The last two of these techniques are used to provide high-speed (> 2400 bits/seconds) data transfer over standard voice-frequency circuits: that is, they are used so as to pack more than one item of information into each information symbol.

Table 7.1 lists some of the modem standards in use today.

7.4   Switching and multiplexing

In general, the transmission of information from one point to another can be considered to require the set of elements shown in Fig. 7.5.

Encoding is the process whereby information is modified to a form suitable for transmission; clearly, *decoding* is the reverse process. Some common examples of encoding are as follows:

Fig. 7.4   Example of the effect of 'Sub-Nyquist' sampling

Table 7.1 The more commonly used modem standards

Standard	Description
V.21	300 bps full-duplex modem, for use on the public switched network
V.22	1200 bps full-duplex modem, for use on the public switched network
V.22 bis	2400 bps full-duplex modem, for use on the public switched network
V.23	600/1200 bps full-duplex modem, for use on the public switched network
V.26	2400 bps modem for use on four-wire leased telephone circuits
V.26 bis	1200/2400 bps modem for use on the public switched network
V.27	4800 bps modem with manual equaliser, for use on leased telephone circuits
V.27 bis	2400/4800 bps modem with automatic equaliser, for use on leased telephone circuits
V.29	9600 bps modem for use on the public switched network
V.36	Modems for synchronous data transmission, using the 60 to 108 kHz group

Analog
- Analog modulation (AM)
- Frequency modulation (FM)

Digital
- Pulse code modulation (PCM)
- Quadrature amplitude modulation (QAM)
- Frequency shift keying (FSK)
- Phase shift keying (PSK)

Figure 7.6 shows a generalised communication channel, using an encoder and decoder.

Multiplexing is the technique whereby the information transfer process is maximised, and typically it involves the merging of many signal sources

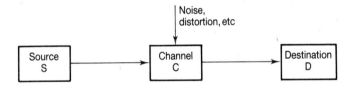

Fig. 7.5 Generalised communication system

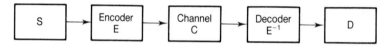

Fig. 7.6   Generalised communication channel involving encoding and decoding

into one communication channel. At the receiving end of the channel, the signal is *demultiplexed* so as to reconstitute the original set of signals. The process is shown in Fig. 7.7.

Multiplexing techniques fall into two general categories: *frequency-division multiplexing (FDM)* and *time-division multiplexing (TDM)*.

FDM is a process in which each information source is allocated a range of frequencies, so that the spectrum is shared between the signal sources. Systems of this type are typically found in high-capacity analog microwave networks, like those used by PTT authorities. The process is shown in Figure 7.8.

The encoder consists of an FM modulator, with carrier frequency f_x, and the decoder is an FM demodulator. Each encoder would be band-limited, so as to prevent contamination from adjacent channels.

TDM is a process similar to that of FDM, except that time is shared between information sources. Usually, this technique is applicable only to digital data – note that analog information is converted to a digital format (usually PCM), for transmission via a TDM channel. TDM works by slotting other signals in between the sampling instants, as shown in Fig. 7.9.

In block diagram format, the generalised TDM system is depicted in Fig. 7.10.

Typical TDM systems are found in high-capacity digital microwave and optical fibre networks, as used by PTTs.

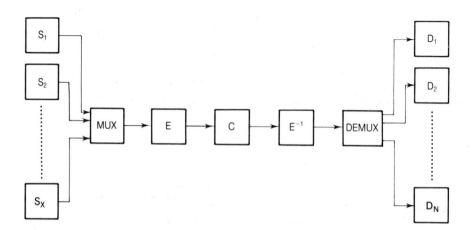

Fig. 7.7   Communication using multiplexers

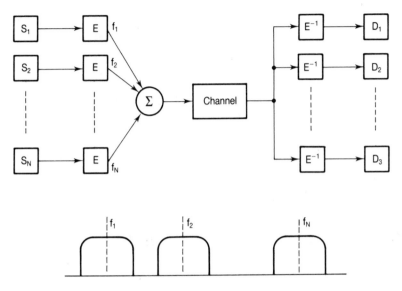

Fig. 7.8   Frequency-division multiplexing

7.5   Serial digital transmission formats and standards

Serial transmission of data involves sending data one bit at a time, from one digital device to a second digital device. There are three alternative modes for sending serial information:

- simplex
- half-duplex
- full-duplex.

Simplex data transmission is defined as data flow in one direction only, along a channel linking two digital devices.

Half-duplex is defined as data flow in both directions along a channel, but in only one direction at a time.

Full-duplex is defined as data flow in both directions simultaneously.

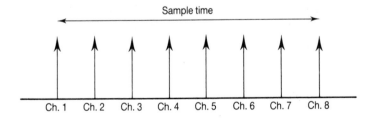

Fig. 7.9   The TDM sampling process

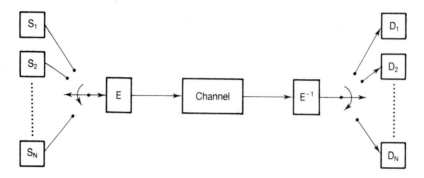

Fig. 7.10   Generalised time-division multiplexing

In any of these three modes, data may be transmitted either *asynchronously* or *synchronously*.

7.5.1   Asynchronous data transmission

With asynchronous data transmission, a data word is sent as a burst of pulses occurring at a predetermined bit rate, but commencing at no particular moment in time. Each data word is sent in a binary bit frame as shown in Fig. 7.11.

The start bit marks the beginning of the frame, and tells the receiving device that data follows. The data following this start bit may be 5, 6, 7 or 8 bits in length: note that the transmitting and receiving devices must have their data length parameters each set to the same value. Following the stream of data bits is the parity bit. *Parity* is a simple form of error correction, and can assume any of the following formats:

- *odd parity* – the parity bit is a logical 0 if the number of logical 1s in the preceding set of data bits is odd; otherwise, it is set to logical 1
- *even parity* – the parity bit is a logical 0 if the number of logical 1s in the preceding set of data bits is even; otherwise, it is set to logical 1.
- *mark parity* – the parity bit is always a logical 1
- *space parity* – the parity bit is always a logical 0
- *nil parity* – the parity bit is unused.

The final bit in the asynchronous data frame is the *stop bit*, which in fact may be either 1, $1\frac{1}{2}$ or 2 bits in length. The stop bit is used to mark the end of the frame.

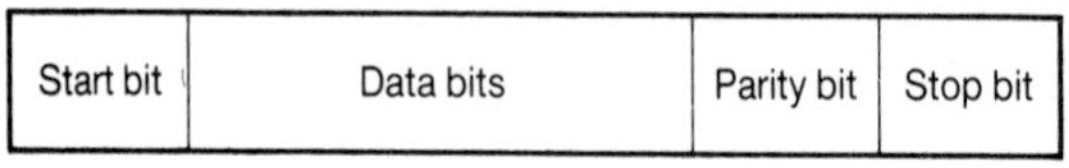

Start bit	Data bits	Parity bit	Stop bit

Fig. 7.11   Binary bit frame for asynchronous serial data transmission

Link control determines how two digital devices negotiate the transmission of data from one to the other. The first parameter which must be set commonly for communication to occur is the transmission speed. Serial data transmission speed is measured in *bits per second* (bps), with values ranging from 110 bps to 115 kbps. Most commonplace speeds are 1200, 2400, 9600 pbs and 19.2 kbps.

Note, however, that it is possible for two devices set for different speeds to communicate, provided that an intermediate device, known as a *baud rate changer*, is interposed: this is shown in Fig. 7.12. In a situation such as this, *buffering* of data is required, whereby data words are temporarily stored until the slower device, when receiving, is ready to accept a new data word.

Three alternative types of link control are used in asynchronous communication:

- hardware flow control
- XON/XOFF flow control
- ENQ/ACK flow control.

Hardware flow control uses the control signals provided at the communications interface – refer to Section 7.6.

XON/XOFF flow control uses single characters to start and stop transmission: the XON character is ASCII 17_{10}, whilst the XOFF character is ASCII 19_{10}. A typical scenario involves one device transmitting to a second device, which is capturing the data into a buffer. As the memory capacity of the buffer is approached, the receiving device will send an XOFF character, which will halt transmission of data; once the buffer has been (either completely or almost) emptied, the receiving device will send an XON character, which will cause transmission to resume.

The *ENQ/ACK flow control* method is block orientated, in that the transmitting device sends one block of data in a burst. A typical scenario has the transmitting device send an ENQ character (ASCII 5_{10}) when it is ready to commence transmission; if the receiving device is ready to accept data, it will return an ACK character (ASCII 6_{10}). Upon receipt of the ACK character, the transmitting device will send the block of data.

7.5.2 Synchronous data transmission

With synchronous data transmission, data are transmitted with special characters which synchronise the transmitter and receiver of the communications equipment. The use of *synch bits* facilitates the transmission of data, without the overhead of start and stop bits for each character.

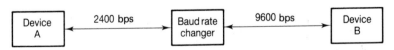

Fig. 7.12 Communication via a baud-rate changer

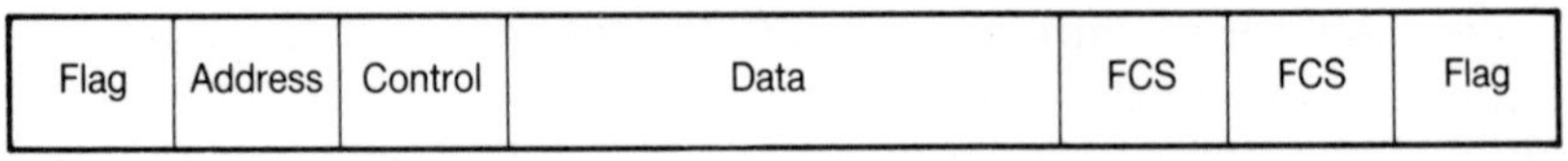

Flag	Address	Control	Data	FCS	FCS	Flag

Fig. 7.13   SDLC frame structure

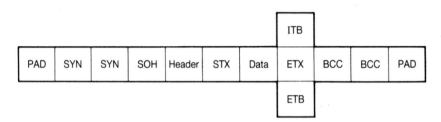

Fig. 7.14   Bisynch frame structure

There are two principal categories of synchronous data transmission:

- bit orientated
- character orientated.

Bit orientated synchronous mode is one which does not specify character boundaries, whereas each boundary is clearly defined with *character orientated synchronous mode*.

One of the most commonplace bit orientated protocols is *Synchronous Data Link Control (SDLC)*. The structure of an SDLC frame is as shown in Fig. 7.13.

The fields within this frame structure are defined as follows:

- *Flag* – this is the frame start/stop byte, and has the value $7E_{16}$.
- *Address* – for a *Control* frame, this field contains the address of the receiving device; for a *Response* frame, this field contains the address of the transmitting device.
- *Control* – this field indicates the function and purpose of each frame.
- *Data* – this field contains the data to be transmitted, and typically but not essentially is divided into groups of eight bits.
- FCS – this field contains a *Frame Check Sequence*, and is used for error control. Typically, this field will contain a *cyclic redundancy check* value.

A typical character-orientated protocol is *Binary Synchronous Communications (Bisynch)*. The structure of a Bisynch frame is shown in Fig. 7.14, whilst the hexadecimal value and description of each Bisynch character are given in Table 7.2.

The fields in a Bisynch frame are defined as follows.

- SYN – this field is used to establish and maintain synchronisation between the transmitting and receiving devices.
- SOH – this field indicates the start of an (optional) header block.
- Header – this field is user-definable, but is typically used for:
  - source or destination identification
  - priority
  - date
  - message type
- STX – this field both terminates the header and indicates the beginning of the data field.
- DATA – this field contains the information component to be transmitted.
- ITB/ETB/ETX – this field indicates:
  1. End of Intermediate Block, and more data will follow in the next frame.
  2. End of Transmission Block, and no more data will follow for this message; however, more messages may follow.
  3. End of Transmission Block, and no more data will follow.
- BCC – this field is a block check character, and is used for error control purposes.

7.5.3   Binary data encoding

The data to be transferred over any communications channel must first be encoded in a form which is recognisable by both the transmitting and

Table 7.2   Table of Bisynch control codes

Bisynch character	Hexadecimal Value	Description
SYN	32	Synchronous Idle
PAD	55	Start of Frame Pad
PAD	FF	End of Frame Pad
DLE	10	Data Line Escape
ENQ	2D	Enquiry
SOH	01	Start of Heading
STX	02	Start of Text
ITB	1F	End of Intermediate Block
ETB	26	End of Transmission
ETX	03	End of Text

Table 7.3   ASCII and EBCDIC codes for data communications

Character	ASCII base 10	EBCDIC base 16	Character	ASCII base 10	EBCDIC base 16
A	65	C1	q	113	98
B	66	C2	r	114	99
C	67	C3	s	115	A2
D	68	C4	t	116	A3
E	69	C5	u	117	A4
F	70	C6	v	118	A5
G	71	C7	w	119	A6
H	72	C8	x	120	A7
I	73	C9	y	121	A8
J	74	D1	z	122	A9
K	75	D2	0	48	F0
L	76	D3	1	49	F1
M	77	D4	2	50	F2
N	78	D5	3	51	F3
O	79	D6	4	52	F4
P	80	D7	5	53	F5
Q	81	D8	6	54	F6
R	82	D9	7	55	F7
S	83	E2	8	56	F8
T	84	E3	9	57	F9
U	85	E4			
V	86	E5	SP	32	40
W	87	E6	!	33	5A
X	88	E7	"	34	7F
Y	89	E8	#	35	7B
Z	90	E9	$	36	5B
a	97	81	%	37	6C
b	98	82	&	38	5D
c	99	83	'	39	7D
d	100	84	(	40	4D
e	101	85	)	41	5D
f	102	86	*	42	5C
g	103	87	+	43	4E
h	104	88	'	44	65
i	105	89	—	45	60
j	106	91	.	46	4B
k	107	92	/	47	61
l	108	93	:	58	7A
m	109	94	;	59	5E
n	110	95	<	60	4C
o	111	96	=	61	7E
p	112	97	>	62	6E

Table 7.3   *Continued*

Character	ASCII base 10	EBCDIC base 16	Character	ASCII base 10	EBCDIC base 16
?	63	6F	FF	12	0C
@	64	7C	FS	28	22
[	91		GS	29	
\	92	E0	HT	9	05
]	93		IFS		1C
^	94		IGS		1D
—	95	6D	IL		17
`	96		IRS		1E
{	123	C0	IUS		1F
		124	6A	LC		06
}	125	D0	LF	10	25
~	126	A1	NAK	21	3D
ACK	6	2E	NL		15
BEL	7	2F	NUL	0	00
BS	8	16	PF		04
BYP		24	PN		34
CAN	24	98	PRE		27
CC		1A	RES		14
CR	13	0D	RLF		09
DC1	17	11	RS	30	35
DC2	18	12	SI	15	0F
DC3	19	13	SM		2A
DC4	20	3C	SMM		0A
DEL	127	07	SO	14	0E
DLE	16	10	SOH	1	01
DS		20	SOS		21
EM	25	19	STX	2	02
ENQ	5	2D	SUB	26	3F
EOB		26	SYN	22	32
EOT	4	37	UC		36
ESC	27	27	US	31	
ETB	23	26	VT	11	
ETX	3	03			

Note that the ASCII codes have been listed in a decimal-valued representation of the first seven bits for the character: the value of the eighth bit, which is the parity bit, depends upon the type of parity selected for each application. On the other hand, the EBCDIC codes have been listed in a hexadecimal-valued representation for the full set of eight bits for the character, and remain valid independently of parity considerations.

receiving devices. There are two commonplace encoding formats available for data communications: *American Standard Code for Information Interchange (ASCII)* and *Extended Binary-coded Decimal Interchange code (EBCDIC)*, which is an IBM standard. These codes are set out in Table 7.3.

7.6   Serial communications interfaces

Communications interfaces have been standardised by the Electronics Industries Association (EIA) and the Consultative Committee for International Telegraph and Telephone (CCITT). Three commonplace data interfaces have been developed: RS 232C/V.24, RS 449 and V.35.

7.6.1   RS 232C/V.24 data interface

This is probably the most commonplace data interface available at the present time. The RS 232/V.24 standard is applicable to the 25-pin interconnection of Data Terminal Equipment (DTE) and Data Communication Equipment (DCE). Table 7.4 shows the standard pin allocations.

The RS 232C/V.24 interface defines the transmission of serial data up to 19.2 kbps over a distance of 15 metres.

7.6.2   RS 449C data interface

To provide for communications over longer distances, the RS 449 standard was developed. This interface is applicable to 37-pin and 9-pin interconnection of DTE and DCE. Table 7.5 shows the standard pin allocations.

The B pins provide the return paths for the RS 449 interface.

The electrical characteristics of RS 449 are defined in RS 422 and RS 423. RS 422 specifies a method of transmission balanced about signal common potential, with differential voltages between 2 and 6 volts. Signalling rates are up to 100 kbps, over distances up to 1000 metres. Rates of 10 Mbps are also possible, over distances up to 10 metres.

RS 423 specifies an unbalanced method of transmission, with voltage levels between 4 and 6 volts with respect to signal common. Signalling rates are up to 3 kpbs, over distances up to 1000 metres. Higher data rates may be accommodated over shorter distances.

7.6.3   V.35 data interface

The V.35 interface was developed to cater for the need for high-speed communications, and was defined by the CCITT for data transmission at 48 kbps. Table 7.6 lists the signal descriptions for this interface.

7.7   Serial communications devices

This section presents examples of dedicated communications devices, which typically are tailored to work with a specific family of microprocessors.

Table 7.4   RS 232C/V.24 Signal descriptions

Pin number	Description
1	Protective Ground
2	Transmitted Data
3	Received Data
4	Request to Send
5	Clear to Send
6	Data Set Ready
7	Signal Ground
8	Carrier Detect
9	
10	
11	
12	Secondary Carrier Detect
13	Secondary Clear to Send
14	Secondary Transmitted Data
15	Transmitter Signal Element Timing
16	Secondary Received Data
17	Received Signal Element Timing
18	
19	Secondary Request to Send
20	Data Terminal Ready
21	Signal Quality Detector
22	Ring Indicator
23	Data Signal Rate Selector (DCE/DTE)
24	Transmitter Signal Element Timing (DTE)
25	

7.7.1   MC6850 asynchronous interface adaptor (ACIA)

The MC6850 ACIA is manufactured by Motorola, as part of its 68xx family. This device provides the data formatting and control functions necessary to interface serial asynchronous data to bus-organised micro-processors. Features of this device are:

- 8- or 9-bit data format
- even or odd parity generation
- parity, overrun and framing error checking
- up to 1 Mbps data rates
- modem control functions.

Table 7.5   RS 449 Signal descriptions

9-Pin number	37-Pin number A	B	Description
1	1		Shield
5	19		Signal Ground
9	37		Signal Common
6	20		Receiver Common
	4	22	Send Data
	6	24	Receive Data
	7	25	Request to Send
	9	27	Clear to Send
	11	29	Data Mode
	12	30	Terminal Ready
	15		Incoming Call
	13	31	Receiver Ready
	33		Signal Quality
	16		Signalling Rate Selector
	2		Signalling Rate Indicator
	17	35	Terminal Timing
	5	23	Send Timing
	8	26	Receiver Timing
3			Secondary Send Data
4			Secondary Receive data
7			Secondary Request to Send
8			Secondary Clear to Send
2			Secondary Receiver Ready
	10		Local Loopback
	14		Remote Loopback
	18		Test Mode
	32		Select Standby
	36		Standby Indicator
	16		Select Frequency
	28		Terminal in Service
	34		New Signal

The internal structure of the MC6850 is shown in Fig. 7.15.

To the microprocessor, the MC6850 ACIA appears as two addressable registers. In fact, there are four internal registers: the Transmit Data and Control registers are write-only, whilst the Receive Data and Status registers are read-only. The serial interface signals of this device comprise:

Table 7.6 V.35 Signal descriptions

Pin identification	Description
A	Protective Ground
B	Signal Ground
C	Request to Send
D	Clear to Send
E	Data Set Ready
F	Received Line Signal
H	Data Terminal Ready
J	Ring Indicator
K	Local Test
R	Received Data (Signal A)
T	Received Data (Signal B)
V	Serial Clock Receive (Signal A)
X	Serial Clock Receive (Signal B)
P	Send Data (Signal A)
S	Send Data (Signal B)
U	Serial Clock Transmit Ext (Signal A)
W	Serial Clock Transmit Ext (Signal B)
Y	Serial Clock Transmit (Signal A)
a	Serial Clock Transmit (Signal B)

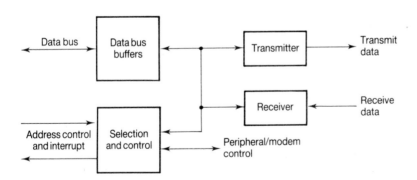

Fig. 7.15 MC6850 asynchronous interface adaptor block diagram

- Transmit Data
- Receive Data
- Data Carrier Detect
- Request to Send
- Clear to Send.

It can be seen that it is a simple matter to implement an RS 232C-compatible serial interface. In fact, this device has been used in many personal computers to perform this very function.

7.7.2 MC6852 synchronous interface adaptor

The MC6852 is also manufactured by Motorola, as part of its 68xx family. This device provides all of the functions necessary to implement a synchronous data interface for bus-orientated microprocessors. Features of this device are:

- 7-, 8- or 9-bit data format
- optical parity generation
- up to 1.5 Mbps data rates
- modem control functions
- character synchronisation, or one or two Synch codes.

The internal structure of the MC6852 is shown in Fig. 7.16.

   To the microprocessor, the MC6852 appears as two addressable registers. The MC6852 differs from the MC6850 in that two of the internal registers are read-only and five are write-only as follows:

- Status Register – read-only
- Receive Data Register – read-only
- Control 1 Register – write-only
- Control 2 Register – write-only
- Control 3 Register – write-only

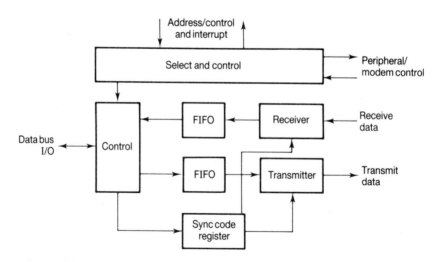

Fig. 7.16 MC6852 synchronous interface adaptor block diagram

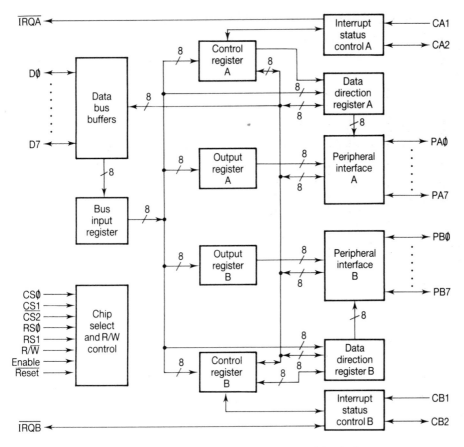

Fig. 7.17 MC6821 peripheral interface adaptor block diagram

- Synch Code Register – write-only
- Transmit Data Register – write-only.

The control of synchronous data transmission and reception is far more complex than that of asynchronous data transmission, as is evidenced by the additional control registers required by this device.

7.7.3 MC6821 peripheral interface adaptor (PIA)

The MC6821 is also manufactured by Motorola as part of the 68xx family. It is not strictly a pure serial-communications device. Primarily, it provides two 8-bit bidirectional data lines and four control lines. The MC6821 has great flexibility: it can be used to implement both serial and parallel communications interfaces. The internal structure of the MC6821 is shown in Fig. 7.17.

7.8   Parallel digital transmission formats and standards

Parallel communications are typified by high speed and short transmission distance; the latter is due to the increased capacitance of the transmission medium.

7.8.1   IEEE–488 programmable interface standard

This Standard was first known as the *General-Purpose Interface Bus (GPIB),* which was pioneered by Hewlett Packard as a means to enable intelligent instruments to communicate with each other. The GPIB gained so much acceptance that it became a standard ratified by the IEEE.

The IEEE–488 interface transmits data in a byte-serial/bit-parallel format between devices. Such devices can be classified into three distinct groups:

- talkers
- listeners
- controllers.

Talkers are devices which are configured to allow them to place data on the bus. Normally, only one talker is active at any one time.

Listeners are devices which can accept data from the bus, via a defined handshake sequence. More than one listener may accept data from the bus, at any one time.

Controllers are devices which assume control of the bus, via the *ATN* line. Only one controller can be active, at any one time.

The IEEE–488 standard is for a bus-based architecture consisting of sixteen conductors, which comprise data lines, control lines and handshake lines. The standard uses *negative logic*: that is, a line is held at $+5$ V DC for a logical 0, and at 0 V for a logical 1. Table 7.7 lists the signal descriptions for this interface.

The data lines are labelled DIO1 through DIO8, and are used for both data and control words. The data may be in either natural binary or ASCII format.

The control lines used by the bus are:

- IFC – *Interface Clear*
- SRQ – *Service Request* – used to indicate a need for a service sequence
- REN – *Remote Enable* – used to disable the *front panel* of an instrument; that is, all actions to be performed by the instrument will be controlled by a remote device
- EOI – *End of Identify* – used for dual purposes:

  - by a talker, to indicate the end of a multi-byte data transfer

Table 7.7 IEEE–488 Signal descriptions

Pin number	Description
1	DIO1
2	DIO2
3	DIO3
4	DIO4
5	EOI
6	DAV
7	NRFD
8	NDAC
9	IFC
10	SRQ
11	ATN
12	Shield
13	DIO5
14	DIO6
15	DIO7
16	DIO8
17	REN
18–23	Ground
24	Logic Ground

- by a controller, in conjunction with the ATN line, when executing a parallel-poll sequence.

The handshaking lines used by the IEEE–488 bus are:

- DAV – *Data Valid* – used by the controller to indicate when a control byte has been placed on the data lines
- NRFD – *Not Ready for Data* – used by a listener to stop further action by a controller or talker
- NDAC – *No Data Accepted* – used by a listener to indicate that it has not yet accepted the last byte placed upon the bus.

7.8.2 CAMAC interface standard

The *CAMAC*, or *Computer Automated Measurement and Control* standard was developed in Europe to facilitate electrical and mechanical interfacing of previously incompatible devices. The development is now covered by the IEEE–583 Modular Instrumentation and Digital Interface System standard. The system incorporates a data highway (*Dataway*), together with modular functional elements which are completely compatible. Additional levels of compatibility can be achieved, through the use of standardised parallel highways (covered by IEEE–595) and serial highways (covered by IEEE–596). IEEE–683 is an additional standard which covers the specification of *block transfer* software commands, whereby the execution of short program statements can facilitate the transmission of large quantities of data.

The system requires all hardware modules to be designed for a common enclosure, called a *crate*. The signal interface is achieved by means of the bused Dataway at the rear of the crate. The Dataway cycle time is 1 μs, and the device addressing capability is very high. The Dataway specification includes provision for:

- 24 Read lines (buses)
- 24 Write lines (buses)
- 24 Station Addresses (dedicated lines)
- 24 Station Demand lines
- 16 Substation Addresses (binary coded) per Station Address
- 32 Functions (binary coded).

A crate can have up to 24 devices mounted within it, each one occupying a *station*, except that the *crate controller* will normally occupy two stations.

The standard parallel highway system enables up to seven such crates to be networked. The parallel highway is capable of very high data transfer rates over short distances. Each parallel port consists of a 132-contact connector accommodating 65 signals and their return lines, plus a cable shield. Parallel-data ports are interconnected electrically in parallel, in *daisy-chain* fashion. A number of these parallel highways may be used in order to assemble a large system, in which case one such parallel highway, together with its interconnected devices, is called a *CAMAC Branch*.

The standard serial highway systems enable up to 62 crates to be networked. The serial highways have lower data transfer rates, but can transmit over long distances. The serial highway may be either *bit serial*, using one data signal and a bit-clock signal, or *byte serial*, using eight data signals and a byte-clock signal. Thus, data may be transmitted either as bit streams or as byte streams, as the case may be. Clock rates up to 5 MHz may be used. Serial-data ports are interconnected electrically in series, to form current loops.

Power bus connections to each crate station make provision for -6, -12, -24, $+6$, $+12$, $+24$ and $+200$ V DC, together with 117 V AC and various power neutral connections.

The advantage of CAMAC systems are considered to be:

- flexibility
- interchangeability
- ease of restructuring

- delayed obsolescence
- high degree of computer independence
- reduction of interfaces
- easy interchange between installations
- reduction of inventory
- ease of serviceability
- reduction of design effort
- wide range of supplier sources
- software economies.

CAMAC systems have been installed in Europe and the USA, in industrial, laboratory, medical and aerospace applications.

7.8.3   Centronics interface standard

One of the most popular methods for implementing computer-to-printer communications is the *Centronics* standard interface. The standard uses a 36-pin byte-wide interface, which has eight data lines carrying the respective bits in parallel. Transmission of the data is controlled by the *STROBE* pulse. Flow control across the interface is achieved by use of the *ACKNLG* and *BUSY* lines. Table 7.8 lists the signal descriptions for the Centronics standard bus.

The IBM PC implements a slightly modified version of the Centronics standard, as part of its parallel port. The printer interface is implemented on a female DB 25 connector, with the signal descriptions as given in Table 7.9.

7.9   Parallel communications devices

7.9.1   MC146823 parallel interface

The MC146823 is a CMOS-based parallel interface device, manufactured by Motorola. Essentially, it consists of twenty-four data lines, implemented as three 8-bit ports, and fifteen control registers. The internal structure of the MC146823 is shown in Fig. 7.18.

Using either Port A or Port B and bits 0 to 3 of Port C, it is a relatively straightforward exercise to implement a parallel (Centronics-compatible) printer port.

7.9.2   The 8255 programmable peripheral interface

The 8255 is a parallel interface manufactured by Intel. This device is compatible with the Intel microprocessor families, and is used as a general-purpose peripheral interface. The internal structure of the 8255 is shown in Fig. 7.19.

Table 7.8 Signal descriptions for the Centronics standard bus

Signal pin number	Return pin number	Signal description
1	19	STROBE—strobe pulse
2	20	DATA 1
3	21	DATA 2
4	22	DATA 3
5	23	DATA 4
6	24	DATA 5
7	25	DATA 6
8	26	DATA 7
9	27	DATA 8
10	28	ACKNLG — data received OK
11	29	BUSY—not ready
12		PAPER OUT
13		SELECT
14		SUPPLY GND
15		OSCXT—external clock
16		LOGIC GND
17		CHASSIS GND
18		+ 5 V
31	30	INPUT PRIME
32		FAULT

Table 7.9 Signal descriptions for the IBM PC parallel port

Pin number	Signal description	Pin number	Signal description
1	STROBE	10	ACKNLG
2	DATA 1	11	BUSY
3	DATA 2	12	PAPER OUT
4	DATA 3	13	SELECT
5	DATA 4	14	AUTO FEED
6	DATA 5	15	ERROR
7	DATA 6	16	RESET
8	DATA 7	17	SELECT INPUT
9	DATA 8		

This device offers features similar to those of the MC6821 and MC146823: that is, it is a general-purpose device, which can be used in order to implement parallel communications interfaces.

7.10 Local area networks

The use of Local Area Networks (LANs) is increasing as the need for management information increases. The following descriptions deal with those LANs which may be found in industrial situations.

7.10.1 Ethernet and IEEE–802.3 LANs

Ethernet is a generic term which has come to describe any LAN based upon *CSMA/CD* techniques. The original term 'Ethernet' was coined by Xerox Research Laboratories to describe a 10 Mbps network employing co-axial cable. This type of network uses Carrier-Sense Multiple-Access/Collision Detection (CSMA/CD) techniques to arbitrate bandwidth on the LAN.

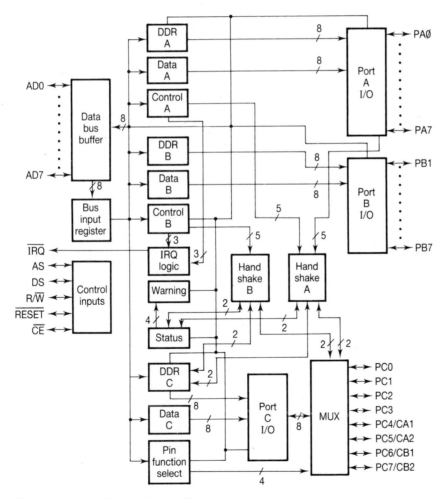

Fig. 7.18 MC146823 parallel interface block diagram

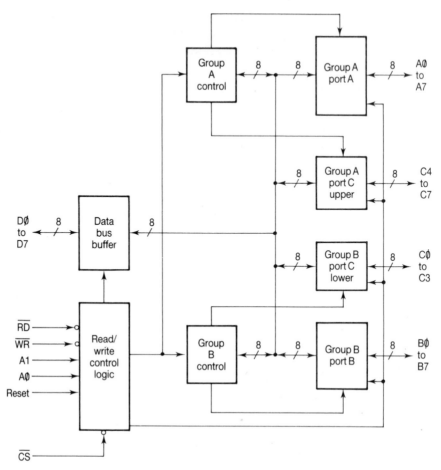

Fig. 7.19   8255 programmable peripheral interface block diagram

The term 'CSMA/CD' simply defines the manner in which many computers can share a common communications medium, by sensing (via the carrier) whether other stations are currently transmitting. The collision detection comes into effect when two stations happen to transmit simultaneously. In this situation, both of the transmitting stations will postpone transmission by different random values of time delay.

Figure 7.20 shows a typical Ethernet system of transmitting and receiving stations.

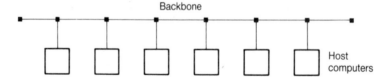

Fig. 7.20   The Ethernet local area network

The other type of CSMA/CD LAN in common use is based upon the IEEE–802.3 Standard: this is very similar to Ethernet and, in fact, the two systems can co-exist on the same cabling scheme, if necessary. The IEEE–802.3 Standard Committee defined:

- the cabling schemes
- the media access unit
- the data packet structures
- the time delay algorithm
- the transmission rate (10 MBytes/s).

IEEE–802.3 LANs may be implemented on the following cable types:

- thick co-axial cable (10 Base 5)
- thin co-axial cable (10 Base 2).
- twisted-pair cable (10 Base T).

A typical arrangement of CSMA/CD LANs in an industrial application is shown in Fig. 7.21.

The mathematical evaluation of a CSMA/CD-based LAN can be extremely rigorous, unless the analysis is restricted to certain limiting cases: the point to be made here is that it is very difficult to predict the performance of such a LAN. Thus:

- typical average utilisation can range from 1 percent to 10 percent peak
- 1-second utilisation may be as high as 60 percent to 70 percent
- average utilisation figures in excess of 20 percent indicate a LAN which is extremely busy and could create delays.

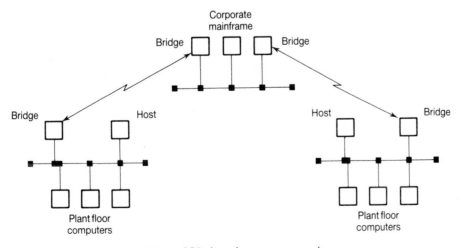

Fig. 7.21   A typical CSMA/CD local area network

7.10.2   Token-ring LANs

Token-ring LANs are not as common as other types of LAN in the industrial arena. This is due mainly to the fact that IBM, as the developer of the system, has been mainly concerned with the commercial market-place. This situation is likely to change in the future, as the AS/400 Series (IBM's small to medium-sized minicomputer range) becomes more popular in industrial applications.

The Token-ring LAN has also been adopted as a standard by the IEEE, and granted the number IEEE–802.5. LANs of this type are typically implemented on IBM Types 1, 2 or 3 twisted-pair cable. There are two alternative speeds for operation of a Token-ring LAN: 4 MBytes/s and 16 MBytes/s.

Figure 7.22 represents a typical Token-ring network. A *token*, which is a block of specially formatted data, is passed from station to station. The station which is currently passing the token has control of the network at that moment in time, and may thereby transmit data to any other station. The token may only be retained by any one station for a specific interval of time, after which it must be passed on to the next station in the ring.

Because the value of the retention time is known, it is possible to determine exactly the response time of a Token-ring LAN, irrespective of loading upon the network.

Typically, the use of Token-ring LANs in industrial applications is very similar to that of CSMA/CD LANs. In fact, the only difference is the architecture of the network. An example of a typical industrial application of a Token-ring LAN is shown in Fig. 7.23.

7.10.3   Token-bus LANs

The IEEE–802 series of standards also defines another type of network structure, which is similar to that of the Token-ring except that a bus structure is used. The standard is known as IEEE–802.4, and is used

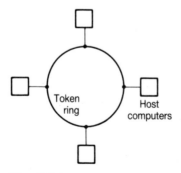

Fig. 7.22   A token-ring local area network

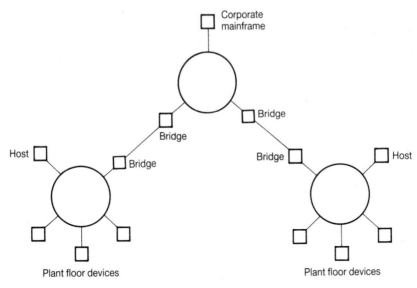

Fig. 7.23   Industrial application of a token-ring LAN

predominantly in the *Manufacturing Automation Protocol (MAP)* for interconnecting industrial computers in a plant environment.

7.11   The OSI standard

In 1983, the International Standards Organisation (ISO) approved the *Reference Model for Open Systems Interconnection (OSI)* as an international standard. This standard was developed in order to allow multi-vendor communications capability. It will be many years before truly 'open' communications exist between disparate system vendors, but the OSI Model will provide the basis under which it will occur. The Model can act as a bridge between two proprietary networks which previously could not communicate. Members of different computer families may communicate and share workload and data files, without the need for special *gateways* for the translation of signals passing between incompatible protocols. With this technology, disparate operating systems, application programs, databases, programming languages and communication networks may interact. Public and commercial networks may be linked.

The OSI Reference Model is an abstraction, not a single tangible entity. It is broken into seven *layers*, which represent a group of related functions or tasks, defined so as to make their interfaces and roles easily understood. The roles which control communication between two devices (for example, two computers) are the *OSI Protocols*, and these define the functions represented by the layers, but do not define *how* the functions are implemented.

Any hardware/software product need not provide all of the services of any one layer, nor just the services of only one layer. The functions within a layer can vary, so long as the communication process implicit in the inter-layer connections remains intact.

The OSI Model is shown symbolically in Fig. 7.24. The communication message path between devices enters at the Application Layer of the originating system, travels via a physical interconnection channel from the Physical Layer of the originating system to the Physical Layer of the receiving system, and is passed to the latter for processing, via *its* Application Layer.

In the originating system, each layer adds a *header* to the message before sending it to the receiving system, in which each layer in turn reads and interprets the information, and strips off the header added by its counter-part in the originating system.

The lower four layers provide *Network Management Services*: that is, the means by which data from one user are directed to another user across the network. The upper three layer protocols provide *Application Management Services*, and ensure that the data generated by another user on the network are presented in a usable format; these services include directory services, security management and application configuration management.

Note that the OSI Model does not refer to any one type of protocol or network. The main purpose is to provide the architecture by which vendor-independent networks may be implemented. The functions of the seven layers can be summarised as follows:

- *Application Layer.* This is the highest-order layer in the Model, and its purpose is to ensure that two applications (that is, programs) co-operate in order to carry out the interchange of information. It serves as a window through which application programs access the OSI Model's communication services.

- *Presentation Layer.* This layer isolates the applications from differences in representation of the data being transmitted; for example, an enquiry into a Digital Equipment Corporation database from a Unisys Manu-facturing Planning program. It allows an application to properly interpret the data being communicated. This layer also handles the set-up and termination of transactions.

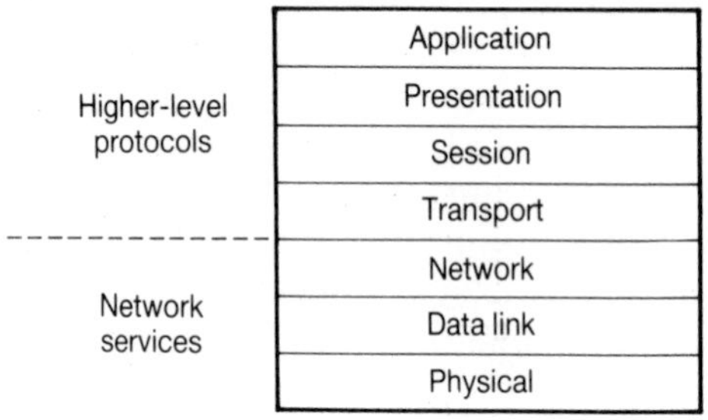

	Application
Higher-level protocols	Presentation
	Session
	Transport
	Network
Network services	Data link
	Physical

Fig. 7.24   The OSI reference model

- *Session Layer.* This layer, as the name suggests, is responsible for handling communications sessions. It co-ordinates the interaction between end-user application programs.
- *Transport Layer.* This layer provides those layers above it with a reliable data-transfer mechanism. It shields the Session Layer from the particular network implementation. Two types of data transmission are available within the Transport Layer: connection-orientated and connectionless transmissions.
- *Network Layer.* This layer provides the actual communication service to the Transport Layer, which is thus shielded from the communications network architecture. It controls communication functions such as routing, relaying and data link connection.
- *Data Link Layer.* This layer controls the transfer of data between two physical layers, and thereby provides error-free, sequential transmission of data over a link in a network.
- *Physical Layer.* This layer defines the physical requirements for attaching to, or for accessing, the network. It determines hardware interconnection details and controls byte-stream encoding for transmission purposes.

7.12 Wide–area networks

Wide-area networks (WANs) are becoming increasingly important for providing plant-status reporting, to centralised management, from a set of widely geographically-dispersed sites. The typical use of a wide-area network involves the linking of multiple local-area networks (of any of the types already discussed), so as to provide the capability of information sharing. Typically, WANs have much lower transmission rates than those of LANs: speeds between 9.6 kbps and 2.048 Mbps would be the normal range. Figure 7.25 shows a typical WAN configuration.

Typically, WANs use the following alternative communication modes:

- X.25-leased links
- TI/CEPT-leased links
- leased point-to-point synchronous links.

X.25 is a standard developed by the CCITT to define packet-switched networks. In networks of this type, all of the data are packetised. Such networks can be very economical for applications involving low traffic volumes. Data rates in the range from 9.6 kbps to 64 kbps are commonplace. Further information on this standard is beyond the scope of this volume, but is available from many reference works.

The TI standard was developed, in the US, to define a digital data stream of 1.544 Mbps. Thus, links of this type are reserved for high-volume applications. Generally, TI is used by telecommunications authorities to provide twenty-four 'time-slots', where each time slot provides either a

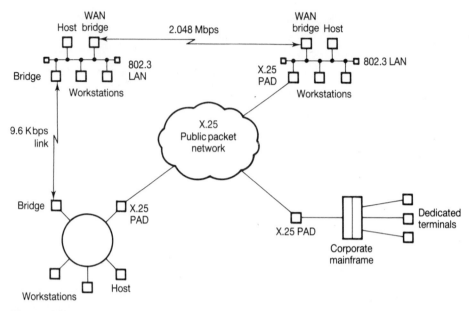

Fig. 7.25   A typical wide-area network configuration

PCM-coded voice channel (8-bit width after sampling at 8 kHz) or a 64 kbps data stream. The CEPT standard is very similar to TI, except that it was developed in Europe. The base CEPT standard – *DI* – involves 2.048 Mbps data rates and thirty time-slots.

Leased point-to-point synchronous services are used where it is uneconomic to use X.25: that is, where the traffic volume is too high for X.25 but insufficient to justify the expense of TI/CEPT. Typical data rates for these services range between 9.6 kbps and 128 kbps.

8

Computer Operating Systems for Control

8.1 Introduction

An operating system is a comprehensive program, normally supplied by the vendor concurrently with a computer hardware installation. It characterises a *virtual machine*: that is, all installations using the same operating system appear to the user to be the same machine, in terms of features and 'feel'. The operating system provides a high-level interface between the user and the computer, such that the hardware becomes readily and cost-effectively usable.

Modern operating systems are considered to have a layered (*onion*) structure with a nucleus of basic features, together with additional layers to provide more sophisticated features.

Non-realtime operating systems, such as UNIX, DOS, and so on, are generally industry-standards, in terms of the commands and features provided, although the code for each has to be customised to suit the requirements of the specific computer hardware platform on which it is to reside. They are usually developed by independent software houses, rather than by computer vendors, although exceptions do exist.

Realtime operating systems, on the other hand, are highly specific and are normally produced by the hardware vendor: no industry-standard systems exist at the present time. The realtime executable programs are generally embedded in the platform in which they reside, and in this role must run under the control of the operating system.

Realtime systems respond to external stimuli generated by sensors, and as a consequence generate signals to energise actuators, either at regular intervals or in direct response to the sensor data. Realtime processes are either *cyclic*, where actuation is to occur at regular intervals in time, or *event-driven*, where actuation is to be a direct and immediate consequence of the external sensor stimulus. In both cases, *interrupts* cause execution: the cyclic process initiates a new actuation cycle following a timer-driven interrupt, whilst the event-driven process initiates a new cycle following a sensor-driven interrupt.

The realtime system must perform to precise deadlines. Often, the consequence of missing a deadline can be catastrophic; in some cases involving cyclic processes, however, missing a deadline may cause actuation to be delayed to a later cycle, without disastrous consequences. *Interrupt latency* is the time elapsing between the arrival of sensor or timer data and the generation of the actuating signal; it will depend upon a combination of the time required to access the interrupt routine, plus the time required to execute that routine. Access time to an interrupt routine will be heavily dependent upon the level of *priority* assigned beforehand to that specific interrupt.

Normally, the features required for a non-realtime operating system are also present in realtime operating systems, and these will be reviewed first in this chapter. However, it is the additional features, present in the *realtime executive*, which distinguish the realtime operating system: these will be given special treatment towards the end of the chapter.

8.2   General requirements for a non-realtime operating system

These requirements can be considered on the basis of such features as:

- general features
- performance
- CPU utilisation
- I/O system
- interrupts
- buffering
- time-sharing techniques
- job scheduling and management
- memory management
- memory protection
- addressing
- file management
- resource management
- process synchronisation.

8.2.1   General features

The user should not be required to have detailed knowledge of the input/output peripheral hardware. Simple input/output software commands should be sufficient to execute data transfers between the processor and the peripherals.

The operating system must be capable of detecting run-time software errors, and provide failsafe reaction together with constructive error messages.

Where appropriate, the oeprating system must allow apparently-simultaneous multiple access by different users.

The operating system must maintain directory structures and file contents, whilst preserving their security against either accidental or deliberate corruption or illegal interrogation.

The operating system must provide software utilities: for example, a text editor.

8.2.2 Performance

The operating system should try to optimise the use of available resources. Some tasks, such as those requiring considerable mathematical computation, will tend to monopolise the CPU. Tasks which require frequent reading from, and writing to, disc memory can cause the system to become *disc-bound*. Yet again, tasks which require frequent I/O data transfers can reduce the CPU utilisation factor to close to 0 percent.

Ideally, the operating system should attempt to keep the utilisation factors of all elements of the system to as high a value as possible. This is more easily attained in practice if there is a good mix of job types, and this can be achieved with *multiprogramming*; in this situation, several programs are interleaved so that they appear to be executing simultaneously (this implies *multitasking* but does not necessarily imply a multiuser environment), although they may then end up in contention for the same system components.

With multiprogramming, completion of a peripheral data transfer must be detected: this can be achieved either by *polling*, in which a *status register* is interrogated, or by the use of interrupts, whereby the status of a peripheral is signalled by its hardware.

The mix of jobs can be controlled by making the I/O-limited job control the peripheral, thereby halting the job during data transfer, so that the CPU-limited job can be run during that phase and can switch back to the I/O-limited job when completion of data transfer is indicated. This process can be assisted by the allocation of *buffer storage* to the peripheral, so that operation of the peripheral can be almost independent of the CPU.

With multiprogramming, the operating system can be used to rapidly shift data in and out of disc store, as required, to suit the current demands of the user programs, using *direct memory access (DMA)* techniques. This makes for efficient use of both CPU and peripherals, whilst the random-access nature of disc storage ensures fast access to data and facilitates prioritisation of tasks, since disc accesses can be readily scheduled by operating system commands.

8.2.3 Job scheduling and management

The *Job Scheduler* keeps a list of all tasks currently in some stage of execution by the computer, together with the status of those tasks and their current I/O requirements.

The *Job Processor* is instructed which task is to be next executed, and the location on disc of the program and data associated with that task.

The *Output Scheduler* keeps a list of all data to be output. When the output peripheral is free, this scheduler selects the next job in the output queue and identifies the location, on the disc, of the data to be output.

The *Disc Manager*:

- reads/writes blocks of data from/to disc storage
- allocates free blocks of storage on disc
- returns surrendered blocks to a pool, and reorganises them so as to maximise disc utilisation.

The *Co-ordinator* schedules the various system processes, and synchronises their execution.

The *Terminal Manager* loads into buffer memory a single character or a single line of input data. The first item of input data from a keyboard will be sent by the Terminal Manager to the Job Scheduler: normally, this will be an ID, password or accounting information. The Job Scheduler assigns a Job Processor to that terminal, and informs the Terminal Manager. All subsequent input and output operations are then made directly between the Terminal Manager and the Job Processor.

8.2.4   Buffers

A simple buffer scheme is inadequate in many cases. For example:

- external devices may generate data continuously
- magnetic tape drives cannot stop dead, but have to back up in order to be able to run up to speed
- multiple terminals are required to service a multiuser environment.

One possible solution is to use multiple buffers, as appropriate; another is to use a large *circular buffer* with *pointers*, with some checks that the pointers do not overtake the data. Such buffers should accommodate the following requirements:

- the user process can be initiated only when there is a complete line of characters to be processed
- the user must be able to type in at least 80 characters at a time
- the user must be able to correct typing errors
- the user should be able to type in multiple lines, if the system is too busy to process the data immediately
- a *prompt* should be output if the system is waiting for data.

A circular buffer with pointers can satisfy these requirements.

8.2.5   Job scheduling principles

The objective with job scheduling is the allocation of CPU time to the execution of tasks, so as to optimise some aspect of system performance:

- to achieve a short response time
- to achieve high CPU utilisation
- to achieve high resource utilisation.

Normally, some degree of compromise is involved.

The Job Scheduler determines which task to allocate to the Job Processor, taking into consideration job priority, foreground/background operation, special user requirements, and so on.

The Co-ordinator selects the process to be re-entered, loads the register for that process, and as a consequence causes the process to be re-entered. The process is then executed until it stops to wait for an event. Since there must be some interaction between processes (as distinct from tasks), a facility is required for passing use of the CPU to another process currently waiting to execute.

The processes are usually queued in order of priority. Once a process commences executing, it may be permitted to run to completion (signified by a *wait* command): this is called *non-preemptive* scheduling.

While a process is running, completion in a reasonable time may not be possible, in which case the Co-ordinator needs to take appropriate action. In a time-sharing system, there will be several processes concurrently requiring execution, so that all need access to the CPU in turn, during any short time interval. A task requiring much CPU use must be *preempted* – that is, interrupted – every few milliseconds, so that the other tasks may be serviced.

This requires an interrupting clock and dynamic adjustment of priority, in a process known as *time slicing*. This function may be performed either by the Co-ordinator or by another module known as a *Process Scheduler*.

In summary, there are three possible levels of scheduling:

- *high-level scheduling*, by the Job Scheduler, which determines the order in which tasks are to be serviced
- *intermediate-level scheduling*, by the Process Scheduler, which adjusts priority of processes and applies time slicing
- *low-level scheduling*, by the Co-ordinator, which applies logical synchronisation of processes.

Often, there are conflicting factors influencing the optimum value for the time slice interval, which may need to be determined probabilistically.

8.2.6   Memory management

Memory, as well as the CPU, must be shared between processes, in a time-sharing system. If memory were to be allocated proportionally to processes, and many of those were inactive for significant time intervals, memory use would be very inefficient. Thus, a memory swapping strategy is required.

The code of inactive processes is temporarily stored on disc, in the form of *core images*. When a process is activated, it is allocated a time slice and the code is loaded (using DMA) back into memory before the process is re-started. On completion of the time slice, the code is transferred (using DMA again) back to disc which, for this purpose, may be on a specially allocated fast-access drive.

One of the principal tasks of the Memory Management system is to optimise both the allocation of memory space and the timing of the DMA transfers. For example, it will need to determine how many processes should reside in memory concurrently. It will also need to protect program code from both accidental and malicious corruption.

8.2.7   Virtual addressing

When time slicing is being used, it is rarely feasible to load the same process repeatedly into the same area of memory. Thus, program modules need to be written in relocatable code, with the address of the current base memory location being held in a *base* or *offset register*.

The relocation mechanism may be assigned to the disc (store) accessing hardware. Each process therefore 'sees' a *virtual store*, which appears (to the process) to occupy the same memory space but which, in fact, frequently relocates to different areas in memory as well as back to disc. This methodology can provide a successful basis for multi-access, but some problems remain to be solved:

- fragmentation
- memory utilisation
- sharing of code and data.

8.2.8   Fragmentation

Processes in any one system differ in size, so that, as they are swapped into and out of memory, vacant memory areas too small to accept new processes can remain. (Similar problems can arise with discs.)

Processes could be 'shuffled' down through memory, as part of the swapping procedure, so as to fill up vacant areas, but this may be unacceptably time consuming.

Alternatively, a list of free memory blocks and their sizes could be maintained, so that the next process to be transferred-in may be inserted into the next available block of sufficient size.

A third alternative is to postpone swapping-in new processes until a sufficient number of processes have been swapped-out for adequate compacted memory space to be available. However, the extended time now needed for swapping could exceed the response time required from the system.

8.2.9 Memory utilisation

Memory itself is uniform, but its usage will not be so. *Static sparseness* arises as a result of the common practice among both programmers and compilers to allocate the lower end of memory address space to program code, and the upper end to data, thereby leaving a potentially large unused area in the intermediate address range. *Dynamic sparseness* arises from the fact that, in any one time slice, only a fraction of the address space storing the related program code and data will actually be accessed. Both types of sparseness have the potential to minimise memory utilisation.

8.2.10 Sharing of code and data

In many cases, it will be advantageous for programs to share code; for example, programs will all need to access compiler functions, and it would be impractical to store a separate copy of the compiler for each program. Similarly, multiple programs will often need access to common databases.
    For code to be shared in this way, it must be *pure* code; that is:

- it must not be self-modifying
- data must be stored separately from programs
- programs must be *re-entrant* – meaning that it must be possible to cease execution at any convenient stage, following an interrupt, and recommence execution at the same place in the process, at a later instant in time.

Modern operating systems have multiple base limit registers, facilitating multiple distinct areas in memory for code and data; the virtual memory then is divided into a number of *segments*, each of which has its own base limit register. The efficiency of the time-sharing process is considerably enhanced as a consequence. *Segmentation* can be regarded as the division of virtual memory into a set of logical areas. Fragmentation can then still occur, but can be alleviated by dividing the user's address space into a number of equal *pages*. Pages are of fixed size and usually are smaller than segments. Fragmentation can still occur, but on a smaller scale, and in this context is called *internal fragmentation*.

8.2.11 File management

The operating system must manage files such that:

- files can be created and deleted, as user requirements change
- access to files can be controlled, and reserved for users having the appropriate *privileges*
- reference to files by symbolic name must be possible, without knowledge of their physical location
- files can be shared, as appropriate

- file contents can be listed
- files can be protected against software and hardware failures.

Each directory entry should have provision for indicating the symbolic name of the file, the position and size of the file on disc, and those users permitted access to the file. Directory structures may be single-level or hierarchical, involving levels of sub-directories in the latter case.

Provision needs to be made for generating frequent *back-ups* of all relevant files, so that in the event of failure the system can be restored without loss of integrity.

8.2.12   Resource management

Contention can arise when two processes are active, and are such that they each require access to the same peripheral device. Each enters a *wait* state, and *deadlock* occurs. If the situation can be foreseen, then the programs can be written so as to avoid its occurrence. In addition, the operating system must be capable of recognising the deadlock situation when it does occur, and of applying the appropriate remedy.

The operating system must also apply a hierarchical structure of privileges to different classes of user, in order to protect files from accidental or malicious corruption by non-privileged users.

8.2.13   Process synchronisation

The elements of the operating system considered up to this point have been presented as largely independent processes. With parallel processing, many could be run concurrently but, in the case of a single processor at least, they have all to be scheduled to be executed sequentially, although interleaving is possible. In any case, data will need to be passed between the processes at appropriate instants, and this is achieved by using shared data structures.

Obvious rules for such structures are:

- data cannot be read or removed from a list before they have been placed there
- data placed on a list must not be overwritten before they cease to be needed.

Processes must be able to inform each other that a particular action has been completed; processes which share access to a resource must be prevented from gaining access concurrently.

8.2.14   Examples of non-realtime operating systems

The most popular operating system in use today is MS-DOS (PC-DOS), produced by Microsoft and IBM respectively for personal computers built around the Intel 8088, 8086, 80286 etc. family of microprocessors. This operating system is very basic, in terms of the range of features offered: for example, it supports only a single user running a single task at any one time. The reason that this operating system is so popular is its ease of use. MS-DOS finds wide use in the Control field predominately in supervisory roles, for which application programs have been written to provide user interfaces, process monitoring, process optimisation and process modelling.

UNIX, first developed by A T & T, has found wide acceptance in the scientific and engineering communities as the operating system of first choice. This operating system is multi-user and multi-tasking, but does *not* offer realtime features. UNIX provides its users with great power and flexibility. Traditionally, it has suffered from its poor operator interface and from its not being particularly easy to use. These drawbacks are being overcome by the use of the X-WINDOWS graphical interface, which is far more easily used. Typically, UNIX is used in a host computer, which gathers process information and subsequently manipulates and analyses it for consumption by management.

OS/2 is being represented by its proponents as the logical evolution of MS-DOS. In simple terms, OS/2 is a single-user multi-tasking operating system, developed again by Microsoft and IBM. This operating system is far more complex, but offers many more features than standard MS-DOS. The two operating systems address the same types of user application.

Table 8.1 presents a short comparison of commands for the three operating systems described so far.

Graphical User Interfaces (GUI) are becoming increasingly important for DOS, UNIX and OS/2. In the DOS environment, the GUI 'standard' is

Table 8.1   Command comparison–DOS and OS/2 versus UNIX

Command	DOS and OS/2	UNIX
Directory	DIR	ls
Wide directory	DIR/W	ls–lf
Copy	COPY	cp
Move	COPY	mv
Print	PRINT	lp
Find	FIND *or* SEARCH	find
Format floppy disc	FORMAT	format
Change directory	CD	cd
Make directory	MD	mkdir
List file	TYPE	cat *or* more
Delete file	DEL *or* ERASE	rm–f–l
Delete directory	RD *or* RMDIR	rm–l
Power off	*none*	powerdown *or* shutdown

Microsoft WINDOWS: this GUI sits on top of DOS, since it has not been crafted as part of DOS. WINDOWS offers pull-down menus together with multiple, scaleable windows.

In the UNIX environment, the issue of GUIs is complicated by the existence of two systems – OPEN-LOOK and MOTIF – fighting for dominance. These graphical interfaces offer similar features, the only significant differences being the manner in which the commands are implemented.

For OS/2, the standard GUI environment is known as *Presentation Manager*: this graphical interface has been crafted to be an essential part of the operating system, and it provides features similar to those offered by Microsoft WINDOWS.

All of the current trends indicate that operating systems must now begin to offer more than the standard text-based *command line*. Today's users demand ease of use and graphical interfaces from their computers.

CONCURRENT-DOS, developed by Digital Research, is almost as old as MS-DOS and, as its name implies, it offers multi-user multi-tasking capability for computer systems built around the Intel family of microprocessors. Applications for CONCURRENT-DOS are restricted mainly to supervisory computers applied to process monitoring.

It will have been noted that the operating systems considered thus far have found little application within process control elements and process instrumentation: this is due to the fact that these devices normally require realtime operating systems in order to guarantee response times.

8.3   Realtime operating systems

8.3.1   General

In the specific case of a realtime operating system (*RTOS*), only two of the features already discussed for non-realtime operating systems are absolutely essential: these are the generation and handling of interrupts, and the assignment of priorities. Applying these features judiciously produces the best approximation to truly realtime processing that can be achieved with a single serial processor.

In practice, interrupts can be recognised and used to generate a responsive action in about 10 ms, at best. In that timeframe, no more than about 10 000 object code instructions can be executed, most of which in any case need to be devoted to general housekeeping activities. It is therefore impractical to cause interrupts to occur more frequently than at 10 ms intervals, since, if they were to do so, they could not be adequately serviced.

Interrupts need to be *vectored*: that is, any particular interrupt needs to generate a responsive action directly targeting the relevant process required to respond. As a consequence of an interrupt, the RTOS needs to set up a decision procedure involving a combination of scheduling, re-starting and stopping processes, and branching. Interrupts, users and tasks need to be prioritised; the time slicing alternative is generally impractical in realtime situations.

Most RTOSs refer to every software/hardware module as a *resource*. The RTOS then becomes a set of programs to manage a full set of resources.

8.3.2 Resource allocation

The RTOS determines the nature, timing, priority, sequence and duration of all resource allocation. When two or more users require access to the same file, the file is allocated a *resource number (RN)*, which is communicated to the users; the user programs then request access to that RN and, if one user has already been granted access, the other has to wait its turn, irrespective of its priority. Typically, RNs are *not* allocated to tasks; otherwise, priority and interrupt signals could not be used to achieve their proper objectives.

RNs are stored in memory tables and, where one needs to be shared between many users, it is designated a *global RN*. If a program currently has control over a particular global RN and the program fails to execute properly, the RTOS must be capable of recognising the situation, and restore the RN to the memory table, so that other users can have subsequent access to it.

The RTOS also determines the allocation of memory space to each user and each task, and controls all memory input/output transactions. Access to memory is at two levels:

- *interactive* – using a *Command Interpreter*, the user is provided with program access to a wide range of commands for actuating external hardware; execution of such commands, in fact, may temporarily bypass the RTOS

- *programmatic* – the user is provided with program access to a library of software routines; in this case, the RTOS is not bypassed. (As an example, a FORTRAN *write* command might be equivalent to 18 000 bytes of RTOS object code.)

8.3.3 Interrupts

To the *Realtime Executive (RTE)* of the RTOS, any external event generates an interrupt. Such an event, for example, could be:

- a keystroke produced by an operator
- a disc drive action
- a printer action
- an event detected by external field equipment.

The purpose of the interrupt is to cause the RTOS to recognise that an event has, in fact, occurred. The RTE interrupt process is a sub-module of the RTOS: it must be able to identify the interrupt, determine its relative priority, and initiate the relevant response process. An *interrupt table* is included in the RTE, for this purpose.

If the interrupt time-resolution is 10 ms, the software can resolve interrupts only when the trigger events occur more than 10 ms apart. If two events occur within the same 10 ms interval, they must be serviced in order of priority which, in this case, can only be resolved by hardware.

Interrupt software routines need to be kept as short as possible, to ensure that low-priority interrupts are eventually serviced. The routines can be designed to pass parameters quickly to other software processes, so that the routines are soon freed-up.

8.3.4   User environment

In some RTOSs, an extra section of the system creates, for each of several concurrent users, the appearance of a single-user environment. This is achieved by a *Session Monitor*, which generates for each user a *logical environment*, rather than a physical one. Each input/output device is allocated a logical unit (*LU*) number, and the Session Monitor sets up a table which relates actual system devices to user LUs, before the user gains access to the RTOS: the process is completely transparent to the user. The Session Monitor will also be responsible for logging CPU usage and user connect times.

8.3.5   Memory management

Normally, realtime programs should be memory resident, since they will require frequent and rapid execution. They may be stacked together at one end of the user area by causing them to be so loaded from disc immediately the system is powered-up; in effect, this part of memory then becomes *RTOS area*. The remainder of the user area is then allocated to swappable programs; this part of memory can then be regarded as *background* area.

8.3.6   System clock

The RTE requires that an accurate record be kept of actual elapsed time. This may well necessitate the use of an external *realtime* (hardware) *clock*, with the record of time surviving any system power failure. Typical accuracy for such a clock would be in the order of 10 seconds per year.

Typical use for the clock would be for:

- running a program at a specific time of day
- delaying execution of a program, irrespective of interrupt status, for a specific time interval
- executing a program at regular time intervals – for example, a *watchdog* routine to check for system failure
- for regular sampling of external sensor signals.

Sampling frequencies for sensor signals should be high enough to avoid missing fast changes in signal, but low enough to avoid sampling random noise superimposed on 'useful' signals. It is often necessary to pre-condition sensor signals before they are sampled, using frequency-sensitive and/or amplitude-sensitive (*front-end*) hardware networks. Alternatively, pre-conditioning of sensor data may also be undertaken using software routines.

8.3.7 Computations

Generally, most realtime computations are performed in integer arithmetic. With 16-bit processors, this means that numbers in the (unsigned) range of 0 to 65535_{10} can be accommodated. Many variations are possible: for example, one bit may be allocated to denote sign, and/or some bits may be allocated to fractional parts of numbers. Again, some realtime computations may be performed on floating-point numbers, but with the ranges for the mantissa and exponent severely limited.

User programs will often involve the use of floating-point representations of *real* numbers, in which case translation routines for converting *reals* in the user programs to *integers* in the realtime processes, and vice-versa, could be required from the RTOS.

8.3.8 REAL/IX

Most realtime operating systems are proprietary, with little information being published about them. A possible exception is REAL/IX, which is the first realtime A T & T Version V UNIX with a fully *preemptive kernel*. Such a kernel ensures that:

- any realtime task gains the direct attention of the CPU within a specified time limit
- high-priority tasks will be executed in realtime mode
- tasks with lower priorities will not be executed until the realtime tasks have been completed
- any high-priority task or interrupt will preempt the CPU, irrespective of whether it is currently in *user* or *kernel* mode, thereby ensuring complete realtime responsiveness.

REAL/IX uses both priority and time-slice scheduling, during the sequence of task handling. The scheduler handles a combination of 256 realtime and time-slice priority levels, with realtime tasks being allocated the highest priority.

The I/O subsystem supports:

- asynchronous I/O operations
- prioritised I/O queuing
- direct I/O between peripheral devices and user-level programs
- connected interrupts.

Realtime tasks are memory-resident, thereby reducing the amount of disc access which would otherwise be necessary. Both realtime and standard UNIX file systems are provided, so that existing UNIX files can be readily ported into REAL/IX, with full compatibility assured.

Typically, interrupts are enabled within a few microseconds, in contrast to the several milliseconds achieved with standard UNIX: a performance improvement of 25 times would be typical.

9

Computer Languages for Controlling Plant

9.1 Introduction

The choice of a computer language is a very personal thing. People become surprisingly attached to a particular language and often will defend the use of it despite powerful countervailing arguments. Becoming fluent in a new language requires the investment of considerable intellectual effort and time; so, using a particular language, although it may be less than optimum, could still be the correct decision economically. The project may not justify the extra expense of learning a new language.

Nevertheless, selection of a particular language is still part of engineering design, and should be approached as dispassionately as the choice of any component of hardware.

In this chapter, a number of programming languages are discussed with regard to their potential as tools for the development of software for engineering applications in general, and control engineering applications in particular.

Computer programming languages are dynamic and evolutionary, although, in the interests of portability, there are always attempts to standardise, and language standards committees proliferate. Often, the standards committees reach agreement long after several dialects of a language have come into use, and there arise unresolvable differences. In addition, there are always vendors who will produce a compiler with enhancements. Many of the comments in this chapter are generalisations, and they are intended to be a guide only. The engineer must decide whether or not a particular version of a compiler is appropriate for the software for the intended application.

As an example, standard PASCAL does not really support concurrency, realtime operation, or external hardware drivers. However, there are texts available describing the use of PASCAL in process control applications, but the language used was an enhanced, concurrent, realtime PASCAL. It should be pointed out that such versions are not generally available for all processors.

9.2   Factors which influence the choice of a language

There are many factors which must be considered when selecting a computer language: which factors are the most important will depend upon the particular project. Some factors may be beyond the control of the engineer. For instance, there may not be a programmer available with fluency in the language of first choice.

9.2.1   Cost

Cost of writing code, and the cost of maintaining the code over the lifetime of the program, must be high on the list of factors. Programs are more quickly written in some languages than others, but the fluency of the programmer is an important aspect. The cost often will depend also upon the size of the project. A language such as FORTH may be the cheapest solution for a very small microprocessor project, whereas, for a similar but much larger project, C might be the most cost effective.

9.2.2   Portability

The ability of the software to be transported from one computer system to another is also important, since it is likely that the computer hardware will become obsolete before the software. It is important to note that, in most projects, the cost of software usually outstrips the cost of hardware by an order of magnitude. Usually, the better specified and standardised a language is, the more portable it is. The most rigidly standardised language discussed here is ADA and probably the least portable of the higher level languages is BASIC. Code written in ASSEMBLY language normally would have to be scrapped with the machine unless the replacement machine fully supports the instruction set of its predecessor (which most personal computer upgrades will).

9.2.3   Type declarations

The ability of a programmer to specify the *type* of a variable within a program is usually considered to be important. Strong *typing* imposes structure upon the program and enables the compiler to detect a number of common programming errors.

   However, in dedicated engineering applications, such as the control of plant, the programmer may legitimately wish to perform operations normally considered to be bad practice; for example, adding a character to an absolute address, such as with a look-up table. In a language such as FORTH, there are no variable types and it will happily perform the unusual. With PASCAL, such operations are strictly limited, but with C it is possible to change the type of a variable with more ease.

9.2.4   Modularity

Most languages support the declaration of subprograms such as functions, procedures, processes and the like. By writing small sections of code and calling them later on in further modules, the overall program assumes a dendritic structure. On the other hand, in assembly language, the use of instructured *GOTO* statements or *BRANCH* instructions results in an amorphous body of code. The latter type of program is exceedingly difficult for another programmer to interpret and, as a result, debugging and maintenance costs are likely to soar. To keep costs down, programs should be modular and well-ordered.

For large software projects, the modules are often written by different programmers and *imported* into the final product. Different languages handle the importation of modules in different ways. Of the languages discussed here, C, MODULA-2, ADA and OCCAM handle module importation well.

Most large projects (and now even small ones) are software engineered. The project begins with a *Software Performance Specification* (often called by a different name). The performance specification outlines in detail all of the work the software must perform, together with its presentation, input and output, timing constraints and so on. From the performance specification, a *Functional Specification* is produced, and this lists all of the subprograms by name, their particular tasks, their formal parameters, and their use of global parameters. Finally, the functions are encoded and assembled into the product. This approach to programming reduces errors, reduces costs and generates proper documentation as the project proceeds.

9.2.5   Recursion

Recursion is the ability of a subprogram to call itself. The technique is useful in some kinds of arithmetical algorithm, such as those implementing Maclaurin Series, and in parsing. However, the importance of recursion is often overrated and is of little value in control engineering work.

9.2.6   Re-entrancy

Programs which run different tasks concurrently, or which employ interrupts, may call a particular subprogram while it is already in the middle of a call from elsewhere. Repeated interrupts could cause repeated nested execution of the same subprogram. If the subprogram uses a different storage area of local variables upon each call, it will be *re-entrant*. The use of stacks for local storage usually ensures re-entrancy.

Unlike much business software (with the exception of multi-user operating systems), control engineering software often employs many interrupts and also multitasking. Re-entrancy can therefore be an important aspect of an engineering language.

9.2.7   Concurrency

Concurrency is a term which often causes confusion, because computer users have different points of view. In a multi-user environment, many users share the same computer via a number of different terminals. The users may or may not be sharing the same software and the same data, but there is unlikely to be data passing dynamically between the different users. Typical examples are on-line banking or airline reservation systems. With multitasking, there may be several, or only one, or no users connected to the system but the computer may still be carrying out several tasks simultaneously – such as PID control loops, data acquisition and printing logged information. With this form of concurrency, the various tasks may well be data-interdependent. With parallel computers, multitasking concurrency is a normal part of the manner in which the computer is programmed, since the program is split into many parts, any or all of which may execute simultaneously.

Task switching can be implemented by interrupts, by a simple *round-robin* method, may have a complex internal *priority* arrangement, or may be determined by a realtime clock. In a realtime multitasking environment, certain tasks must take priority if they are to have a guaranteed time-related performance.

9.2.8   External input and output capability

Computers used for the control of plant clearly must be capable of accessing external inputs and outputs. Often, as in *Supervisory Control And Data Acquisition (SCADA)* systems, the I/O is by way of a *Remote Terminal Unit (RTU)* and the only connection to the main computer is via a modem or other serial communications port. In many cases, however, the connections are direct. The physical configuration of the computer will have a direct bearing upon the choice of language, since some languages (for example, standard PASCAL) can handle a serial port but not other, more direct, connections.

Also, there can be a trap with microprocessors, since there are two ways in which external I/O is addressed. Microprocessors that follow the Motorola style of architecture use memory-mapped input and output, whereas followers of the Intel architecture use ported input and output. Users should check the compatibility of the language with the input/output configuration.

9.2.9   Loading code into read-only memory

Software which is for the control of plant is often destined to be installed in Read-Only Memory (ROM) in a target machine. It is necessary, therefore, that the language used be capable of generating ROMable code. Some compilers offer this facility and others do not.

It may also be of importance to know whether the code is *relocatable*: that is, whether the code can be loaded anywhere in memory and still run successfully. With some languages, the compiler generates relocatable code, but this becomes non-relocatable after linking.

It may be necessary to use a *cross-compiler*, whereby the program is compiled on a machine which is different from the machine intended to execute it. Not all languages are available with suitable cross-compilers.

9.3   Assembly language

Assembly language is the language of choice in surprisingly many control engineering projects, and it does have a number of advantages.

The programmer is able to write code which is relocatable, supports interrupts, is re-entrant, concurrent (with a free choice of priority algorithm), ROMable, and will support any kind of input, output or peripheral device. In fact, where unusual external devices are connected, either assembly language or FORTH may be the only options available for the driver routines.

It is generally accepted that code written in assembler executes faster than code written in a high-level language, but this may not always be the case. While a good assembly programmer can apply a trade-off between compactness and speed of execution, the better high-level language compilers produce code which is faster than that written in assembler by the majority of programmers. Conversely, there are some poor compilers on the market which produce very slowly-executing code.

One major disadvantage with assembly language is the lack of portability. Code written in assembler cannot be moved to a machine with a different instruction set.

The other main disadvantage of assembler is the cost. Program code takes about the same time per line to write, debug, test and comment, irrespective of the language. The higher-level languages generally can achieve far more work per line of code than the lower-level ones, and so are much more cost-effective.

9.4   Special languages

For the control of plant, there exists a large number of special languages and programming environments to serve a variety of needs. Many of these are unique, and are important enough to deserve mention here; others are too numerous to be included. There are, for instance, between ten and twenty commonly used languages for robot programming, of which one is discussed briefly.

9.4.1   Ladder logic

About eighty percent of programming for programmable controllers (PCs) and programmable logic controllers (PLCs) is carried out in *ladder logic*.

Since the PC/PLC is the commonest type of embedded control computer, there are many programmers using it.

Ladder logic is based upon the American standard for drawings for electrical discrete control equipment. The active power line is drawn vertically on the left, and the zero volts, or the power neutral, is drawn vertically on the right. Control connections are drawn horizontally between the two. The logic is implemented by the use of imaginary relays, with coils, normally-open contacts and normally-closed contacts. There are also timers, counters and, very often, digital inputs and outputs, analog-digital conversion, digital-analog conversion, arithmetic and proportional control.

The programming method is graphical, and the ladder is converted into machine instructions for storage in the PC/PLC (either in battery-backed RAM or in ROM). The ladder usually is viewed graphically on the screen, or can be printed in graphical form. Refer to Chapter 11 in the companion volume for further information on these machines.

9.4.2 Mnemonic languages

Some programmable (logic) controllers offer other programming options, including a form of assembly language. Some of the assembly language mnemonics are made available to the programmer, who may use them in preference to ladder logic. Other offerings comprise sets of mnemonics that are virtually macros, and which allow the programmer to convert logical sequences into code.

9.4.3 G-code

The programming of numerically-controlled machine tools and, now, *computer-numerically-controlled (CNC)* machine tools, poses special programming difficulties. The user must be able to define trajectories in two- or three-dimensional space, and make allowances for tool size and shape, spindle speed, step-and-repeat of subprograms, mirror-imaging of subprograms and so on.

There are several different programming languages for dealing with these problems, and many of them are similar to *G-code*. The name derives from the use of the letter *G*, followed by a number, to indicate a process or procedure. The language and its variants are derived from several American standards for punched paper tape and numerically-controlled machine tools standard numbers EIA RS-227A, EIA RS-267A, EIA RS-274C, EIA RS-281A, EIA RS-358, and USAS X3.4-1967.

G-code is an interpreter similar to BASIC, and stores the text of the program and interprets it instruction-by-instruction right through the program. The instructions usually are very simple, consisting of single letters followed by a number.

Typical letters are:

- N sequence number
- G process identifier
- M miscellaneous function
- T tool select
- F feedrate
- X x axis command
- Y y axis command
- Z z axis command
- I x axis arc centre
- J y axis arc centre
- K z axis arc centre.

The *sequence number* is a numerical line label (similar to those used in BASIC and FORTRAN), and usually conditional statements and *GOTO* or *GOSUB* statements are available. Programming in these codes can be tedious at first and prone to mistakes, because the instructions are difficult to relate to the action of the machine, and, hence, difficult to remember. However, regular programmers rapidly become proficient.

Traditionally, G-code is made and stored on punched paper tape, which many computer people regard as quaint. Punched tape is used because of its robustness. In a machine shop, the tape may become exposed to magnetic fields, be contaminated by oil, water, dirt and swarf – all of which it can survive. Torn tape can even be repaired. Nowadays, it is more common to generate the code on an ordinary text editor, store it on disc, and download to the machine through a serial communications port. However, the computers and disc storage must be provided with a clean environment.

There are now translation packages available which attempt to convert the files produced by *computer-aided design and drafting (CADD)* software into G-code. It is expected that such packages will become more commonplace.

9.4.4   VAL

The problems with the programming of robots are similar to those with computer-numerically-controlled machine tools, although there are added complexities, such as the need to fully accommodate three dimensions, more degrees of freedom, and reference point and reference plane changes.

Many of the robot programming languages are similar to G-code. VAL is possibly the best known robot language, since it is the language of one of the largest robot manufacturers – Unimation. VAL is, again, an interpreter rather like BASIC, and it uses numerical line labels, unstructured *GOTO* statements and *GOSUB* statements. VAL has interpreter primitives, however, which differ considerably from those of BASIC, and some of which carry out quite complex calculations, such as the mirror-imaging of specified routines.

Some other robot programming languages are: T3 (Cincinnati Milacron), AML (IBM), AL (Stanford University), HELP (GE), MCL (McDonnell Douglas), RAIL (Automatix), and KAREL (GMF Robotics). Some of these languages more closely resemble PASCAL than BASIC.

9.5 BASIC

BASIC was written at Dartmouth College in the USA, and throughout the 1970s was installed on mainframe and minicomputers on campuses around the world. Because it was so simple to learn, it became the language of choice for people without formal training in mathematics, engineering or computer science. Thus, it was learnt by biologists, chemists, social scientists, medical, legal and arts people. When the first home and personal computers appeared, BASIC was the language supplied, and it also became the hobbyist's language. BASIC also was supplied in alphanumeric pocket calculators, and gradually found its way into schools.

There are so many widely divergent forms of BASIC that making generalisations about it is difficult. The abundance of dialects of the language appeared for two main reasons. Firstly, the shortcomings of the early versions soon became clear to users and sales-people alike, and there was an obvious demand for improvements. Secondly, although it was necessary for computer sales personnel to supply a BASIC with the machine, they often tried to outdo each other by offering better enhancements than those of their competitors.

The outcome from this was that some of the later forms of BASIC became as difficult as PASCAL or FORTRAN to learn, and programmers themselves had difficulty transferring from one package to another. Naturally, programs written in one form of BASIC would not run with other dialects, and so, with the exception of assembly language, it is the least portable of all the languages discussed here.

The original BASIC was an *interpreter*. The program was stored in its text or source code form and never converted into machine code. Each time the program ran, the interpreter had to parse the text, word by word and line by line. There are obvious advantages to this approach. There is no difficulty with relocatable code, and the user does not have to understand about *compilers, linkers, executable code* and so on. BASIC programs are quite like FORTRAN, since the same numerical line labels are used.

The main difficulty with interpreter BASIC is that it runs exceedingly slowly, and there are documented (and embarrassing) cases of programs written in BASIC having to be rewritten in assembly language in order to obtain the necessary speed of execution.

A further difficulty with the early BASIC is the absence of programming structures, since the programmer must rely on a *conditional* plus a *GOTO* statement, or a *GOSUB* (jump to subroutine) statement, for branching and looping. Thus, even short programs become amorphous and difficult to follow. The writing of large programs, the use of subprogram libraries, the modification of programs and general maintenance (especially by other programmers) are all extremely time-consuming. It should also be stated that BASIC teaches people bad programming technique, and now is generally avoided as a teaching language.

The BASIC interpreter is quite different in operation from the FORTH interpreter, since the program is stored in memory, and interpreted from there. With FORTH, the text is interpreted directly from the keyboard, and only compiled if compiler words are invoked. The two languages share one attribute, however (and for different reasons): the program source code can easily be plagiarised.

A high proportion of modern BASICs are compiled languages, providing many of the high-level structures found in other languages. These forms of BASIC are so different from the original that they should have been given different names, or at least standardised version numbers, as has occurred with FORTRAN.

Despite all the foregoing, BASIC is still often supplied as the fundamental programming tool with a variety of control equipment, such as Programmable Logic Controllers, three-term controllers, distributed control systems, digital controllers, backplane bus cards for embedded computers, machine tools and robots. BASIC has always had instructions for absolute memory addressing, and some of the enhanced versions are usually well endowed with instructions and routines for handling input and output, files, graphics, external hardware devices and realtime programming.

For the control of plant, a modern enhanced *compiler* BASIC is no less appropriate than, say, an exhanced version of PASCAL, or even a version of C, but neither is it necessarily more easily learnt.

9.6 FORTRAN

FORTRAN is the 'grand old man' of programming languages, since (with COBOL and ALGOL) it was one of the first three important high-level languages to be written. There have been successive improvements to the language over the years (FORTRAN 4 and FORTRAN 77) to bring it into line with programming developments. During the 1980s, there have been attempts to upgrade it once again to FORTRAN 8x, but the standard has not, to date, been properly formulated. It should be emphasised that FORTRAN 8x is markedly different from FORTRAN 77, since some of the traditional methods are to be scrapped. The implication is that there will not be full upward portability and essentially 8x will be a different language.

Until FORTRAN 8x is standardised, FORTRAN 77 will remain the most common version in use. While FORTRAN has developed considerably since its inception, FORTRAN 77 is still a programming tool of the same technical level as PASCAL.

One of the strengths of FORTRAN is its ability to execute a very wide range of mathematical operations including complex numbers, multidimensional arrays, and matrices. The other principal strength is its ability to handle the character strings and printed output in complicated formats for the display of computed data. Nevertheless, FORTRAN was always regarded as a language for scientific calculations, and was never intended for system programming or hardware- and machine-related work.

Naturally, being the language for calculations, FORTRAN is the language most commonly used on supercomputers and, to some extent, parallel computers. There is a version available, for example, for the Transputer chip. Also, very many engineers and scientists have received training in FORTRAN, and so it is quite popular for engineering applications for that reason alone. Because it has been the most widely used and supported high-level language (with the possible exception of BASIC), there is an immense body of established programs in use and in storage round the world, and FORTRAN continues to receive widespread support from many computer manufacturers. FORTRAN 77 is, naturally, highly portable.

FORTRAN 77 can sometimes support re-entrancy, depending upon the operating system of the host computer. In general, there is no concurrency, no realtime programming, no records or structure declarations, limited type-changing capability (usually integer-character only), no access to absolute addresses or external hardware, and no capability of creating ROMable code.

Because FORTRAN is so widely used, there are several dialects on the market which address some of the deficiencies. There are versions which support concurrency and parallel programming, and there are powerful vectorising compilers for the supercomputers. There are also FORTRAN compilers which offer realtime programming and external input/output capability. Caution is advised, though, because these versions are machine specific, and may not be portable. Whereas there are popular and well supported enhanced dialects of PASCAL, C and BASIC for personal computers, there appears to be no counterpart in FORTRAN.

9.7 PASCAL

PASCAL was developed by Niklaus Wirth and released in about 1970. It was always intended to be used as a teaching language, and is excellent for the purpose. The structure is based on ALGOL 68, which many regarded superior to FORTRAN and COBOL, its competitors at the time.

PASCAL is a high-level language which imposes a strict discipline upon the programmer. Apart from variations in style, there are few possible variations in structure, and PASCAL programmers are able to read and interpret, with ease, even lightly-commented source code produced by other programmers. For teaching purposes, this is of considerable benefit, since the correction of student assignments usually is a major problem. The reserved words of PASCAL lend an English-like readability to the code, especially if meaningful variable and procedure names are declared.

The principal advantage of PASCAL is its discipline and structure. The programmer is forced into preparing a program structurally, by declaring all types, variables and constants early in the program. Because forward declaration of procedures and functions is uncommon, the source code usually begins with the simplest structures, and these increase in complexity as the program develops. The main part of the program is usually quite short, and is placed at the very end. The result is a modular and well-presented program.

Usually, a well-ordered modular piece of software is easily debugged and easily maintained, and for software which has a long projected lifetime, PASCAL programs can keep software maintenance costs down considerably.

Many programmers find PASCAL compilers to be overly pedantic, which, for teaching students, is an advantage. There is a bonus, however, in that many software errors are discovered by the compiler, and any program which compiles successfully has a good chance of executing reasonably well, more so than with many other languages.

PASCAL has some very useful attributes, of which *pointer type* and *record type* declarations are particularly noteworthy. The lack of type-changing facilities in standard PASCAL can be such a problem as to render it unusuable by computer engineers, who often need to change variable types.

PASCAL handles all input and output information by way of files. This is fine for student use, and, to a large extent, for office use, but for engineering applications is a serious disadvantage. Most PASCAL compilers simply refer to keyboard, screen and printer as special 'files' and provide no facilities for handling other types of input and output data. There are no facilities, either, for handling direct memory access or memory-mapped input and output.

Not surprisingly, regular PASCAL is rarely used in control engineering applications, but because of its structure and ease of use, it is often used in calculation and data processing work. A number of dialects have arisen in order to provide extra facilities, such as type-changing, direct memory access, input/output functions, graphics and so on.

There is at least one PASCAL dialect which runs on personal computers, and is so well enhanced that it is more like a version of MODULA-2 than PASCAL; it comes with a whole range of graphics, input/output functions and direct memory access. For fluent PASCAL programmers, such a dialect is a valid choice for dedicated control applications even though the programs may not be very portable.

9.8   MODULA-2

The shortcomings of PASCAL quickly became obvious to programmers attempting to write systems software, and so in 1975 Wirth produced a new language – MODULA. The language was revised to MODULA-2, which was first implemented on a PDP-11 computer in 1979. The first compiler was released in 1981.

MODULA-2 is a modern very-high-level programming language, widely used for large software projects, for systems software (operating systems for larger computers), and for professional level embedded computers. It supports recursion, re-entrancy, concurrency, realtime operations, absolute addressing, type changing (even lower/upper case character changes), external hardware interfacing and the importation of modules.

The syntax of MODULA-2 is so like that of PASCAL that programmers trained in either can understand most of any program written in the other. The underlying philosophies of the two are quite different, however, since MODULA-2 is based upon the concept of a *module*, which is a programming unit of two parts – the *definition* part, and the *implementation* part.

MODULA-2 is a well-defined language with few variations, and so is generally very portable. Although it is comprehensive and well-liked by users, it is not always available for specific computers.

9.9   FORTH

FORTH was written in the early 1970s by Charles Moore, who was working at Kitt Peak Observatory, USA. Most work at the observatory was being written in FORTRAN, including the control of the telescopes, and Moore thought that there could be a better approach. FORTH is the one example of a language which was written specially, in response to control engineering requirements.

FORTH is such an unusual language that programmers usually have strong feelings about it one way or the other. Many programmers (and many companies) either dislike it, or will not publicly admit to using it. However, there have been some remarkably successful pieces of software written in FORTH. On one occasion, the computer network for an entire airport was programmed in FORTH, including all the digital communications software. Several United States Government military and space industry contractors are known to use FORTH in dedicated processors, especially for debugging and diagnostic work. FORTH also is popular in low-level industrial control projects involving microprocessors.

The reason why FORTH is so unusual is that it is an *interpreter*, but permits the user to compile code directly through the interpreter. The user simply types instructions into the computer in the form of a succession of words delineated by spaces. When the carriage-return is pressed, the computer executes the instruction. Naturally, the instruction is never stored, so to repeat the action it must be typed in over again. There are, however, a number of defining words, which allow the user to create new words and add them to the end of the dictionary. The new words can then be used many times. Further new words can call previously defined words, so that large and elaborate programs can be built quite quickly.

The computer user can always mix simple instructions and define words in a way that is impossible with other languages. The FORTH editor permits FORTH text to be stored on disc from where it can be loaded. The text either simply executes or compiles as it is loaded.

Although FORTH is exceedingly easy to learn, there are some complex advanced concepts which may take years to master, and to which many FORTH programmers never progress. For example, there are words which permit the creation of new defining words, and it is possible to write new FORTH compilers or to re-model the old ones. It is also possible to create recursive definitions. Many users complain about the use of cryptic symbols, and the stack-orientated (reverse Polish notation) arithmetic and logic.

FORTH is a low-level language, since it operates close to the instruction set of the machine. Direct memory access is the norm and the only data structure used is the stack, or *Last-In/First-Out (LIFO)* buffer. The user may create new data structures without difficulty, but there is not the use of a *heap*, or of easy type declarations that are available with higher-level languages. In fact, there is no *typing*, since FORTH does not distinguish between characters, bytes or words: everything is stored as sets of sixteen-bit integers. FORTH does not have any floating-point arithmetic, unless a floating-point library can be purchased or written.

The lack of typing and data structures means that it is not possible to have anything but rudimentary error checking, and FORTH is notorious for allowing the most outrageous examples of programming practice.

The lack of typing, the low-level and the interactive interpreter are precisely what makes FORTH so useful as an engineer's tool. There is no better medium for debugging microprocessor hardware, especially in the field, since it is possible to write and run simple test routines immediately, and to use meters and oscilloscopes to search for output pulses, memory write cycles or information from transducers. It is possible to write directly to output ports from the keyboard, a capability which is particularly valuable in robotics. However the same features render the code very easy to break into and copy, since the source is actually compiled into the vocabulary. This language is not recommended where software security is important.

It is often tempting to use FORTH for larger projects and where time is a constraint. This is attractive, since FORTH code is fast to write. However, FORTH code is so modular that large programs become almost homogeneous in their structure, and very difficult to maintain. In addition, there can be difficulty with preparing relocatable code for a target machine, unless a target compiler is purchased. FORTH is not really appropriate for large software projects.

9.10   C

C has become not only the de-facto standard language for embedded systems and control applications, it has also become just about the most popular language for writing all kinds of general-purpose and systems software. There is even a MODULA-2 compiler on the market which converts source code into C.

C was developed at Bell Telephone Laboratories by Dennis Richie and Brian Kernighan in the mid 1970s. It is a relatively low-level language, since it operates fairly close to the machine instructions and provides no keyboard-, screen- or file-handling instructions; nor does it have any high-level arithmetic-, string- or array-handling operators. The core of the C language is compact, and it is relatively easy to write a C compiler for a new machine.

The original Kernighan and Ritchie standard has been updated by the ANSI standard C, which provides for a wide range of standard library functions. It is the library functions which provide all of the higher-level facilities, such as keyboard-, screen- and file-handling. Most suppliers of C also include a plethora of libraries for handling graphics, external input and output, interrupts, strings, arithmetic and so on.

Although the library functions can vary from one supplier to another, in a sense the portability is not compromised, because the heart of the compiler is usually ANSI standard. In reality, programs are not so portable if they call non-standard functions, although replacement functions are relatively easy to write.

It is claimed that C is a very cost-effective language, partly because of the small number of reserved words, and it is easy to learn. Some users have, however, found it quite difficult to learn for three reasons. Firstly, it is quite cryptic, since it employs a number of symbols (as does FORTH) instead of words. Secondly, variations in legal expressions, designed to allow compactness in the source, give rise to *clever* programming which can be difficult for beginners to interpret. Thirdly, there are so many standard functions that the number of effective words the beginner must remember is quite large.

C permits only a few data types, although there are several sizes of integers and floating-point numbers. There are several derived data types which can be created with pointers, arrays, structures and unions. It permits type *coercion* (or type conversion for compatible types), and type *casting* (bit-for-bit conversion of non-compatible types). The normal compilers do not have the rigorous type checking of languages such as PASCAL, but a separate programme, *LINT*, can be used to catch errors which might pass through the compiler.

Like most modern languages, C is highly modular and its simple structure uses only functions as subprograms (the *void* function type acts like a procedure), and a final function, *main*, acts as the main program. C lends itself quite well to software engineering techniques for both large and small applications, and it is generally a straightforward matter to import code from other sources, especially assembly language code and separately-compiled functions.

C permits recursive functions as standard, although functions cannot be nested. Re-entrancy is not a problem with C. External interrupts can be handled without difficulty, since functions can be written in assembly language and included in the program.

For the same reason, there is no difficulty with the handling of absolute addresses or any external input and output, which normally are handled using specially-written driver functions.

To produce ROMable code, it is necessary to take certain precautions. C initialises declared variables at load time, and, if the code is in ROM, the initialisation will not occur. The programmer therefore must take care that initialisation-sensitive variables are initialised by the program. It is also necessary to purchase a cross-compiler, which allows the user to specify absolute address ranges for data, program code, stacks and so on. Fortunately, there are many good cross-compilers available for embedded systems.

Standard C does not support parallel programming or concurrency in any way, but, because it is quite low-level, prioritised concurrency can be built into a large program as can some realtime operation. There are available several variations of C which do support concurrency, including one for the Transputer, but this does not provide the explicit parallelism of OCCAM.

9.11 ADA

The US Department of Defense is the largest single user of computer power in the world, and in 1975 its High Order Language Working Group (HOLWG) determined that the widespread use of dozens of languages was creating a compatibility nightmare. The decision was made to commission a special language for all the military subcontractors to use for general work, systems work, and in particular for embedded systems work.

Tenders were called, and a team was put together, and by 1980 work had started on the new language – ADA. All the tenderers had suggested that the PASCAL syntax was the most appropriate starting point. Compilers for ADA have to meet very strict criteria in all aspects of compiler operation, so that the software is as portable as it is possible to make it. Any ADA program must compile on any compiler and run on the particular target machine.

Unfortunately, this strictness has been the downfall of the language because few software houses could afford the investment needed to get a new compiler past the standards committees. Accordingly, certified ADA compilers did not appear for some time, and even now there are not enough available to serve the needs of general industry. ADA is, therefore, almost entirely restricted to US military contracts, and there is no question that it is a superb tool for embedded computers in really advanced control systems such as spacecraft, military aircraft, missiles, and servosystems for gun, missile launch and antenna platforms.

The syntax of ADA is very like that of PASCAL (and hence MODULA-2), with procedures, functions, type declarations, records, pointers and so on.

It supports recursion, re-entrancy, many kinds of interrupt and exception handling, parallel processing and concurrency, realtime programming and absolute addressing. In addition, it has a wide range of general input/output facilities and external hardware interface features.

Although it is certainly the best language for advanced control system engineering work ever produced, ADA is really too comprehensive (and unwieldy) for general use and has lost ground in the marketplace to its lighter-weight relative, MODULA-2.

9.12   OCCAM

OCCAM was developed in the early 1980s by Inmos Ltd in the UK. It is a development of the language CSP (or Concurrent Sequential Processes), which was written by Tony Hoare.

OCCAM was always intended to be the 'assembly language for the transputer', although it is significantly higher in level than the instruction set of the machine.

OCCAM is somewhat like PASCAL in its syntax, but it has many of the features of C. It is actually quite a low-level language, since it does not possess anything but the most rudimentary arithmetical operators and no input/output routines. However, when OCCAM is purchased, it comes with a very large library of support software which contains all the keyboard, screen and disc processes, as well as floating-point to ASCII conversions and so on.

The Transputer Development System is also equipped with all the tools necessary for the preparation of code for target processors, including program loaders and ROMable code generators.

OCCAM is set up to handle very large software projects and the unusual folding editor is popular with those who take the trouble to learn it, but it does have a few unfortunate habits, such as saving code which is better scrapped. The ability to separately compile subprograms and to import processes and library routines from elsewhere is handled very efficiently. Some users complain about the compiler because it only identifies one error at a time.

The principal advantage with OCCAM is that it is designed for multitasking across either one or an indefinite number of processor chips. Programs are written in the form of concurrently-executing processes, which are declared as either *sequential, parallel* or *alternate*. The parallelism can be nested as many times as are desired, and parallel processes can be split for eventual execution on different processors. The main difficulties which programmers new to OCCAM experience are not with the syntax, which is straightforward, but with the concepts and rules of writing code *in parallel*.

For control engineering, the concurrency makes writing operational software quite easy since it is more like returning to the days of programming analog computers, except that processes start up when required and shut down again when not used. The Transputer has an on-chip timer accessible through OCCAM, and this renders the writing of realtime programs a simple matter.

10

Algorithm and Strategy Development for Computer Control

10.1 Introduction

Many texts providing mathematical treatment of control systems cover the same field as this chapter, and similarly many texts deal with the algorithms and strategies in more detail. Often, however, the engineer is left with a gap between the theory and programming practice. This chapter is an attempt to bridge that gap by describing the more common algorithms and strategies in computer control, some practical methods, and, where it helps to clarify a point, some fragments of code.

10.2 Scaling data

The problem of scaling data is closely related to the resolution of the number representation within the computer. Table 10.1 gives the resolution of some typical representations.

Table 10.1 Resolution of some typical number representations

	Resolution		Smallest number
8-bit integer	1:256	$= 0.391\%$	
16-bit integer	1:65 536	$= 15.259$ ppm	
32-bit integer	1:4 294 967 296	$= 0.0002328$ ppm	
IEEE 32-bit floating-point	1:8 388 608 (23 bits)		$\sim 1.1755 \times 10^{-38}$
IEEE 64-bit floating-point	1: $\sim 4.5 \times 10^{15}$ (52 bits)		2^{-1022}

10.2.1   Amplitude scaling

If resolution is important to an understanding of scaling, so is the concept of *headroom*. Any representation of a variable implies a maximum value and often a minimum value. If a 4 to 20 mA process control signal exceeds its maximum value, it limits at 20 mA. If an 8-bit integer exceeds its maximum value, it will either limit at 255 or it will overflow to 0, depending upon the implementation.

In computational work, it is generally undesirable to allow a variable to reach its maximum value, and *headroom* is the amount of leeway between the normal operating range of a variable and the maximum value. It is also undesirable to allow a variable to become so small that it approaches the order of the resolution, because the round-off error can cause an increasing inaccuracy, particularly in the absence of feedback.

It is necessary to strike a balance between keeping variables well above the resolution and yet still leaving adequate headroom. Normally, this is not a problem with 32-bit integer or with floating-point arithmetic, but it can be serious with 16- and 8-bit integer representations.

There are several simple techniques that can be employed to optimise the operating range of variables in integer computations. It is important to choose an optimum operating point for variables according to the balance described: the optimum can vary according to the type of operation. For addition and subtraction, an arithmetic midpoint is appropriate, say between 4096_{10} and 32768_{10} for 16-bit arithmetic. For multiplication and division, a geometric midpoint, say 256_{10} to 1024_{10}, is more suitable.

Where there are operations involving widely divergent numbers, it may be better to split the operations by using localised scaling constants. For example, in 16-bit integer arithmetic, computing

$(20*a)/b,$

where a is about 10000_{10} and b is about 1000_{10}, would cause saturation after the multiplication. This would be better computed as:

$((20*(a/100))/(b/10))*10$

Localised scaling is accomplished by the insertion of dummy multipliers or divisors. Care must be taken to ensure that any dummy multipliers are re-scaled at an appropriate point in the computation, and that mis-scaled variables are not inadvertently added or subtracted.

Another method for maintaining the balance between headroom and roundoff error is to rearrange the order of computation. In the previous example where $a = 10000_{10}$ and $b = 1000_{10}$:

$(20*a)/b$ will limit, but computing

$(a/b)*20$ will give a more accurate answer

Even so, the quotient a/b may easily have a roundoff error of up to 10 percent if the remainder is neglected. A yet more accurate answer may be obtained if the solution for the quotient and the remainder are computed in a different order. A typical piece of pseudocode to illustrate this is:

```
u := (a DIV b)*20;
w := ((a MOD b)*20)/b;
x := u + w;
```

DIV is an integer division which retains the quotient and throws away the remainder. *MOD* is an integer division which keeps the remainder and throws away the quotient.

In FORTH this would appear as:

a b / MOD 20 * SWAP 20 * b / +

The advantages with using integer arithmetic are that software is more easily written, especially in lower-level languages, and the computation proceeds much more quickly than otherwise (assuming the absence of an arithmetic co-processor). The disadvantages are that it is possible to make mistakes with the techniques just described, and it is necessary to know in advance what the range of a particular variable is likely to be.

10.2.2 Time scaling

Time scaling is often necessary for mechanising computer models of systems. In most cases, it is desirable to speed up the simulation of a slow process: a model of the evolution of the solar system is an extreme example. Occasionally, there are models which must be slowed down in order to examine the minutiae of change. In control engineering, computer models are most widely used in predictive control and process optimisation, and an increase in speed of one or two orders of magnitude may be required.

For a model to be time scaled properly, it is necessary to alter the time variable by the same amount wherever it occurs throughout the entire simulation. In linear dynamic models the only time-dependent functions are integration and differentiation, with integration being the more common. In these cases every integrator and differentiator must be scaled by the same scaling factor.

Time scaling of integrators is relatively simple. For example, the trapezoidal integrator

$$C_n = C_{n-1} + h/2*(R_n + R_{n-1})$$

may be time scaled by 60 (to render it 60 times faster) by multiplying the gain by a factor of 60; thus:

$$C_n = C_{n-1} + 60*h/2*(R_n + R_{n-1})$$

To slow down the time frame by (say) 15, that is, to time scale by 1/15, we get:

$$C_n = C_{n-1} + [h/(15*2)]*(R_n + R_{n-1})$$

Time scaling of differentiators is similar to that of integrators. For example, to time scale the differentiator

$C_n = (R_n - R_{n-1})/h$ by a factor of 60, one obtains:

$C_n = (R_n - R_{n-1})/(60 * h)$

10.2.3 Scaling of output data

Section 10.2.1 deals with the problems of roundoff error and headroom within a computer itself. However, usually the hardware attached to the analog outputs from the computer, such as power amplifiers or final control elements, imposes different constraints upon the variables. It may be necessary to compensate for a nonlinearity, although the system feedback will usually suffice for this.

Often, the digital-analog converter will have a resolution of 8, 10 or 12 bits, despite 16- or 32-bit integer or floating-point resolution within the computer. If this is the case, a floating-point value must be converted to integer, which is thereby truncated before being output. Normally, only the requisite number of more-significant bits would be used and the remainder would be discarded.

Usually, signals fed to final control elements must be calibrated for both range (or span or gain) and zero (or offset). Often, this is best accomplished in the computer algorithm and via the keyboard, rather than by making adjustments at the output amplifier or elsewhere in the hardware. Such span and zero adjustments often interact, and it is easier to prevent this within software.

10.2.4 Typical mathematical constants

Often, it is necessary, even with integer arithmetic, to use physical and mathematical constants. Most constants can be implemented using an approximation involving two integers. Some typical constants are given in Table 10.2.

10.3 Linearisation of input data

Nonlinearities present in the forward path of a closed loop control system are usually of little consequence, since negative feedback has the consequence of reducing the overall effect of the nonlinearity. Nevertheless, feedback cannot reduce the effects of nonlinearities outside the forward path of the loop, and no component is more sensitive than the feedback transducer. Process variable and other computer input signals are often artificially linearised by applying a cancelling static characteristic, and this is a method used for thermocouples, RTDs, pressure, humidity, flowrate and other types of transducer.

Another, less common, method of linearisation, and one which is generally found in final control elements, is the application of *local* negative feedback. Examples of this are flapper-nozzle amplifiers, force feedback differential-pressure transducers, valve positioners and hydraulic servo

Table 10.2   Integer equivalents of some mathematical constants

Constant	Approximation	Error
π	22/7	0.04%
π	355/113	0.09 ppm
$\sqrt{2}$	41/29	0.03%
$\sqrt{2}$	239/169	8.8 ppm
$\sqrt{3}$	26/15	0.075%
$\sqrt{3}$	97/56	53 ppm
$\sqrt{3}$	362/209	3.8 ppm
e	106/39	0.013%
e	193/71	0.001%
e	1457/536	0·65 ppm
ln 10	175/76	0.002%
ln 10	624/271	0.89 ppm
sin 45°	70/99	51 ppm
sin 45°	408/577	1.5 ppm
sin 60°	13/5	0.074%
sin 60°	181/209	3.8 ppm
degrees/radian	573/10	74 ppm
degrees/radian	4068/71	0.085 ppm

valves. This technique is covered in the companion to this volume, and only the method of applying a compensating nonlinearity is described here.

The deliberate introduction of nonlinearities is also used for simulating nonlinear systems, and the same methods may be applied.

10.3.1   Look-up tables

The simplest and best known method of linearising a known nonlinearity, or deliberately introducing a nonlinearity, is to use a *look-up table*. The name is descriptive of the technique. A table is kept in memory, or as a file, and for every possible input value there is tabulated a corresponding output value. When the nonlinear function is to be invoked, the input value is used to refer to the table and a corresponding output value is generated.

There are two ways to implement a look-up table: using an *array* and using a *reserved block of memory*.

An array would normally be used in a high-level language, and could easily be initialised by loading from a file. For (say) an 8-bit integer, the array would contain 256 elements, each generating an output value. The input value is interpreted as the identifying address of the array element. To use the table, the following typical instruction could be used:

```
output_value := lookup_table [input_value];
```

A reserved block of memory would be used in a low-level situation, such as with assembly language or possibly FORTH. The table would begin at a start address and would occupy, say, 256 bytes of memory. To access the table, the program would add the input value to the table starting address, and fetch the contents of the new address so generated.

There are two main advantages with a look-up table. Firstly, it is much quicker to access than computation of a nonlinear function, and, secondly, any function can be implemented, even a non-analytic one.

The main disadvantages with look-up tables are that the method can consume much memory space (and inefficiently so), or, if space is limited, the resolution is very coarse.

One compromise solution is to use a look-up table for a coarse approximation (say, to within 8 bits) of the function, and then to use a computation method such as Newton-Raphson to obtain a final accurate result. Sometimes, a combination of a coarse look-up table and a linear or polynomial interpolation algorithm is used to give a relatively accurate fine result, and at least one high-performance 32-bit microcontroller chip on the market includes this method in its instruction set.

10.3.2   Normalisation of functions

Sometimes, an input variable may be defined by a known continuous mathematical function and this can be linearised by applying a cancelling function. The commonest example is the use of a *square-root extractor*, which is used with differential-pressure types of flowrate transducer.

Rather than use a look-up table, it may be preferable to compute the new function using one of the methods described in Section 10.4.

Unfortunately, the range of a variable may be considerably altered by the function. For example, a 4 to 20 mA signal fed to a square-root extractor would become a 2 to 4.472 mA signal, and an 8-bit integer would saturate at 4 bits, giving a resolution of 1:16.

In order to retain the operational range or headroom of a variable, it is usual to normalise it. The unscaled variable is divided by its maximum value on both sides of the mathematical operation, so that the scaled variable is always referenced to a maximum value of 1, which represents the maximum value of the carrier signal – say 20 mA or 8 bits, and so on.

Table 10.3 gives normalised values of the square-root function, for 4 to 20 mA signals.

10.4   Arithmetical operations and functions

Most software packages are available with floating-point mathematical functions. Some languages, such as FORTRAN and PASCAL, have the functions built into the compiler. Other languages need library routines, which are either supplied or can be purchased separately.

Table 10.3   Square-root function for 4 to 20 mA signals

INPUT SIGNAL mA	OUTPUT SIGNAL (square-root) mA
20	20.000
19	19.492
18	18.967
17	18.422
16	17.856
15	17.266
14	16.649
13	16.000
12	15.314
11	14.583
10	13.789
9	12.944
8	12.000
7	10.928
6	9.657
5	8.000
4	4.000

There are instances, however, when it is necessary to write mathematical routines in assembly language or FORTH, for example, and this is especially true if there is a problem with resolution and headroom.

There are two principal methods for obtaining a mathematical function: the *method of iteration* and the *power series method*. In both cases, the algorithms derive from Taylor series representations of functions.

The general form of the Taylor series is:

$$f(x) = f(a) + \frac{(x - a)f'(a)}{1!} + \frac{(x - a)^2 f''(a)}{2!} + \cdots$$

Taylor series usually converge rapidly and can often be truncated to yield an acceptable result. There are many possible Taylor series, since a function f(x) is described in terms of a function f(a) and its derivatives, and f(a) can be any other differentiable function.

10.4.1   Iteration methods

In an iteration method, a first approximation is made to find a starting value. A formula is then applied to compute a correction factor, which is added to the initial value. The process is repeated until the correction factor is small enough to be acceptable.

The commonest example of an iteration method is the Newton-Raphson technique for finding the roots of an algebraic equation. The Newton-Raphson method converges very quickly, once the answer is within a few percent of the correct value. The truncated Taylor series yields

$$f(x_0 + \delta x) \cong f(x_0) + f'(x_0)\, \delta x$$

and one can obtain a formula for the correction factor δx as

$$\delta x = -\frac{f(x_0)}{f'(x_0)}$$

In the iteration, the new value x_1 of x is:

$$x_1 = x_0 + \delta x, \text{ where } x_0 \text{ is the previous value of } x.$$

The formula appears to work for any algebraic function, but is particularly useful for square-roots, cube-roots, and so on.

To find a square-root $x = \sqrt{A}$, the equation is:

$$f(x) = x^2 - A = 0, \text{ and the derivative } f'(x) \text{ is } 2x.$$

The correction factor $\delta x = \dfrac{A - x_0^2}{2x_0}$

The iteration formula is:

$$x_1 = x_0 + (A - x_0^2)/2x_0$$

or $$x_1 = \frac{1}{2} \cdot \left\{ \frac{A}{x_0} + x_0 \right\}$$

To find the *n*th root x of a number B, where $x^n = B$, the iteration formula is:

$$x_1 = x_0 + \delta x = \frac{1}{n} \left\{ \frac{B}{x_0^{n-1}} + (n-1)x_0 \right\}$$

10.4.2   Power series methods

Since there are many possible Taylor series for any function, academics and private companies often develop their own proprietary algorithms for obtaining mathematical functions. One straightforward method is to use the well-known Maclaurin series, which is obtained from the Taylor series as follows. By making the value a = 0 in the Taylor series, one obtains:

$$f(x) = f(0) + \frac{x \cdot f'(0)}{1!} + \frac{x^2 \cdot f''(0)}{2!} + \frac{x^3 \cdot f'''(0)}{3!} + \cdots$$

so that:

$$e_x = 1 + x + \frac{x^2}{2!} + \frac{x^3}{3!} + \cdots$$

$$\sin x = x - \frac{x^3}{3!} + \frac{x^5}{5!} - \frac{x^7}{7!} + \cdots$$

Naturally, other trigonometrical functions can be obtained from various identities such as:

$$\cos(x) = \sin(\pi/2 - x)$$
$$\tan(x) = \sin(x)/\cos(x)$$

10.5 Integration with respect to time

By definition, integration is a mathematical operation which involves continuous variables (either absolutely or piecewise continuous). Integration is, after all, the continuous summation of a dependent variable over infinitely small intervals of the independent variable (often *time*). The process of sampling the data (which is a necessary part of digital computation) means that variables are no longer continuous and differentiable. Because the intervals are not infinitely small, it follows that any digital implementation of integration must imply an approximation and an error in the result.

The need to treat sampled data has given rise to a special branch of mathematics – the theory of sampled data systems and the use of z transforms. The search for greater accuracy with sampled data has also generated a large number of innovative integration algorithms, although there are always penalties (three penalties) for improved accuracy.

The first of the penalties is that the more-complex algorithms have a greater tendency towards instability. With the digital mechanisation (implementation) of a continuous function – say a filter or other transfer function – the program itself can become unstable even if the analog version of the function is stable. The point of instability occurs at a particular value of sampling interval (h), called the critical sampling interval, and the ratio of critical sampling frequency $1/h_{crit}$ to the highest natural analogue frequency f_{nat} is important. This ratio, $1/h_{crit}:f_{nat}$, which must always be greater than 2_{10} and would generally be less than 10_{10}, is greater for a more-complex integration algorithm. The implication is that inputs must be sampled more often when the more complex integration algorithms are used, in order to ensure the stability of the software.

The second penalty is that the more-complex algorithms are badly behaved in the region of signal discontinuities and other transitory phenomena. The reason for this is that the more-complex algorithms are generally of higher order and thus require a longer sequence of data histories in the computation. Higher-order algorithms take longer to settle after start-up (since histories would all be initialised to zero) and can fluctuate wildly following a step change of input signal – both situations which are common in control systems.

The third penalty is simply one of computation time. The more-complex algorithms tend to consume more processing time for each computation.

The algorithm designer must take several factors into account when selecting an appropriate integration algorithm. There is, however, one overriding rule of thumb: select the simplest and most general algorithm which will give sufficient accuracy.

The first major factor is the question of sampling interval, which may be fixed or may vary. The case of a fixed sampling interval generally applies to data which have previously been sampled and then stored for subsequent processing, but can also apply to systems with hardware constraints. A good common example of fixed sampling interval data is weather (rainfall) figures, which are generally available with a sampling interval of 1 day. More common in the control arena, however, are systems with a variable sampling interval. Usually, in computer control systems the sampling interval is determined by the time taken to carry out all of the requisite computations for the control algorithms, and, in a non-synchronised system, this may vary from one sample to the next. It will almost certainly depend upon the efficiency of the code. The programmer should optimise the code for speed but, despite this, the system accuracy may actually be improved by the use of a simpler (and less accurate) algorithm, which computes more quickly and thereby permits a higher sampling frequency.

In the absence of any other guide, a useful rule for the sampling frequency is to make it at least ten times the maximum signal frequency of any significance. The program will then be stable for most lower-order integration algorithms.

A second factor is whether the system is open loop or closed loop. If the integrator is in the forward path of a closed loop system, the presence of feedback around the integrator and associated components will tend to reduce any nonlinearities and inaccuracies therein. For example, there would be little point in using anything other than a very simple integration algorithm in a 3-term PID controller. However, in an open loop system such as an inertial navigation system, where two cascaded integrations are performed to obtain position, there is a case for using the most accurate algorithm possible. In such a case there is, generally, no way to apply absolute position feedback to the vehicle.

A further factor is the predictability of the input signal to the integrator. Most so-called integration algorithms are in fact numerical methods of solving differential equations, and the Euler-Cauchy method, Simpson's rule and the Runge-Kutta algorithm presuppose a knowledge of the function f in, say,

$$dx/dt = f(x, t) \qquad\qquad 10.1$$

In control engineering and other technical fields, one is not usually solving a known analytic differential equation, or, if there is a linear differential equation, it is non-homogeneous and has an unpredictable driving function. For digital mechanisation of dynamic systems, a general-purpose integration algorithm is required rather than a method of solving a differential equation.

Many of the higher-order algorithms are based on a truncated Taylor series representation of the function f, and one attempts to improve accuracy in two ways. An attempt is made to fit a curve to the independent variable, in order to fill in some of the lost data between the samples. Equation 10.1 possesses inherent feedback, which normally is delayed by one sampling interval, and some algorithms try to predict the integrator output in order to remove the inaccuracy caused by this delay.

The question of the existance of time delays through digital integration algorithms is a subtle point and gives rise to some confusion over the application and even the naming of algorithms. The confusion arises because the time delay is of vital importance for theoretical analysis but is often inconsequential for the programmer. This point is explored in more detail later. Nevertheless, the timing of the samples determines the type of algorithm. An algorithm which uses an input sample, coinciding with the last output sample generated by the integrator, computes a future value of integral. Such an algorithm is said to be a *predictor* or *explicit*. An algorithm which uses a new input sample and a previous integral to compute a new integral is said to be a *corrector* or *implicit*.

The convention for the algorithms, here, is to use R for the input signal, C for the output signal, and h for sampling interval (or *step length*) in seconds – equivalent to the parameter T in Section 3.7. The histories are denoted by subscripts: thus, R_n is the most recent sample of input, R_{n-1} is the previous sample, and R_{n-2} is the sample before that. C_n is the latest output sample, C_{n-1} denotes the previous output sample, and so on.

10.5.1 First-order algorithms

There are two first-order algorithms:

the RECTANGULAR algorithm: $C_n = C_{n-1} + h \cdot R_n$
the EULER algorithm: $C_n = C_{n-1} + h \cdot R_{n-1}$

The only apparent difference between these two is the time delay of one sampling interval in the input signal. This results in a time shift of one sampling interval in the output waveform. This time delay causes h_{crit} for the Euler algorithm to be roughly one quarter of that for the Rectangular algorithm for the simulation of a typical second-order transfer function. Normally, the programmer would use the Rectangular algorithm for programs, but the Euler algorithm may be necessary in a rigorous mathematical treatment. In fact, the Rectangular algorithm is a corrector algorithm and the Euler is a predictor.

10.5.2 Second-order algorithms

There are many possible second-order algorithms, but two are commonplace. The first, the *trapezoidal rule of integration*, conservatively takes the mid-point of a straight line between the latest value of input and the previous value; thus, it is a corrector. The straight line approximation is often accurate enough for oscillatory systems, since the errors may cancel in opposite half-cycles. This rule is very popular, and is often seen as the well-known *bilinear* substitution.

The second rule, the *Adams rule*, is a predictor method and uses a straight line continuation of the variable, obtained by extrapolation of the line.

The two algorithms are:

the TRAPEZOIDAL algorithm: $C_n = C_{n-1} + h/2 . (R_n + R_{n-1})$

the ADAMS algorithm: $\qquad C_n = C_{n-1} + h/2 . (3R_{n-1} - R_{n-2})$

With both of these algorithms, there is some debate over the incorporation of a unit delay into the formula, and the trapezoidal algorithm is occasionally written as:

$$C_n = C_{n-1} + h/2 . (R_{n-1} + R_{n-2})$$

It is interesting to note that the first-order algorithms can be viewed as members of the family of second-order algorithms; thus:

Rectangular: $C_n = C_{n-1} + h/2 . (2 . R_n + 0 . R_{n-1})$

Euler: $\qquad C_n = C_{n-1} + h/2 . (0 . R_n + 2 . R_{n-1})$

Output waveforms for four integration algorithms for a typical square-wave input waveform are shown in Fig. 10.1. It is assumed that the integrator has no stored energy, so that initial values are all zero. Stored energy will give a DC bias to the waveform.

10.5.3   Higher-order algorithms

There are very many higher-order polynomial expansion algorithms. The *Runge-Kutta* method is very accurate, giving an error in the order of the 5th power of h, but it requires a knowledge of the complete differential equation, which is applied four times in the algorithm. More useful in control systems are algorithms which use polynomials involving past histories of the driving variable.

All third- and higher-order algorithms are only accurate with systems of a fixed sampling interval h. They should not be used wherever h varies.

The *quadratic Adams-Bashforth* algorithm is a predictor which under-values the integral somewhat. The formula tends to suit coarsely sampled sinusoidal waveforms.

The QUADRATIC ADAMS-BASHFORTH predictor algorithm:

$$C_n = C_{n-1} + h/12 . (23R_{n-1} - 16R_{n-2} + 5R_{n-3})$$

The 4-POINT ADAMS-BASHFORTH predictor algorithm:

$$C_n = C_{n-1} + h/24 . (55R_{n-1} - 59R_{n-2} + 37R_{n-3} - 9R_{n-4})$$

The QUADRATIC ADAMS-MOULTON corrector algorithm:

$$C_n = C_{n-1} + h/12 . (5R_n - 8R_{n-1} - R_{n-2})$$

The 4-POINT ADAMS-MOULTON corrector algorithm:

$$C_n = C_{n-1} + h/24 . (9R_n - 19R_{n-1} + 5R_{n-2} + R_{n-3})$$

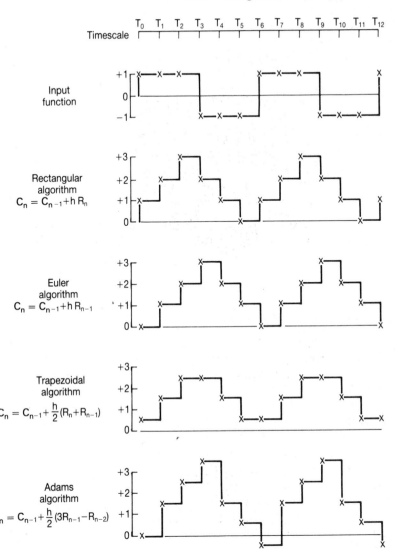

Fig. 10.1 Output waveforms for four integration algorithms

10.5.4 Unit delays in feedback

A first-order linear differential equation may be solved using the simulation shown in Fig. 10.2. In this case, the diagram represents the differential equation

$$r = dc/dt + a.c$$

Conversely, $r - a.c = dc/dt$

The a.c term is a negative feedback variable and c is produced as a result of a previous integration. The output of the integrator is not available for

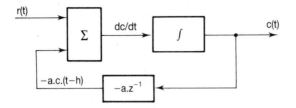

Fig. 10.2   Simulation of equation $r = \dfrac{dc}{dt} + ac$

use until the integration has been performed. Accordingly, there is a unit delay of one sampling interval, h, in the feedback path (shown as the z^{-1} multiplier) which occurs even if the computation time is very much smaller than h. The unit delay must be included in any rigorous analysis of the simulation.

10.5.5   Predictor-corrector algorithms

In the simulation of Fig. 10.2, it is possible to perform the integration twice. The input to the integrator is made up of a newly sampled driving variable and a previously calculated integral, which is fed back and multiplied by a coefficient. Computation of an interim integral c_n^* can be performed and the value of c_n^* (multiplied by the feedback coefficient) is then used to derive a new input to the integrator. A second integration calculation obtains a more accurate value for c_n. This is the predictor-corrector method for solving differential equations.

Sometimes, different integration formulae are used for the prediction (the first computation) and the correction (the second computation).

The first term of the input variable R for predictor algorithms is given here as R_{n-1}. Whether the input terms truly possess a unit time delay or a very small delay depends upon the synchronisation of sampling to computation within the program. If sampling takes place immediately before computation of C_n, generally the first input term could be R_n and not R_{n-1}. The user must decide which representation is appropriate.

10.5.6   Checking integration accuracy

There is one generally accepted method used for checking the accuracy of integration algorithms and which works equally well for simulation.

Select a typical, coarse, value of sampling interval, h, and integrate or simulate for a standard precise driving function: say, a constant, ramp, exponential or sinusoidal function. The computed output should be tabulated. Repeat the simulation for a new value of sampling interval of h/100 and tabulate the results for every 100th computation. Comparison of the two tables will yield an error function to around 1 percent.

10.6   Control law mechanisation

There are many methods for simulating dynamic systems or for solving differential equations by computer. Of these, Taylor series methods and matrix methods are well known. However, control engineers need to simulate dynamic systems in realtime or scaled realtime, as an essential part of the control function, not in order to investigate the behaviour of the equations. For this reason, the flow of data through the simulation is continuous and unceasing, and methods which require a fixed-size array of data are unsuitable.

There are three suitable practical methods for preparing equations for realtime simulation of control laws:

1 the intuitive construction method
2 the z-transform method
3 the state variable method (with or without matrices).

10.6.1   The intutitive construction method

This approach to algorithm design is less formal and less rigorous than the other two, but, as with most design, can be efficient and effective if tempered with experience.

The method consists of manipulating the system block diagram using classical control theory. The overall transfer function is separated into its component parts by factorising polynomials, expanding by partial fractions, separating cascaded or parallel blocks, or identifying feedback transfer functions, if possible.

Each small block is simulated independently, using either one of a library of functions or one of the two other methods discussed.

By way of example, a 3-term (PID) controller algorithm can be simulated easily by this method. As seen in Fig. 10.3 the PID controller consists of four elements: an integrator, a differentiator, a summation and a multiplication by the gain.

The integrator and differentiator can be implemented using methods described in Section 10.5. The outputs from the integrator and the differentiator are to be summed with the error and the result is to be multiplied by the gain.

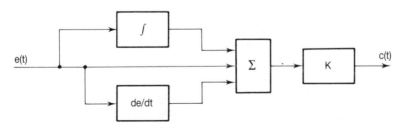

Fig. 10.3   PID controller algorithm

A fragment of code to implement this, written in a 'C' style, would look as follows:

```
#  define h 0.1 /* sampling interval is 100 millisec */
void pidcontrol(float gain, float Ti, float Td);
{
        float integrator_gain;
        float newerror;
        float olderror;
        float newintegral;
        float oldintegral;
        float derivative;
        float summation;
    integrator_gain = 1/Ti;
    while (do_not_stop)          /* do-forever loop */
        {
        newerror     = setpoint − process_variable;
        newintegral  = oldintegral + (h*integrator_gain *newerror);
        derivative   = (newerror − olderror)*Td/h;
        summation    = derivative + newintegral + newerror;
        output       = summation*gain;
        olderror     = newerror;
        oldintegral  = newintegral;
        }
}
```

10.6.2   The z-transform method

The z-transform method has become the classical method for writing software to simulate dynamic linear functions. As indicated in Section 3.7, the procedure consists of manipulation of the transfer function to be simulated, substitution by functions of z^{-1} for functions of s^{-1}, further manipulation, and derivation of the difference equations. The difference equations then are converted directly to program code.

Step 1. All algebraic terms in the transfer function must be converted to terms in $1/s$, so that both numerator and denominator polynomials would normally be divided by s^n, where n is the order of the system. This yields terms in $1/s^n$, $1/s^{n-1}$, $1/s^{n-2}$, and so on.

Step 2. The terms in $1/s$ ... $1/s^n$ are replaced directly by appropriate functions of z^{-1}. Several typical substitutions are given in Table 10.4. The substitution is directly related to the integration algorithm, and a discussion of integration algorithm selection was presented in the preceding section. One popular substitution is the *bilinear* (or *Tustin*) substitution, which corresponds to the Trapezoidal Integration Algorithm. Substitution formulae may easily be derived from the difference equations for integration algorithms, by replacing each unit time delay by z^{-1}, and manipulating to give $C(z)/R(z)$.

Table 10.4 Alternative substitutions for $1/s^n$, where $\Delta = z^{-1}$

Method	Substitution for: $1/s$	$1/s^2$	$1/s^3$	$1/s^4$
First difference	$\dfrac{h}{1-\Delta}$	$\left(\dfrac{h}{1-\Delta}\right)^2$	$\left(\dfrac{h}{1-\Delta}\right)^3$	$\left(\dfrac{h}{1-\Delta}\right)^4$
z Transform	$\dfrac{h}{1-\Delta}$	$h^2\dfrac{\Delta}{(1-\Delta)^2}$	$\dfrac{h^3}{2}\dfrac{\Delta(1+\Delta)}{(1-\Delta)^3}$	$\dfrac{h^4}{6}\dfrac{\Delta(1+4\Delta+\Delta^2)}{(1-\Delta)^4}$
Tustin	$\dfrac{h}{2}\dfrac{1+\Delta}{1-\Delta}$	$\left(\dfrac{h}{2}\dfrac{1+\Delta}{1-\Delta}\right)^2$	$\left(\dfrac{h}{2}\dfrac{1+\Delta}{1-\Delta}\right)^3$	$\left(\dfrac{h}{2}\dfrac{1+\Delta}{1-\Delta}\right)^4$
Boxer-Thaler	$\dfrac{h}{2}\dfrac{1+\Delta}{1-\Delta}$	$\dfrac{h^2}{12}\dfrac{1+10\Delta+\Delta^2}{(1-\Delta)^2}$	$\dfrac{h^3}{2}\dfrac{\Delta(1+\Delta)}{(1-\Delta)^3}$	$\dfrac{h^4}{6}\dfrac{\Delta(1+4\Delta+\Delta^2)}{(1-\Delta)^4}-\dfrac{h^4}{720}$
Madwed-Truxal	$\dfrac{h}{2}\dfrac{1+\Delta}{1-\Delta}$	$\dfrac{h^2}{6}\dfrac{1+4\Delta+\Delta^2}{(1-\Delta)^2}$	$\dfrac{h^3}{24}\dfrac{1+11\Delta+11\Delta^2+\Delta^3}{(1-\Delta)^3}$	$\dfrac{h^4}{120}\dfrac{1+26\Delta+66\Delta^2+26\Delta^3+\Delta^4}{(1-\Delta)^4}$

Thus, the substitution for the rectangular integration algorithm $C_n = C_{n-1} + hR_n$ becomes,

$$\frac{C(z)}{R(z)} = \frac{h}{(1 - z^{-1})}$$

Step 3. The equation is manipulated algebraically to obtain a form with $C(z)$ on the left-hand side and terms in $C.z^{-m}$ and $R.z^{-m}$ on the right-hand side.

Step 4. Terms in $C.z^{-m}$ are replaced by C_{n-m}, and terms in $R.z^{-m}$ are replaced by R_{n-m}. This yields a difference equation for the system and which then is coded directly, in an appropriate language.

As an example, consider the transfer function:

$$\frac{C(s)}{R(s)} = \frac{25}{(s^2 + 5s + 25)}$$

Thus, $C(s^2 + 5s + 25) = 25R$

or, $C + \dfrac{5C}{s} + \dfrac{25C}{s^2} = \dfrac{25R}{s^2}$

Applying the *bilinear* substitution of $1/s$ replaced by

$$\frac{h(1 + z^{-1})}{2(1 - z^{-1})}$$

yields

$$4C(1 - z^{-1})^2 + 10hC(1 - z^{-1})(1 + z^{-1}) + 25h^2C(1 + z^{-1})^2$$
$$= 25h^2R(1 + z^{-1})^2$$

or $4C(1 - 2z^{-1} + z^{-2}) + 10hC(1 - z^{-2}) + 25hC(1 + 2z^{-1} + z^{-2})$
$$= 25h^2R(1 + 2z^{-1} + z^{-2})$$

or $C(4 + 10h + 25h^2) + z^{-1}C(-8 + 50h^2) + z^{-2}C(4 - 10h + 25h^2)$
$$= R.25h^2 + z^{-1}R.50h^2 + Rz^{-2}$$

Finally,

$$C_n = \frac{(8 - 50h^2)C_{n-1} + (-4 + 10h - 25h^2)C_{n-2} + 25h^2R_n + 50h^2R_{n-1} + 25h^2R_{n-2}}{(4 + 10h + 25h^2)}$$

This difference equation can now be coded directly in a high-level language.

10.6.3   The state variable method

The *state variable method* derives from a technique of representing the system differential equation by a series of integrations. There is one integration per degree of system order, so that a fourth-order system, for example, would require four integrations. The modern approach is to represent the system mathematically by a set of single-order differential equations, and to incorporate these in two groups of matrix equations. There is a system state equation, and a set of output equations – one for each output. The mathematical treatment of this material is covered by many excellent texts.

For relatively simple systems, however, an algorithm may easily be obtained by using a state variable approach very similar to that originally used for programming analog computers.

The algorithm is developed in two stages, corresponding to the two groups of matrix equations. The first step is to generate a state model defining the dynamic characteristics of the system, and the second step is to generate the output models defining the amplitudes and phases of the various outputs.

By way of example consider the system:-

$$\frac{C(s)}{R(s)} = \frac{3s + 2}{s^2 + 5s + 25}$$

To obtain the state model, one must use the characteristic equation as follows:

$$X(s^2 + 5s + 25) = \text{input-function (s)}$$

or $$\frac{d^2x}{dt^2} + \frac{5dx}{dt} + 25x = \text{input-function (t)}$$

or $$s^2X = -5sX - 25X + \text{input-function (s)}$$

or $$\frac{d^2x}{dt^2} = -\frac{5dx}{dt} - 25x + \text{input-function (t)}$$

The model uses one summation and two integrations. The summation generates the state variable d^2x/dt^2 from the addition of the three other terms. The first integrator generates the state variable dx/dt from d^2x/dt^2, and the second integrator generates the state variable x from dx/dt.

This model is shown in Fig. 10.4.

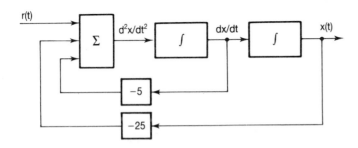

r(t)

d^2x/dt^2 $\int$ dx/dt $\int$ x(t)

Σ

−5

−25

Fig. 10.4   Generation of state variables for the function
$$\frac{C(s)}{R(s)} = \frac{3s + 2}{s^2 + 5s + 25}$$

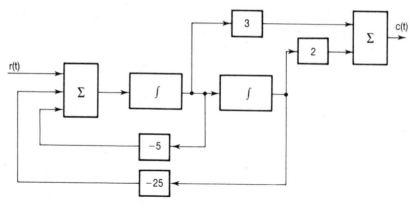

Fig. 10.5 Flow diagram for the simulation of $C(s) = R(s) \times \dfrac{3s + 2}{s^2 + 5s} + 25$

The second step is to create the output models. In this example there is only one output, so that only one output equation exists. It is modelled by a simple summation of terms generated by multiplying the appropriate state variables by their coefficients. The 3s represents 3 × [the variable dx/dt] and the 2 derives from the variable. The final flow diagram is shown in Fig. 10.5.

The model can be encoded in a suitable programming language directly from the flow diagram. A program fragment in 'C' is given for the example. The example uses the rectangular integration algorithm.

```
#  define h 0.1    /* sampling interval in seconds */
void secondorder( )
{
        float firstsv;
        float secondsv_new;
        float seconsv_old;
        float thirdsv_new;
        float thirdsv_old;

    while (do_not_stop       /* do-forever loop */
        {
        firstsv        = input( ) − (5*secondsv_new)
                         − (25*thirdsv_new);
        secondsv_new = secondsv_old + (h*firstsv);
        thirdsv_new  = thirdsv_old + (h*secondsv_new);
        output       = (3*secondsv_new) + (2*thirdsv_new);
        secondsv_old = secondsv_new;
        thirdsv_old  = thirdsv_new;
        }
}
```

The state variable method has two distinct advantages over the z-transform method: less algebra means less chance of making a mistake, and the lines of code tend to be more concise.

10.7 Initialisation

Computer programmers recognise the importance of initialising variables prior to the start of any computation. Whether or not a particular variable must be initialised is largely a matter of commonsense, but in case of any doubt variables should be initialised.

In the same way, control engineers recognise the importance of making initial settings to plant variables prior to startup. In operational control software, it is important not only to initialise variables properly but to select the sequence of initialisation. During the startup of plant, uninitialised variables at a crucial point can give rise to unpredictable and possibly dangerous events, which usually are difficult to reproduce or diagnose. In some cases, variables may have to be re-initialised during plant operation and, in others, variables can be made self-initialising. A good example of the latter case is the PV tracking method described in the next section.

It is common for plant operators to set variables at a nominal 50 percent for startup, and this introduces the idea of a nominal signal value for *zero* – that is, the *datum value* for the variable. The programmer must decide whether the initial value should be at, say, 50 percent of full scale (12 mA in a 4 to 20 mA system) or at the minimum value, or some other arbitrary value.

It is also important to decide whether or not to neutralise outputs before a reasonable value has been computed. This is particularly important with digital outputs, since motors (for example) could start unpredictably while the computer is initialising itself.

Similar decisions must be made concerning the occasion of power failure: should the computer freeze all outputs, or neutralise them, or set them all to some predetermined value? These decisions depend on the particular application.

Selection of initial values in computation will often determine how quickly a system stabilises. In a closed loop system, uninitialised variables have the same effect as external fluctuations and are eventually removed by the feedback. However, they should be minimised for two reasons. Firstly, they represent many disturbances occurring simultaneously, which is abnormal, and may be beyond the capabilities of the control system. Secondly, in nonlinear plant, control loops may pass through an unstable region during startup, requiring manual operation, and unnecessary fluctuations place extra pressure upon the plant operators.

In open loop systems, correct initial settings are a critical part of the operation of the software.

Where integrators and differentiators are simulated in software, initial values of state variables represent *initial conditions*. Initial conditions represent stored energy in capacitive or inductive storage elements. Thus, state variables should always be set to some initial value, which is usually zero in closed loop systems (which generally are switched on in an energy-neutral state). However, there are clearly some dynamic computations (usually open loop) for which the initial stored energy must be carefully evaluated prior to running the computation.

10.8   PID control laws

Although there are many texts which examine the *PID* or *3-term* control law, there are some commonly used variations which deserve mention here. All of the algorithms given may be converted to usable program code by the methods discussed in Section 10.6.

The usual form of the PID algorithm has the transfer function:

$$C(s) = K[1 + s.T_d + 1/(s.T_i)].E(s)$$
where   K   = controller gain
$\quad$ T_d = derivative action time (seconds)
$\quad$ T_i = integral action time (seconds)

In practice, the algorithm is often modified in several ways.

In order to limit the potential for undesirable fluctuations in the manipulated variable, it is common to use *PV tracking*. In *manual* operation, the setpoint should be made equal to the process variable so that, at the point of switching to *automatic* operation, the system deviation is zero. This ensures a *bumpless transfer* from manual to automatic.

Another way to reduce the effect of large fluctuations is to introduce *rate limiting* of the manipulated variable. Here, a history of the PID algorithm output is kept and a differentiation is performed after each computation. If the derivative is too large, the output is reduced to conform to the maximum permitted rate-of-change.

In systems with a fast controller sampling rate and a very long process time constant, the quantisation error can cause difficulty, even with digital resolution of 12, 14 or 16 bits, because the integral action integrates the quantisation error, causing oscillation around the loop. A cure for this is the addition of a few lines of code, which switch off the integral action if the system deviation is close to the quantisation error.

Derivative action can often give an undesirable abrupt response, and it is commonplace to soften this by use of a *pseudo derivative* algorithm. The most usual form assumes the transfer function:

$$\frac{s.T_d}{1 + a.s.T_d}, \text{ where a is usually 10 or 20}$$

The abrupt derivative action is often most pronounced with changes to the setpoint made by plant operators. It is possible to have full PID functions operational on the process variable, but to only have P and I functions on the setpoint. This gives sensitive control to system fluctuations but smoother operation to setpoint alterations. This algorithm is known as *IP-D control*, and the transfer function is:

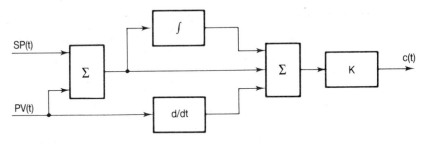

Fig. 10.6   IP-D controller algorithm

$$C(s) = K[1 + 1/(s.T_i)].SP(s) + K[1 + s.T_d + 1/(s.T_i)].PV(s)$$

The block diagram is given in Fig. 10.6.

Yet smoother action on the setpoint is provided by the *I-PD algorithm*. Here, the proportional action is removed from the setpoint, leaving only the integral action. The transfer function for this is:

$$C(s) = K[1/(s.T_i)].SP(s) + K[1 + s.T_d + 1/(s.T_i)].PV(s)$$

The block diagram is shown in Fig. 10.7.

Both IP-D and I-PD algorithms are often used for the second (downstream) controller in a cascaded pair of controllers.

Integrator windup is a problem which occurs in the presence of large disturbances and/or long-time-constant processes. In certain circumstances, the normal operating range of an integrator may be a very small proportion of its total range and, where an integrator output has ramped to saturation point, the settling time of the loop may be unacceptably long.

To prevent integrator windup, it is common practice to introduce limits to the excursion of the output of the integral action, which is a relatively simple matter in software. It is also common to set integrator outputs to zero after some particular event, such as manual/auto transfer.

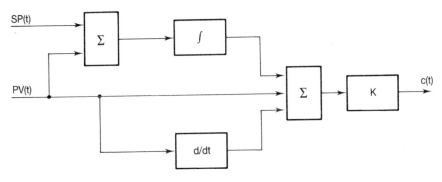

Fig. 10.7   I-PD controller algorithm

10.9   The Smith predictor

The *Smith Predictor* is a specific technique used in process control for handling *dead time* in the forward path of a feedback control system. Dead time is commonly called *pure time delay* or *transport delay* and is quite different from a *time lag*, which is exponential in nature.

Pure time delay in process control usually occurs because of the transport of material, which introduces a period of waiting before the measurement of the process variable is made. The delay can be extremely troublesome, since it can cause the feedback system to be unstable in several modes simultaneously, and normal compensation methods may not remove all of them. PID or feedforward compensation algorithms may simply be inadequate for removing the effects of a pure time delay.

The technique involves adding, around the controller, extra feedback which incorporates another time delay equal in value but opposite in sense to the troublesome one. It is a straightforward matter to build such a loop into the controller itself, and many commercal process controllers offer pure time delay feedback as an option.

10.9.1   Analysis of the Smith Predictor Technique

$$K_m = \text{feedback path and setpoint path gain}$$
$$K_c = \text{controller gain}$$
$$G_c = \text{controller transfer function (excluding the gain)}$$
$$K_v = \text{forward path gain}$$
$$D(s) = \text{disturbance function}$$
$$S(s) = \text{setpoint function}$$
$$P(s) = \text{process variable function}$$
$$B(s) = \text{controller output function}$$
$$e^{-as} = \text{transfer function of the pure time delay}$$

Fig. 10.8 shows a typical system, in block diagram form.

The gain of the process excluding the controller is $K_v.K_m$. The Smith Predictor technique consists of installing feedback around the controller, with a feedback transfer function of the form:

$$H(s) = K_v.K_m.(1 - e^{-as})$$

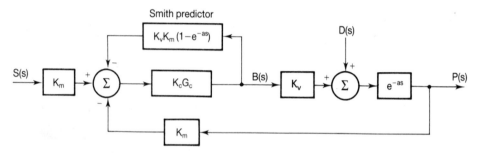

Fig. 10.8   System with Smith predictor incorporated in the controller

For a disturbance change, the overall transfer function of the system

$$\frac{P(s)}{D(s)} = \frac{e^{-as}}{1 + e^{-as}.K_m.K_v.K_c.G_c(s)}$$ in the absence of the Smith Predictor.

The pure time delay e^{-as} appears in the numerator and delays the output function $P(s)$, but it also appears in the denominator, and this is what causes the instability.

By inserting the $H(s) = K_v.K_m.(1 - e^{-as})$ term as feedback around the controller, the new overall transfer function for a disturbance change becomes:

$$\frac{P(s)}{D(s)} = 1 + \frac{\dfrac{e^{-as}}{e^{-as}.K_m.K_v.K_c.G_c(s)}}{1 + K_m.K_v.K_c.G_c(s)(1 - e^{-as})}$$

which reduces to

$$\frac{P(s)}{D(s)} = \frac{e^{-as}(1 + K_m.K_v.K_c.G_c(s)(1 - e^{-as}))}{1 + K_m.K_v.K_c.G_c(s)}$$

$$= \frac{e^{-as} + K_m.K_v.K_c.G_c(s)e^{-as} - K_m.K_v.K_c.G_c(s)e^{-2as}}{1 + K_m.K_v.K_c.G_c(s)}$$

It can be seen that the Smith Predictor removes the pure time delay from the denominator, thus removing the source of the instability. However, a penalty is incurred in the form of a double pure time delay term in the output.

10.9.2   Computer mechanisation of the Smith Predictor

The heart of the algorithm is the pure time delay e^{-as}, and the only way to mechanise this is to use a *delay line*. The simple method of creating a delay line is to use a *queue* or *first-in-first-out (FIFO) buffer*. To generate the delay, incoming data are fed to the end of the queue (input) and outgoing data are taken from the front of the queue (output). All the while, the data inside the queue are moved continuously from the end to the front as space at the front is vacated.

Shuffling data along the queue is inefficient, and the normal approach is to leave the data alone and alter the *pointers* to the head and the tail of the queue. Although the queue occupies a linear memory space, it is circular in shape since the ends must be joined, or *wrapped around*.

As an item of data is added to the tail of the queue, the tail pointer is decremented to the next available address space. As an item of data is removed from the head of the queue, the head pointer is decremented. Either head or tail pointer is decremented to see if it is at the bottom of the allocated address space and, if so, it is wrapped around and made to point to the top of the address space. The arrangement is shown diagrammatically in Fig. 10.9.

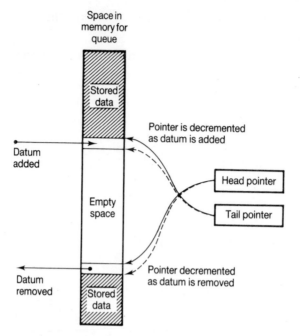

Space in
memory for
queue

Stored
data

Pointer is decremented
as datum is added

Datum
added

Head pointer

Empty
space

Tail pointer

Datum
removed

Pointer decremented
as datum is removed

Stored
data

Fig. 10.9 Implementation of a queue in memory

The function e^{-as} is a pure time delay of time a, and so the queue must be implemented in real time. Any datum inserted into the queue must be retrieved after a seconds. The implication is that the process of inserting and removing data must occur after a specifiable time interval, h, and thus the queue must contain a fixed number a/h of elements.

Generation of the function $K.(1 - e^{-as}).x$ is a simple matter, and the algorithm is as follows:

> variable x arrives
> x is inserted into the tail of the queue
> y is retrieved from the head of the queue
> $z := K*(x - y)$;
> z is the result.

10.10 Self-tuning strategies

10.10.1 Historical background

Research on the subject of adaptive control systems (Fig 10.10) first became prominent in the early 1950s. Prime motivation for the early research effort was the need for autopilots in high performance aircraft. State space and stability theory were introduced in the 1960s, as were important advances in stochastic control theory. Tsyphnin[1] showed that

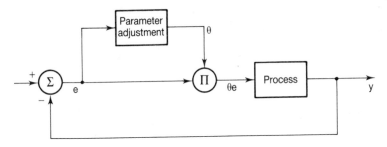

Fig. 10.10   Early adaptive regulator

many schemes for learning and adaptive control could be described in a common framework as recursive equations of the stochastic approximation type. There were also major developments in the area of system identification and parameter identification. Work in the late 1970s and early 1980s yielded valid proofs of the stability of adaptive systems, albeit whilst applying very strict assumptions.

Systems like Fig 10.10 have been successfully flight tested. It is critical to the operation of such a system that it be excited by wide-band disturbances.

Marsik's system (Fig 10.11) also attempts to compensate for variations in loop gain. It is based upon the fact that many frequency responses of closed loop systems have a resonant peak which increases with increasing loop gain. The peak occurs approximately at the open loop gain crossover frequency; that is, where the open loop phase shift is 180°.

Assuming that the command signal has a frequency content which covers a reasonably wide spectrum, the output signal y has a significant component in phase with the error e, if the loop gain is low. Marsik's system sets the gain simply from a measurement of the correlation of the error signal with the output signal.

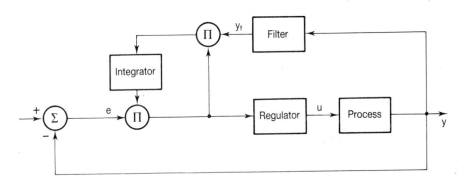

Fig. 10.11   Marsik's adaptive regulator

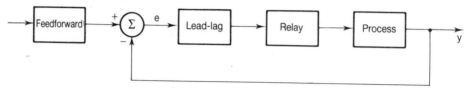

Fig. 10.12   Self-oscillating adaptive system

10.10.2   Measurement of transient response features

Regulator performance parameters can be specified in terms of some features of the closed loop transient response. Zero crossings, overshoot, damping factor and so on are typical features that are considered. The advantage of these schemes is that they are simple to implement. The drawbacks are that they are heavily dependent upon the disturbances having to take the form of isolated steps or impulses.

10.10.3   Self-oscillating adaptive systems

The basic approach with these systems (Fig 10.12) is to create a feedback loop having a gain as high as possible, combined with feedforward compensation to give the desired response to command signals. The high loop gain is maintained by the introduction of a relay in the feedback loop, thus creating a limit cycle oscillation.

10.10.4   Gain scheduling

In some systems, there are auxiliary variables which correlate well with the characteristics of the process dynamics. If these variables can be measured, it becomes possible to use them to change the regulator parameters. This approach is called *gain scheduling*, because the scheme (Fig 10.13) was originally used to accommodate changes in process gain. Gain scheduling is a form of open loop compensation.

The idea of gain scheduling originated in connection with flight control systems. A drawback of gain scheduling is that the design is time consum-

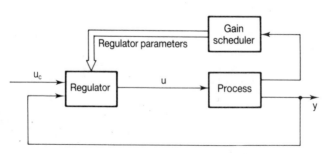

Fig. 10.13   Gain scheduler

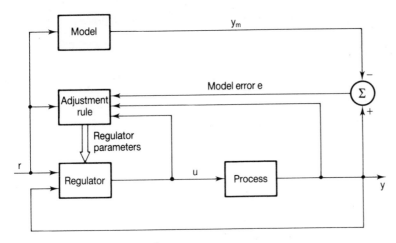

Fig. 10.14   Model reference system

ing, but gain scheduling has the advantage that the parameters can be changed very quickly in response to process changes.

10.10.5   Model reference adaptive systems

These systems were originally proposed by Whitaker, who considered the problem in which the system performance specification was given in terms of the behaviour of a reference model. This model is designed to behave in precisely the way in which the process is ideally required to respond to the command signal.

Referring to Fig 10.14, the parameters of the regulator are adjusted by the outer loop, in such a way that the error e between the model output y_m and the process output y becomes small. The key to the problem is to design the adjustment mechanism so that one obtains a stable system in which the error e converges to zero. The procedure can be applied to nonlinear systems. Some approximations are required in order to make a realisable parameter adjustment control law. It is not possible to guarantee stability with this method.

10.10.6   Self-tuning regulators

The method used with these regulators is different from those discussed previously, in that process parameters are updated and the regulator parameters are obtained from the solution of the design problem.

This type of regulator can be considered to be composed of two stages, as shown in Fig 10.15. The inner loop consists of the process together with a conventional regulator. The parameters of this regulator are adjusted by the outer loop. The self-tuning regulator has received considerable attention because it is flexible, easy to understand, and easy to implement with microprocessors.

10.10.7   Adaptive schemes derived from stochastic control theory

These schemes (Fig 10.16) differ from those previously described, in the fact that the system and its environment are described by a stochastic model. The parameters are introduced as state variables. An unknown constant is modelled by the differential equation $d\theta/dt = 0$, with an initial distribution which reflects the parameter uncertainty. This corresponds to a Bayesian approach, where unknown variables are viewed as random variables. Parameter drift is accommodated by adding random variables to the right-hand side of the equation. A criterion is formulated so as to minimise the expected value of a *loss function*, which is a scalar function of state and control variables.

The problem of finding a control law which minimises the expected loss function is difficult. Making the assumption that a solution exists, a functional equation for the optimal loss function can be derived using dynamic programming. The functional equation, which is called the *Bellman* equation, can be solved numerically only in very simple cases, using a nonlinear estimator and a feedback regulator. The estimator generates the conditional probability distribution of the state from the measurements. The distribution is called the *hyperstate* of the problem. The feedback regulator incorporates a nonlinear function, which maps the hyperstate into the space of control variables.

10.10.8   Industrial products

Adaptive techniques are being used in a number of commercial products. Gain scheduling is the standard method for the design of flight control systems for high-performance aircraft. Self-oscillating adaptive systems are used in missile control. There are several commercial adaptive autopilots used for steering ships. There are adaptive motor drives, and adaptive systems for industrial robots. By far the largest field where adaptive control is enjoying most popularity is in industrial control.

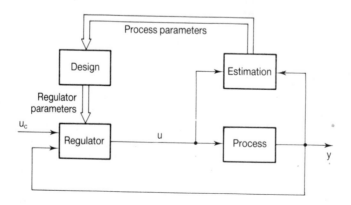

Fig. 10.15   Self-tuning regulator

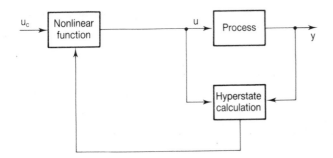

Fig. 10.16   Adaptive regulator derived from stochastic theory

10.10.9   The SattControl auto-tuner

Most industrial processes are controlled by PID regulators. Many instrument engineers and plant personnel are acquainted with the selection, installation, and operation of such regulators. In spite of this, it is common practice that many regulators are poorly tuned. One reason is that simple robust methods for automatic tuning have not been available. A PID regulator can be described by

$$u = K(e + 1/T_i.\int e(t)dt - T_d.dy/dt) \qquad\qquad 10.2$$

where $e = r - y$.

The Swedish company SattControl has developed an auto-tuner (Fig 10.17) which automatically adjusts the parameters of a PID regulator. The auto-tuner is based upon a special system identification technique which automatically generates an appropriate test signal, and a variation of the classical Ziegler-Nichols closed loop method. An interesting feature is that it has the ability to determine whether derivative action is necessary. The Ziegler-Nichols method is based on the observation that the appropriate regulator parameters can be determined from knowledge of one point on the open loop frequency response of the system. This point is the intersection of the polar response curve with the negative real axis: in other words, the gain at phase crossover. It is traditionally described in terms of the ultimate gain K_{pm} and the ultimate period T_u. Refer to Section 14.7.2 for further information on Ziegler-Nichols techniques.

The regulator design is based upon the idea that the critical gain and the critical frequency can be determined from an experiment with relay feedback. When the critical gain K_{pm} and the critical period T_u are known, the parameters of a PID regulator can be determined by the Ziegler-Nichols rule which is expressed as:

$$K = 0.6K_{pm} \qquad T_i = T_u/2 \qquad T_d = T_u/8$$

This rule gives a closed loop system which is sometimes too poorly damped. A major advantage of the auto-tuner is that there are no parameters which have to be set *a priori*. To use the tuner, the process is simply brought to an equilibrium by setting a constant control signal in manual mode. The tuning is then activated by pushing a tuner switch.

10.10.10   The Foxboro Exact adaptive regulator

This regulator is based upon the analysis of the transient response of the closed loop system to set point changes or load disturbances, and traditional tuning methods in the Ziegler-Nichols spirit. Figure 10.18 shows how the control error responds to a step or impulse disturbance.

Heuristic logic is used to detect that a proper disturbance has occurred and to detect the peaks e_1, e_2, e_3 and the period T_p. The control design is based upon specifications on damping, overshoot and the ratios T_i/T_p and T_d/T_p, where T_i is the integral action time, T_d is the derivative action time and T_p the period of oscillation. The damping factor is defined as:

$$d = \frac{e_3 - e_2}{e_1 - e_2}$$

10.3

and the overshoot is

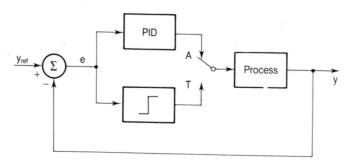

Fig. 10.17   SattControl auto-tuner

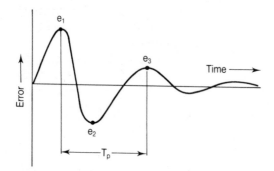

Fig. 10.18   Error response for a step disturbance of a Foxboro adaptive regulator

$$z = -\frac{e_2}{e_1}$$

10.4

In typical cases, it is required that both d and z are less than 0.3. The Ziegler-Nichols tuning rule gives:

$$\frac{T_i}{T_p} = 0.5 \quad \text{and} \quad \frac{T_d}{T_p} = 0\cdot12$$

The numbers 0.5 and 0.12 have been modified, based upon empirical studies. Smaller values are chosen for processes with dominant dead time and larger values are selected for processes with a dominant lag. The tuning procedure requires prior information of the regulator parameters K, T_i, and T_d. It also requires knowledge of the timescale of the process. Some measure of the process noise is also required, in order to set the tolerances of the heuristic range.

10.10.11 The Leeds and Northrup Electromax V adaptive regulator

This regulator is an adaptive single-loop controller based upon the PID structure. The adaptation process is a self-tuning regulator in which a second-order discrete-time model is estimated. The parameters of the PID regulator are then computed from the estimated model, using the *pole placement* method. The regulator can operate in three different modes called *fixed, self-tune*, and *self-adaptive*. In the fixed mode, the regulator operates like an ordinary fixed-parameter PID regulator. In the self-tune mode, a perturbation signal is introduced, a model of the process is estimated, and PID parameters are computed from the model. In the self-adaptive mode, the PID parameters are updated continuously.

10.10.12 The ASEA adaptive regulator (Novatune)

This regulator is based upon least-squares estimation and minimum variance control. The Novatune differs from the previous regulators, because it is not based upon the PID structure. The Novatune is an *implicit* self-tuning regulator. The parameters of a discrete-time model are estimated, using recursive least-squares. The control design is a minimum-variance regulator, which is extended to admit positioning of one pole and a penalty on the control signal.

10.10.13 Industrial experience

In the process control field, the Electromax V was introduced in 1981, the ASEA Novatune in 1982, the SattControl auto-tuner and the Foxboro Exact in 1984. The majority of applications has been in the field of temperature control. The experiences have been generally favourable, although it is noted that adaptive control is not a panacea for every loop. Most of the benefits are derived from self-tuning, although there are some cases where continuous adaptation has been profitable. Difficulties with using these regulators have been observed with processes having an asymmetric process response, rapid parameter variations, or strong non-linearities. Difficulties with regulators used in the self-adaptive mode have also been found under operating conditions in which the measured variable is suddenly disconnected.

10.10.14   Uses of adaptive control

All adaptive techniques can be used to provide automatic tuning. In such applications, the adaptation loop simply is switched on. Perturbation signals may be added to improve the parameter estimation. The adaptive regulator is run until the performance is satisfactory, then the adaptation loop is disconnected and the system left running with fixed regulator parameters. Auto-tuning can be considered as a convenient way to incorporate automatic modelling and design into a regulator. The available industrial experiences indicate that there is a need for automatic tuning of PID controllers. Conventional industrial regulators are often poorly tuned. Derivative action is seldom used, although it can often be beneficial. Some of the available schemes require prior information, which makes them more difficult to use.

10.11   Strategies for decoupling control loops

Control systems with several inputs and outputs are commonplace, and they are referred to as *multivariable processes*. For a system with x inputs and y outputs, there are (x.y) transfer functions, there being one transfer function between each input and each output.

Many of the transfer functions are the normal plant process transfer functions which are to be controlled. Other transfer functions are unexpected, undesirable or inconvenient, and these relationships are often referred to as *cross-coupling* or *interaction* between processes.

Some examples of process interaction may be familiar. The traditional example is the heat exchanger problem, where there is cross-coupling between the flowrate control system and the temperature control system. In high-performance aircraft, there is inertial cross-coupling between the longitudinal dynamic system (elevators) and the transverse dynamic system (ailerons and rudder). In most control systems, there is interaction between load or disturbance input variables and the process variable(s).

Theoretically, it is possible to completely eliminate the interaction between variables by *decoupling*. This should be achieved by introducing another transfer function so as to cancel the effect of the problematic function. In practice, however, it is only possible to obtain a reduction in interaction because no process is ever perfectly linear. There are two possible approaches: the open-loop configuration and the closed-loop configuration. In either configuration a controller is added to the system so

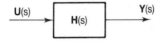

Fig. 10.19   Interacting multivariable system

that there is a controller matrix and a new input matrix, which is a controller setpoint matrix.

The transfer functions obtained for the controller matrix in either configuration may be converted into computer program code by any of the techniques outlined in this chapter.

Figure 10.19 shows a multivariable interacting system where:

$$Y(s) = H(s).U(s)$$
where $Y(s)$ is the process output vector
        $U(s)$ is the process input vector
        $H(s)$ is the process matrix

It is required to control the process by use of a controller to reduce the interaction, giving a new overall process matrix $G(s)$, which, if totally non-interacting, will be a diagonal matrix. The new process is shown in Fig. 10.20.

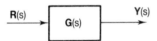

Fig. 10.20   Non-interacting (compensated) system

10.11.1   Open loop configuration

The open loop controller introduces a new matrix $W(s)$ into the process, as shown in Fig. 10.21.

   Now: $Y(s) = G(s).R(s)$
   and $Y(s) = W(s).H(s).R(s)$
   so that $G(s) = W(s).H(s)$
   and $W(s) = [H(s)]^{-1}.G(s)$

The last formula yields a method for obtaining the controller matrix from the desired (diagonal) matrix $G(s)$ and the process matrix $H(s)$.

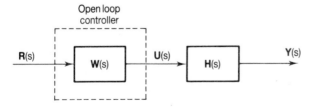

Fig. 10.21   Open loop decoupling controller strategy

10.11.2   Closed loop configuration

The closed loop configuration uses a feedback controller, as shown in Fig. 10.22.

Now: $\qquad$ $\mathbf{Y}(s) = \mathbf{G}(s).\mathbf{R}(s)$
and $\qquad$ $\mathbf{Y}(s) = \mathbf{W}(s).\mathbf{AH}(s).\mathbf{E}(s)$
and $\qquad$ $\mathbf{E}(s) = \mathbf{R}(s) - \mathbf{Y}(s)$
thus yielding $\;\mathbf{W}(s) = [\mathbf{H}(s)]^{-1}.\mathbf{G}(s).[\mathbf{I} - \mathbf{G}(s)]^{-1}$
where $\mathbf{I}$ is a unit matrix.

This yields an approach for obtaining the controller matrix $\mathbf{W}(s)$.

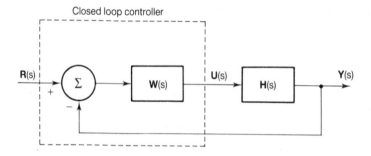

Fig. 10.22   Closed loop decoupling controller strategy

Reference

Tsyphin, Y. Z. (1971). *Adaptation and learning in automatic systems.* Academic Press, New York.

11

Microcomputer Process Control

11.1 Introduction

One of the surprising facts about Control Engineering practice is that the companies which service the needs of process control have developed quite separately from companies which service digital-electrical control. Engineers are thus often faced with a choice of two kinds of hardware, neither of which will completely satisfy their requirements: this choice is between a distributed (continuous) process control system on the one hand and a programmable logic (sequence) control (PLC) system on the other.

Most larger programmable logic control systems additionally offer full analog inputs and outputs and three-term (PID) control. However, the needs of a large process plant cannot be served by the very simple features available on such systems. Modern distributed process control systems are very complex distributed computers with a wide array of features and facilities.

This chapter covers the broad principles and requirements of control for the process industries, and a description of the more commonplace features found on the control systems themselves. The discussion of hierarchical levels of control brings together ideas from many different disciplines, including business and finance, factory management, computing and control engineering. The details of the techniques themselves are explained in other chapters and other works. In particular Chapter 10 should be consulted for details of control algorithms and strategies.

11.2 Hierarchical levels

In order to optimise the operation of a large manufacturing plant, it is necessary to apply a total automation strategy to managing the entire factory as well as to plant-level control. The concept is often termed *strategic process management*.

Strategic process management is based upon a pyramid of hierarchical levels of control, with the more abstract levels higher in the pyramid and the more physical levels lower down. At the tip of the pyramid is a single loop – the profitability control loop. At the base of the pyramid are all of the simple single control loops and simple logic control functions.

The control loops in the *strategic controls* category have humans in the loops, and the control of profitability is usually carried out at boardroom level. Managers can have a considerable range of technologies available to provide feedback information. These technologies are generally computer-based, although other types of reporting technique and management method are also constantly being developed.

Apart from the constant striving for better profitability, other strategic decisions can use control engineering techniques. These included pricing policy, advertising policy, personnel management, interlocking of different processes to maximise use of resources, finding new products to utilise waste and by-products, waste disposal, private power generation and energy management.

It may not be possible to implement fully the ideal solutions found by the application of manufacturing strategies, so that compromises must be made. Finding the best compromises is the province of the next level in the

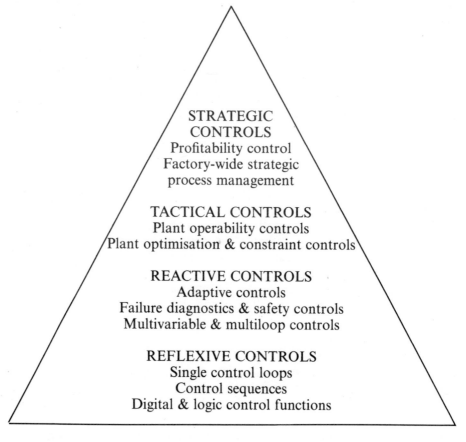

Fig. 11.1   Hierarchical pyramid of strategic process management

hierarchy – plant operability and optimisation controls or *tactical control*. For example, a waste product from one process may be used in the creation of another product, but there may be a conflict between the process conditions for obtaining the best quality of the two products. To what extent the quality of either product can be downgraded to improve the quality of the other is an optimisation problem bounded by the constraints of plant operability. Optimisation techniques are well-known, and computers can be used for solving the difficult sets of equations involved. The two principal optimisation methods are *static optimisation* – also called *linear programming* – and *dynamic optimisation*, which involves variables with a time-domain dependence.

Examples of multiple-product processes abound. In brewing, the by-product from the vats is used to make yeast-extract, which is sold as a foodstuff. In the timber industries, the waste from veneer, plywood and laminated wood products is used for particle-board, hardboard and high-density fibreboard.

In brick-making, a different kind of optimisation is needed: one which can be performed on-line by computer. Different brick products demand different kiln temperatures and different atmospheres. If a tunnel kiln is used, various types of brick product travel through on 'cars'. The kiln temperature profile, the kiln atmosphere and the loading-order of the brick cars have to be carefully calculated in order to maximise the quality of the products, and to minimise the consumption of energy which can constitute sixty per cent of the cost of making bricks.

Plant optimisation strategies are bounded by, and depend for their success on, the way in which the plant itself is controlled. The entire process must be controlled and, since broad areas of processes usually interact, parameters must be coupled, decoupled, referenced and monitored. The control system must be able to handle failure of the plant, together with disturbances such as variability in raw material, environmental conditions and workforce skills. The *reactive control* part of process management must cope with all of these factors.

The interconnecting parts of the process depend, for their operation, upon the effectiveness of the individual control loops, simple logic functions and the reliability and stability of single items of plant. This is the province of the lowest level of the pyramid: the *reflexive controls*.

11.3   Physical and operational distribution of hardware

One distinction between the technologies of distributed process control systems and PLC systems is that the manufacturers of the process control equipment learned very quickly that using a single minicomputer for controlling all of the plant could lead to disaster. This was revealed between 1970 and 1975, when many single-computer process control systems were installed around the world. Unfortunately, the attempt to computerise process control was premature because the minicomputers of the day were quite unreliable and the factories which depended upon them were more often shut down than running. The difficulty was solved when Honeywell introduced the idea of *total distributed control*, which is the modern approach to computerised process control. The technology has developed ever since.

In contrast to the principle of distributed process control, the manufacturers of programmable logic controllers have tended to offer larger and more powerful controllers, with one processor bearing responsibility for a large section of plant. But the main principle of distributed control systems is the provision of the ability of plant to continue running effectively, despite the failure of one or more of the control components.

The first approach for achieving this end is to distribute the responsibility over several (or very many) microprocessors. Each processor is generally responsible for about eight process control loops or equivalent computations such as multiplication functions. Nowadays, this rule is being broken as more powerful and much more reliable processors are being used, but the principle remains.

A second approach is to use the principle of *graceful failure* modes. When a failure does occur, it must be possible to maintain control by some other means. At controller level, this is usually achieved by allowing manual intervention and operation, by virtue of extra (redundant) circuitry. A plant operator can run each loop in manual mode of operation while a technician is replacing the controller, and this implies of course that the manual circuitry is physically separate from the processor. The processors controlling the loops are also independent of any local programming panel, any operator station, or any communications network, so that failure in any one of those components will not prejudice plant operation.

A third approach to the concept is the use of *dual redundancy*. Communications components and power supplies, in particular, are usually duplicated so that failure in one merely registers an alarm and does not halt the plant. Very often, communications cables will follow different routes so as to avoid simultaneous damage to both of the duplicated channels by careless workers or forklift trucks – such an event would be termed a *common failure mode*. For similar reasons, power supplies will often be fed from different switchboards.

It is usual to site the local control processors close to or adjacent to the plant being controlled, and large factories which use such a system have *field stations* which are remote from the main control room. The field stations can be a simple switchboard cubicle or a small, clean, air-conditioned room. A typical field station will contain a rack, power supplies, one or more control processors (each driving several loops, as explained), a local programming or operator's panel and a network communications processor. Usually, the field station is autonomous and each processor in it is autonomous. If it were otherwise, the principle of distributed responsibility would be compromised.

In any medium-to-large plant, there has to be some form of centralised control room and this is usually equipped with one or more *operator stations*. An operator station consists of a large colour video display screen

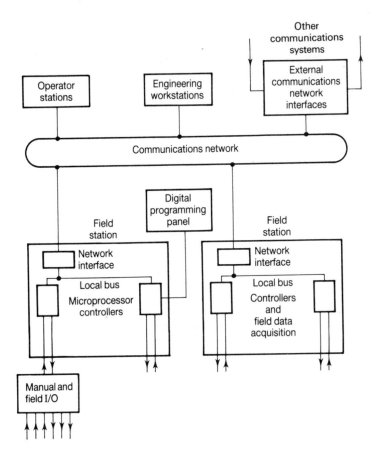

Fig. 11.2 Simplified schematic diagram of a typical distributed control system.

and keyboard, which is normally special-purpose, and not the usual typewriter/computer keyboard. The operator stations are in communication with the field stations, and all control functions can be modified or altered from them. The operator stations can also be used for programming or configuring the controllers. It should be stressed yet again that the operator stations are important for the efficient functioning of the distributed control system, but are not *necessary* for the system to run.

A simplified schematic of a typical system is shown in Fig. 11.2.

11.4 Process interfaces

The massive automation of the process industries brought about by the application of computerised control systems has led to an increase in the number of process transmitters installed. As a result (thankfully), the large range of transmission signal types has been reduced to a small set of *de facto* industry standards, and most manufacturers now adhere to these. Although it is often commercial practice for large companies to sell non-standard interface components so that customers become committed to the product range, most transmitters are now compatible with a wide variety of equipment from different sources.

There are three principal standards accepted for process signals:

1   Electrical – 4 to 20 mA

2   Pneumatic – 3 to 15 psi

3   Electronic – 0 to 600 pulses per second.

Some manufacturers still use voltage standards despite the clear advantages of the 4 to 20 mA current source. Commonly used voltage signals are:

- 0 to 5 volts
- 1 to 5 volts
- 0 to 10 volts.

Power supply voltages vary and the traditional ones are at odds with supplies for computers. Manufacturers have to provide very comprehensive power supply units for both computers and signal interfaces (and often communications systems). Commonplace interface power supplies are:

- 24 volts DC
- 24 volts AC
- 12 volts DC.

The traditional voltages and signals for analog computers are still to be found in instrumentation, but are rarely used for signal interfaces. The power supplies are + 15 volts and − 15 volts, and the signals are generally in a linear regime between + 10 volts and − 10 volts.

Various voltages have been used for logic signals but three standards are common. These are: TTL (0 to + 5 volts), 0 to 24 volts DC, and the AC mains voltage of the country of installation.

The modern trend with inputs and outputs to process control computers is to use serial communications. Increasing use is being made of *smart* (microprocessor based) *transmitters*, which convert data directly into messages suitable for transmission over one of the networks. Also used are *remote terminal units (RTUs)*, which have been used for many years in *supervisory control and data acquisition (SCADA)* systems. An RTU is a small local processor which carries out a minimal amount of control and which is mainly concerned with sampling data and transmitting them back to a main computer, or receiving serial data and converting them to digital or analog form and storing them for local reference.

The advantage with using serial communications is that the cost of installation and, to a large extent, maintenance is much less than the equivalent cost for individually hard-wired signals. The hardware costs of the smart transmitters and receivers are minor in comparison with other costs.

The penalty with using this more advanced technology is that a high level of software is required in order to handle the protocols and the error checking required for data security. Most manufacturers use their own proprietary protocols, which vary from the very simple RS-232C with no handshaking and no error checking (like the one used in weighing machines) to a full seven-layer International Standards Organisation OSI-Model protocol – refer to Chapter 7. There is no single industry standard, although Ethernet is quite commonplace. The MAP standard has considerable potential but has yet to be widely used.

A further penalty with serial communications is that the level of training of installation and maintenance personnel must be considerably higher, and this involves an ongoing cost burden.

One problem which is common to all computers connected to physical hardware is that of electrical interference and noise. Interference is probably the single largest cause of trouble, embarrassment, litigation and even failure in the entire industry, and its importance cannot be stressed too highly. In common with human medical conditions, each case is different and there is no universal cure. There is nowadays a very wide range of products available to reduce the effect of interference, and there is also a body of specialist literature available. Many universities and other institutions run courses on the treatment of interference problems. The subject is complex and severe cases do require the attention of an expert.

There are four causes of electrical interference problems:

1 electrostatic radiation
2 electromagnetic radiation
3 differential-mode noise on power supplies and other physical connections
4 common-mode noise on power supplies and other connections.

Common-mode noise arises with earthing problems, earth loops, and confusion between *earth, chassis, ground* and *signal-common* line connections.

The treatments for each of the four groups are generally different from one another, and all four types of interference should be counteracted – refer to Chapter 2 in the companion volume, *Basic Control System Technology*.

11.5 Operator interfaces

Operator interfaces underwent a dramatic change between 1975 and 1990. The process has been one of quite rapid evolution, so that equipment has become obsolete during this period.

At the start of the period, control rooms were exceptionally large, often with entire walls consisting of rows of three-term controllers, panel meters, chart recorders and mimic diagrams. It required considerable skill on the part of operators just to read the instruments and to keep track of the information. The introduction of the minicomputer was expected to solve the problem, but this idea generally failed for three reasons. Firstly, minicomputers were very unreliable and, although reliability has improved, they are still not reliable enough for the responsibility for controlling an entire plant. Secondly, the computers were not very suitable for the purpose, since they were difficult to initialise – not having automatic boot-up routines – and did not have the advantage of colour graphics. Thirdly, there were no specially-designed methods for programming them for process control applications – the choice of programming language was between FORTRAN and BASIC.

However, there is now a very wide range of operator interface equipment available, and purchasers can select a medium which is suitable for the type of plant and the level of training of the personnel who run it.

Nevertheless, many of the traditional types of controller are still used. Rotary setpoint and process variable indicators are found on lower-cost temperature controllers. Edge-wise meters and bar-graph meters are also available, but these in general have been replaced with LED-type bar-graph indicators, LED or liquid crystal display digital meters, and thumbwheel switches for setting setpoints. Setpoint raise- and lower-pushbuttons are also widely used. The single-loop PID controller will nowadays contain no analog circuitry except for the 4 to 20 mA input and outputs, and the front panel layout merely mimics its fully analog predecessor.

Some controller manufacturers have now abandoned the analog look altogether, and have introduced digital panel meters and keypads for the user interface. The advantage is a saving of space, and the ability to put more than one controller inside a single case with only one set of indicators and buttons. This approach sacrifices ease-of-use, and therefore introduces a penalty in an industry that makes sales on the issue of *user-friendliness*.

Many systems now rely upon the use of personal computers and engineering workstations as the user interface, and this approach possesses one considerable advantage: value for money. By using a computer produced in very large quantities, a high-quality colour graphics terminal can be bought very economically, and both the IBM-style of computer with a VGA resolution screen or the Apple Macintosh are almost as good as the higher-priced engineering workstations.

The Process Engineering industry is still reluctant to use the personal computer as a general operator interface, and there are very good reasons why this is so. Many process plant operators were never trained in the use of computers, and they are generally apprehensive about using the typewriter-style keyboard. In addition, most do not have the necessary typing skills. The keyboard was never designed for use in process control applications, and most of the operations are exceedingly clumsy using that medium. This fact becomes obvious in a large installation when there is an emergency. Because the trend recordings, bar graphs and indicators have to be dialled-up for display on the screen, it is not possible for operators to view several displays simultaneously. It is therefore necessary to have single-keystroke recall of important data, or single-keystroke alteration of settings. Quite simply, the typewriter keyboard is not fast enough.

All of the major manufacturers of distributed process control equipment offer a special-purpose keyboard and screen combination, usually called an *operator station.*

The operator station is usually a workstation dedicated to the business of operating a large process control installation. The workstation consists of a high-resolution (normally 640 × 480 pixel) colour graphics video monitor and a special-purpose keyboard, part of which is user-configurable. Often, the keys are illuminated pushbuttons and have a removable fascia, which the user can inscribe with a legend. Some systems have an additional panel of illuminated pushbuttons, adjacent to the screen itself. Another recent innovation is the use of touch-screen video displays which can, of course, be software-configured with *soft keys*, as opposed to the *hard keys* of the keyboard.

In a process plant, there are different personnel with a legitimate reason for access to the distributed control system, including plant operators, design engineers, maintenance technical officers and strategic process managers. People in each of these groups require access for different purposes, which may conflict. Many manufacturers now supply a system which can accommodate different types of workstation simultaneously, and it is not uncommon for engineers and technicians to have their own PC-based station in a quiet room, away from the plant and away from the control room. In addition, there are usually available several types of software, some of which can be purchased separately, and which have specific uses.

Operators in the control room can supervise process operations without disruption, whilst engineering and technical staff can design, configure, test and maintain systems from a different location. Production managers can examine operational records and summarise plant performance from their desks.

11.6   Software support and configuration of systems

In the early days of distributed control systems, the programming or configuration of the system was a tedious business similar in many ways to hand-assembly of computer machine instructions. Individual bits in a *configuration word* had to be set according to the type of operation required. For example, a set of six bits would define the control algorithm, another bit indicated whether the output was forward or reverse acting, another bit indicated whether displays were in percent or actual values, and so on.

A modern distributed control system offers a wide range of programming media, each of which permits access to the data in the system.

Two programming languages – interpreter BASIC and C – are generally available, and it would normally be engineering personnel who would make use of them.

The principal technique for programming involves a graphical approach. The engineer uses a graphical design editor to assemble building blocks on the screen and to interconnect them so as to form a *schematic block diagram*. The diagram is then converted to a data format before being loaded into the control system. The method is very similar to the ladder logic editor of programmable logic controllers and, indeed, some manufacturers use a very similar layout.

The Bailey Controls distributed control system also offers a natural language configuration package, which allows a configuration program to be written using a normal text editor or wordprocessor. The words allow the program to call the same functional blocks, which are developed using the graphical design method, or using the BASIC or C languages. Some other manufacturers use a version of FORTH, which is quite suitable for this work.

Other software tools are used for configuring the operator workstations and engineering workstation. The larger manufacturers can supply a Computer-Aided Design drafting package to permit the construction of colour mimic diagrams for the screen. Using this tool engineers can, for example, construct a diagram of the actual plant layout complete with parameter values, alarm levels, and even diagrammatic representations of material in hoppers.

Data from the entire distributed process control system can be integrated into files, which can be used by some of the more popular spreadsheet programs. This feature allows managers to work within the higher levels of the factory control hierarchy, particularly the tactical controls of plant optimisation and plant operability. The specific variables which exist in a process control system are commonly referred to as *tags*, one tag being one variable. Tags can normally be analog variables, Booleans, registers (integers), bytes and sometimes characters and floating-point numbers. Apart from permitting the insertion of variables into spreadsheets, some manufacturers also allow manipulation of tag listings by means of a database program.

Another feature found in the larger distributed control systems is the mathematical modelling and simulation feature. This is generally used to predict product quality from the operational parameters being measured, and helps to provide advance warning of any quality problems. This feature is often used in conjunction with product quality monitoring devices such as on-stream analysers, which can also be used to calibrate the model.

The mathematical modelling feature is also useful for operator training, since it can be used to set up typical process models for off-line training simulators. A peripheral advantage is for off-line dynamic simulation for sales demonstration purposes.

A recent innovation in some larger systems is the addition of an expert system inference engine, which can use logged data from the database to make qualitative decisions about plant operation. It can also be used for intelligent diagnosis of operational problems, thereby providing early warning of a reduction in product quality.

11.7   Alarms, data displays and archival records

11.7.1   Alarms

Alarm settings are one of the most important operational features of any process control system. They are used to attract the attention of the operator to any parameter whenever it departs from the normal range, and thereby requires manual intervention and adjustment. Alarms can also be used to generate Boolean variables from analog data, so as to facilitate on/off control or alteration of algorithms, settings or control functions.

The factors behind any variation in product quality are usually examined at a later date, and prevailing alarm conditions are an important part of the data. Alarms are thus routinely logged, either by printer or by storage on magnetic disc. Alarms are also often used simply to alter displays in order to highlight a problematical variable by, for example, displaying it in red, or causing numbers to blink on the screen.

An alarm is a function which gives a Boolean output according to whether a variable is above or below the setting. Alarms can be *low alarms* – *true* if variable less than setting, or *high alarms* – *true* if variable greater than setting. The better systems permit multiple levels of alarm on any variable, each level being more imperative than the previous one. Alarm levels can also be configured to be alterable, so that changing a setpoint, for example, may cause an alarm level on another variable to alter automatically.

Any alarm can be configured so that it must be acknowledged by an operator, and usually an alarm/acknowledgement constitutes an *event* which is logged as such. The larger process control systems can have so many variables that even monitoring alarms can be beyond the capabilities of an operator. In this case, an *alarm management system* is required, and this permits a hierarchical ordering of alarms, which can then be presented in order of priority. The really critical alarms can be distributed plant-wide to all, or a selected group, of operator stations or engineering workstations.

11.7.2   Trends

Trending is the name given to the facility of recording data in graphical format in the same way as with a strip-chart recorder, X–Y plotter or oscilloscope. The operation works very similarly to that of that of a digital storage oscilloscope.

The data from the variables to be trended are sampled, with fixed sampling frequency, and the samples are stored as a file. The files from one or more variables can be recovered and displayed on the screen in graphical form. The better systems permit trends of several variables to be displayed on the same graph in different colours. Also, a range of time-bases is offered, with time-bases of between 15 seconds per screen-width to 24 hours per screen-width being commonplace. For long time-bases, of course, most of the data samples are lost, and only certain ones are displayed. Old data normally have to be disposed of automatically, since normally there is insufficient disc space available to store all of the samples.

Different manufacturers offer a wide variety of different formats for trends. The commonest format appears to be either the half-screen or quarter-screen graph, with the remaining parts of the screen showing bar-graph displays of set point, process variable and deviation for one or two loops. Underneath are written loop parameters such as set point, gain, integral action time, derivative action time, pure time delay (*Smith Predictor*), compensation delay time, alarm levels and alarm status.

Another commonplace format is the full-screen graph with several variables in different colours and/or symbols.

An option offered by some manufacturers is the X–Y plot of one variable against another. This permits the use of *Lissajous figures* for control system analysis. Often, many of the features of digital storage oscilloscopes are also offered, so that movable cursors on the graph can be used to determine either time displacements or the heights of overshoots and undershoots. This is a particularly useful feature for engineers employing one of the formal tuning methods, such as the Zeigler-Nichols methods. Another feature is the ability to display, in detail, a short section of a trend which occurred some time previously, by panning backward along the trace – a natural feature of hard-copy strip-chart recorders.

11.7.3   Logging and archiving

Most large systems can be configured to record, in digital form, data falling into two categories:

1   selected trends (but not all of them), which can be logged over long periods
2   alarms, events and digital changes of state, which are particularly important and all of which usually can be logged.

Often, data from the two categories can be combined so that, for example, trend data immediately before and after a specific alarm or trip condition can be recorded. Also commonplace is the so-called *snapshot log*, which will record the values of a large number of variables at a specific time instant or immediately after an event or alarm.

Digital storage space becomes full very quickly when logging data, and normally an imminent shortage of storage space is indicated to the plant operators. The operators can then choose whether to scrap data or to archive it. The traditional method for archiving data is to print it out, either continuously or in batches. Archiving can also be achieved using removable magnetic media, such as floppy disc or spooler tape.

12

Control of Robots and CNC Machines

12.1 Introduction

One of the difficulties with the realm of robotics is that it possesses glamour. The field has thus invited considerable attention from writers, broadcasters and film makers, and this has led to a plethora of misconceptions from both lay people and technologists alike. This, in turn, has given rise to false expectations and, in some cases, to failed business ventures.

Because of these misconceptions, it has been necessary to devise a definition of a robot so that policy-makers are better able to evaluate the growth of the industry. The definition of a robot, according to the Robot Institute of America, is that:

> *A robot is a reprogrammable, multifunctional manipulator designed to move material, parts, tools or specialised devices through variable programmed motions for the performance of a variety of tasks.*

This definition excludes many specimens which are often described as robots. Excluded are all machines which are designed for one specific task only, although they may meet all other specifications. A domestic dishwasher, for example, is a reprogrammable manipulator, but it only washes dishes. The definition also excludes all devices which require some form of continuous human input, such as theatrical robots, remote controlled manipulators and remotely controlled vehicles like the well-known military bomb disposal vehicles.

In addition there are many so-called *pick-and-place* robots which are not computer or microprocessor controlled, and it is difficult to say whether these fall into the category of true robots.

Computer-numerically-controlled (CNC) machine tools are very similar in both design and application to robots, but they do not have the necessary functional flexibility. Nevertheless, it is appropriate that the two be considered together.

The plain fact is that robotics is a quite ordinary field of endeavour complete with its successes, failures, devotees and critics. A robot is just another machine, and in comparison with many other machines in common use (aircraft, for example), it is not particularly complex. What makes robots so interesting is that they are anthropomorphic, autonomous and are a superb example of the marriage of computing power to mechanisms.

The underlying technology of robotics is not new. The methods for designing servomechanisms were developed for aircraft at about the time of the Second World War. The technology was adapted to other controlled armaments, and, of course, received a considerable boost with the development of rocketry and missile technology from about 1944 onwards.

Much of the technology was not published or described at the time, and was rediscovered for the development of numerically-controlled machine tools in the 1960s by a younger generation of engineers. Servomechanism technology was again invented for robotics in the 1970s, but there is now a wide range of relevant texts available, and much of the basic theory is taught at undergraduate level.

12.2   Robot configurations

Robots can be divided, first of all, into two main types: robot arms and mobile robots.

Most working industrial robots are stationary robot arms or *manipulators*, whereas most mobile robots are experimental. There are exceptions to this generalisation, since there is much experimental work carried out on stationary arms and there are highly automated factories with a form of mobile robot – the *Autonomous Guided Vehicle (AGV)*. Nevertheless, robot arm technology is more mature than mobile robot technology.

All robots have a number of servomechanisms, and each servomechanism is responsible for one mode of movement. The number of independent servosystems on a robot is referred to as the number of *axes* or the number of *degrees-of-freedom*.

Mechanically, there are two types of servosystem: those with rotary motion and those with rectilinear motion. The selection of the number of degrees of freedom, the choice of rotary or rectilinear motion, and the juxtaposition of the two types, plus the choice of limb size, provide for a wide range of possible designs for a robot arm.

Robot arms usually have a minimum of four, and as many as seven, degrees-of-freedom. A commonplace number is six, because six axes are required to specify, fully, any position plus any tool orientation within the workspace: three axes are required for position and three are required for orientation.

It is usual to consider the first three degrees-of-freedom as defining tool position, at least to a first approximation. Three degrees-of-freedom can encompass four geometries, depending upon how many are rotary or rectilinear. The four types are shown in Fig. 12.1 and are:

Architecture	Degrees of Freedom
Cartesian	three rectilinear
Cylindrical	two rectilinear, one rotary
Spherical	one rectilinear, two rotary
Revolute	three rotary

Even within the four geometries, there are many workable variations. Most robots for general work or welding have quite short limbs, but for spray-painting (Fig 12.2.) the requirement is for limbs which are long and slender. The long reach is needed for spray-painting the insides of containers (such as panel vans, for example), which is a very unpleasant task and is commonly undertaken by robots.

Another popular design is the *SCARA* robot or *Selective Compliance Assembly Robot Arm* (Fig 12.3.). This design is a cylindrical geometry machine, but the two rotary movements are in the same (horizontal) plane. The SCARA design is very rigid, and is normally used for fine assembly work such as the insertion of integrated circuits into printed-circuit boards, since the positional accuracy requirement is very high.

12.3   End-effectors and tooling

The provision of grippers and special tools (usually collectively called *end-effectors*) is normally the responsibility of the purchaser of the robot, and most CNC machines and manipulator arms are supplied without them. Tooling for such machines is often available from local business houses, and many users manufacture their own tools.

End-effectors are usually specially designed and made for a particular application, and, naturally, there is a huge range of possibilities limited only by the imagination of the designer. Accordingly, end-effector design is the province of the specialist mechanical engineer.

Since the subject is so specialised, it is not possible to cover the entire range of possibilities here, so that only some of the more commonly encountered techniques are described.

12.3.1   Grippers

The *gripper* is the most fundamental type of tool, and there is an enormous variety of types. The first decision to be made is whether the gripper has only to handle one type of workpiece, or a mutiplicity. If it is the former case, the gripper can be shaped to suit the workpiece. If the latter, it is more usual to manufacture mating, and often locking, parts for the gripper and the tools or workpieces.

Where tools are to be changed, there will usually be a universal gripper or attachment device on the robot arm, and a range of mating tools in a magazine. The attachment device may be virtually unrecognisable as a gripper.

The simplest gripper (Fig 12.4a.) consists of two hinged jaws and an actuator, with the jaws opening along an arc: this may be quite adequate. There are several ways to make parallel-opening jaws (Fig 12.4b.) so as to keep the contact faces of the jaws parallel to each other, although any point on the jaw will still describe an arc. By mounting the jaws on guide rails, true parallel operation can be achieved, much like the action of a vice.

The jaws can be actuated by a rotary electric motor, electric solenoid, or pneumatic or hydraulic actuator. The pneumatic cylinder is the most commonplace. The problem here is that the power supply (or actuating

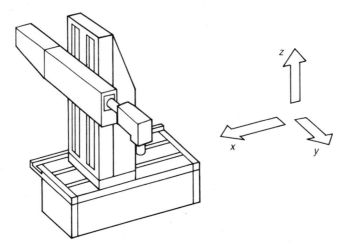

Fig. 12.1a   Cartesian robot arm

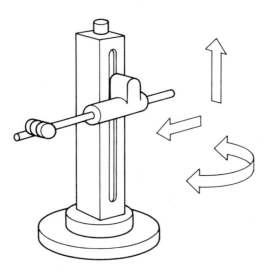

Fig. 12.1b   Cylindrical coordinate robot

signal) must be directly-connected, and this may interfere with the gripper operation, or the connections may chafe and wear. It may also inhibit multiple rotations of the wrist (tool) motion – as is required with a screwdriver, for instance. To solve this problem, a slip-ring assembly must be used for transferring electric power, a rotary union for pneumatic or hydraulic power or a mechanical push-rod may be used. The mechanical push-rod runs through the centre of the wrist-and-gripper attachment, and it has a radial spline so that the tool may rotate, but the forward and backward motion may still be transferred.

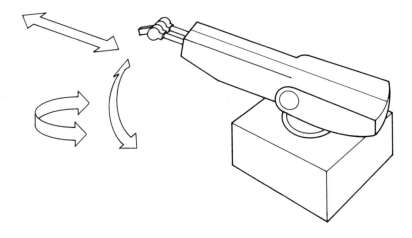

Fig. 12.1c Spherical robot arm

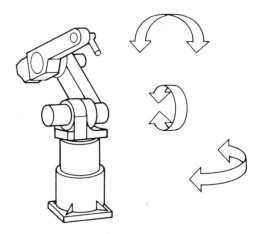

Fig. 12.1d Revolute robot arm

12.3.2 Multiple-jaw grippers

Two-jaw grippers may not be appropriate for awkwardly shaped objects, such as cylinders, spheres and so on, and three jaws or more are often used. There is a variety of designs. Many types are similar to self-centring three- and four-jaw chucks. Other types are similar to a human hand or prosthetic device. Some multiple-jaw grippers have tendons instead of mechanical linkages, and there has been much research carried out in this area.

12.3.3 Special types of gripper

Specially-designed grippers are used for manipulating particular objects.

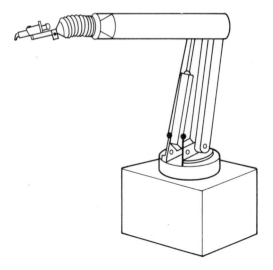

Fig. 12.2   Spray painting robot arm

For flat sheets of glass, board, sheet-metal, plastics and lightweight smooth objects, it is commonplace to use suction pads. The pads are fed by way of electrically-operated valves from a vacuum pump. For flat sheets of steel, electro-magnetic pads can be used instead of suction pads.

Pipes are usually handled by a very wide cradle gripper, with cones which fit into the ends of the pipe. For other hollow cylindrical objects, an

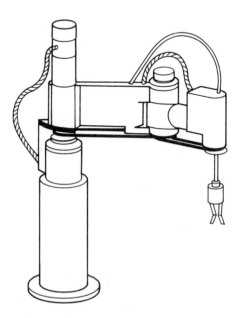

Fig. 12.3   SCARA robot arm

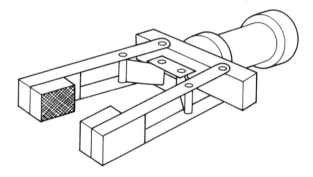

Fig. 12.4a   Simple hinged gripper

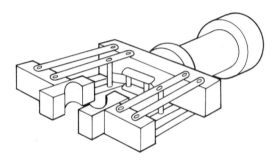

Fig. 12.4b   Parallel opening gripper

internal three-jaw chuck may be used: this expands outwards to grip the inside of the workpiece.

12.3.4   Multiple-operation tools

It is commonplace to maximise the work of robots, so that two or more jobs may be carried out either simultaneously or in sequence. For this, multiple-tool heads can be employed. Provided that jigs are accurately aligned, two entirely different workpieces can be grasped simultaneously and taken to a machining station. At the station, the various manufacturing operations are performed together. It is not uncommon to see an end-effector fitted with three, four, or more different grippers and tools.

Naturally, there are very many different types of tool. These include arc welders and spot welders, spray-painting heads, drills, routers, sanders and polishers, screw-drivers and spanners, compliant assembly grippers, gas torches, sealant applicators and so on. There are some well-known exotic end-effectors. In brick-making, *cartesian gantry robots* use a battery of

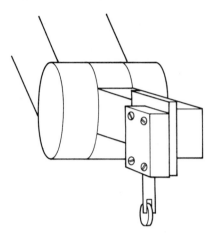

Fig. 12.5   Limit switch fitted to the end of a robot arm

pneumatic bellows on the end-effector to place an entire layer of bricks on a kiln car. The Australian sheep-shearing robot has a complex end-effector with multiple sensors and very sensitive dynamic compliance to follow the contours of the sheep. The principal sensor is an AC voltage source which measures the current, and hence the resistivity between the shears and the skin of the sheep.

12.3.5   Tools with sensors

Sensors, including the problem of compliance, are covered in Section 12.4. Since the end-effector is that part of the robot which comes into contact with the workpiece, many of the extra sensors are fitted there, as with the sheep-shearing robot. Grippers are often fitted with strain gauges (for force feedback), with optical or magnetic scanners, with simple microswitches or elaborate touch sensors. Robots are often used in quality assurance work and all kinds of measuring devices may be fitted, especially electronic dial-gauges for accurate dimension measurement.

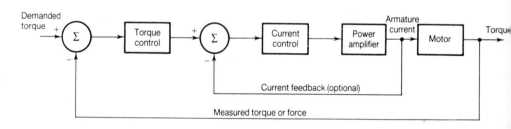

Fig. 12.6   Force feedback control loop

12.4   Robot sensors and measurement strategies

12.4.1   Sensors for servosystems

The most important transducers in robotics are those concerned with the servomechanisms, and are either position transducers or velocity transducers. All types of these transducers are discussed fully in the companion volume to this work. Position (displacement) transducers are obviously used for the measurement of position, but there are techniques for differentiating the measured position data to obtain velocity, and differentiating again to obtain acceleration. Alternatively, velocity can be obtained directly from a velocity transducer. Velocity data may be required in order to control velocity itself, but are more commonly employed for velocity-feedback damping of position control loops.

Some robots do not have any transducers for the axes, but use stepper motors instead in open loop configurations. This method is not recommended, since, if the motor is overloaded and stalls, the count of steps is lost and there is no way to recover the position information.

The most popular position transducers appear to be incremental shaft encoders, because of their linearity and accuracy coupled with fine resolution. The main disadvantage with these is that the measurement is lost if the power to the transducer is interrupted or if the counter information is corrupted by electrical interference. For this reason, many robot designers have returned to synchro-resolver technology. Resolvers possess the ability to generate cartesian information directly from rotary displacement, thus facilitating revolute-cartesian co-ordinate transformations.

Designers should not disdain the use of potentiometers, which are electrically robust, low in cost, and can provide very fine resolution. Some robots on the market use a combination of an incremental encoder and a potentiometer on each axis, in which case the computer control algorithm gives selective weighting to the position information obtained from each, depending upon the circumstance.

12.4.2   Proximity sensors

It is often useful for robots to be able to measure proximity to other objects, be they fixtures, workpieces, tools or even other robots. In most cases, the simplest sensors are the most effective. Limit switches can be implemented in various configurations, either with a simple lever or *cat's whisker* or roller attached. Industrial-quality electromagnetic proximity detectors are very effective and are generally more robust than micro-switches, but are restricted to detecting magnetic materials.

Optical proximity detectors are usually either visible-light or infra-red light types. If the light transmitter and receiver are mounted elsewhere other than on the robot, the sensor detects the presence of the robot when it traverses a particular fixed line (and hence interrupts the beam). If the transmitter and receiver are mounted side-by-side on the robot tool, it is able to detect the presence of a moderately reflective surface at a fixed distance from the sensor. With such sensors, the surface need not be shiny, and often the sensitivity (and thus the distance away) can be adjusted.

Ultrasonic transmitters and receivers are often used, mounted on robots, to sense proximity. They can work extremely well but they are difficult to set up, because of the problem with multiple reflections. Often (particularly with mobile robots), a combination of several types of proximity sensor can give a more comprehensive impression of the surroundings. The difficulty then becomes that of writing software smart enough and robust enough to interpret the information correctly.

12.4.3   Force sensors

Force is one of the more important variables to be measured since it can be used to infer other key variables. Force feedback often is used for grippers in order to close on an object with a determinable amount of pressure. The force used for gripping an egg would be different from that for gripping a golf ball, for example. Force feedback is also used on tools for tightening screws with a fixed amount of torque, or for applying a defined pressure to adhesives, and so on. A force feedback control loop is shown in Fig. 12.6.

Force measurements can also be used to determine slippage of a workpiece within a gripper, or overloading of an arm, or an out-of-balance condition. Very importantly, force can be used to determine moment-of-inertia and gravitational loads. The same techniques can also often be used to measure bending of a robot arm.

The measurement of force is usually made by measuring displacement and inferring force from the mechanical properties of the displaced body. The usual method is to use strain gauges mounted on a metal plate or directly on the robot gripper or limb. The gauges measure bending strain, and normally there is a linear relationship between strain and force.

A second method for measuring force is to mount the core of a linear variable differential transformer (LVDT) on a spring. This method is also used to measure acceleration, since the mass of the LVDT core will be displaced when accelerated.

A novel, and as yet experimental, method is to measure the resistivity of pads of conductive neoprene foam: the electrical resistance across each pad varies as the foam is compressed, the objective being to give the robot a sense of touch.

12.4.4   Measurement of moment-of-inertia

The moment-of-inertia of a robot arm can vary by a factor of up to 4:1. When retracted, an unloaded arm has a much lower moment-of-inertia than when extended and loaded. This property has important consequences for the servosystem dynamics. Rise time, degree of overshoot and natural frequency all depend directly upon the moment-of-inertia, and they will vary if the moment-of-inertia changes. Furthermore, prediction of servosystem behaviour depends upon a linearised approximation of the dynamics, in transfer function form. A system with a variable moment-of-inertia is mathematically nonlinear, which makes the task of the designer exceedingly difficult.

The solution to this problem is to employ an adaptive control algorithm, whereby the gain of the velocity-feedback damping is varied according to the known value of the moment-of-inertia. If the moment-of-inertia is computed and the velocity-feedback gain calculated according to a mathematical rule or a look-up table, the adaptive algorithm is a *feedforward* type, and is *open loop*. If the effect of the moment-of-inertia upon the system behaviour is measured and the gain adjusted accordingly, the adaptive algorithm is a *feedback* type, and is *closed loop*.

The design of such a system is not trivial, since the moment-of-inertia varies with two independent variables: load mass, which usually remains constant for periods of time and then changes suddenly, and arm position, which is always changing. The adaptive algorithm must be very responsive.

There are three principal methods for measuring moment-of-inertia:

Method 1 Since the position of the arm is always known from the position transducer data, it only remains to measure the load mass: this can be effected using strain gauges. The arm can be held in a roughly horizontal position, and strain gauges used to measure the arm deflection due to gravity. Alternatively, the arm can be motored with constant acceleration, and the horizontal deflection of the arm measured. The disadvantage of this method is that it is difficult to measure the moment-of-inertia when the arm is in certain positions.

Method 2 If the arm is fitted with a DC servomotor, it can be motored under a constant-acceleration condition by controlling the second-derivative of position. The moment-of-inertia can then be inferred by measuring the motor armature current.

Method 3 Again using a DC servomotor, the motor can be fed with a constant armature current – thereby generating constant torque – and the acceleration measured, so yielding a knowledge of the moment-of-inertia.

It should be emphasised that *adaptive control of moment-of-inertia* is still very much an experimental technology.

12.4.5   Gravity compensation

There are many causes of inaccuracy in the control of absolute position of robot arms, and a commonly encountered one is bending due to heavy loads. One method for dealing with this is to use an accurate (rigid) geometry such as the SCARA design.

For a robot arm which does bend under load, the error in absolute position can be estimated either by calculation – if the mass of the load is known – or by measuring the deflection of each limb using strain gauges.

The position of the arm can then be adjusted to compensate. This is an *open loop* compensation method, since there is no really effective way to measure absolute tool position, and hence no way to apply absolute position feedback.

Normally, bending due to gravity is not important if the robot trajectory between endpoints is non-critical, in which case the compensation need only be applied as the final position is approached.

Gravity compensation is also an experimental technology at present.

12.4.6   Passive and active compliance

The position error in a robot arm can be troublesome if the robot is being used for accurate work such as assembly of components. There is a variety of causes of position error. The most commonly cited cause is the dynamic position error in servosystems undergoing constant-velocity slewing: all other causes are mechanical in nature. Offsets in the position transducer data can be caused by wear or by misalignment of the shaft. Backlash in gears or in a chain drive can cause a significant increase in position error as the robot becomes worn, and drive chains can stretch quite soon after the delivery of a new robot. Moreover, in many designs, the position transducer is situated between the motor and drive chain or gearbox, in which case the backlash occurs outside the loop and cannot be corrected by feedback. Fluctuations in the output levels of poorly-designed power supplies can cause offsets. Position error resulting from the bending of robot limbs has already been discussed.

Another main cause of position error is misalignment or random fluctuation in the position of the workpiece. This is particularly common-place when jigs are poorly made or not properly maintained, or where the workpiece itself is flexible.

Where there is a difference between the tool position and the position of the work, attempting to carry out parts assembly can cause damage to the workpieces or even to the jig or the robot. Thus in order to minimise damage or wear, the robot is made to be *compliant*. *Compliance* is the ability of the robot to adjust itself in order to compensate for position errors – to comply with variations in position requirement.

There are three approaches to the design of a compliant robot. These are the proper design of the workpiece, the use of passive compliance, and the use of active compliance. Proper workpiece design should be undertaken as a matter of course, but either or both of the other two approaches can be used as well.

Where component parts must be assembled, the parts should be designed so that mating faces automatically guide themselves into position. If necessary, lugs and indents can be moulded into the parts to provide accurate relative positioning. If the robot gripper is to release the workpiece, spring clips can be fitted or moulded so that, after a press fit, the parts remain locked together. For example, where a bearing is to be fitted to a shaft, the end of the shaft and the inside edge of the bearing can be chamfered to assist assembling.

With *passive compliance,* the robot tool or gripper is spring-loaded so that, if the tool is required to bend as component parts are fitted together, it can accommodate this without causing damage. In most cases, a *compliant wrist* is fitted to the robot and some companies that specialise in making compliant wrists have developed a range of excellent designs. The springs can be made in a variety of nonlinear configurations and even have different degrees of stiffness for different directions of movement. Compliant wrists can also be fitted with strain gauges for the application of active compliance.

With *active compliance*, force feedback is used to reposition the robot arm. Force sensors (usually strain gauges mounted on the arm, tool or wrist) are used to determine the direction and degree of misalignment as component parts are brought together. The information generated is used as feedback data and causes the servosystems to move so as to minimise the distortion of the tool position.

12.4.7   Robot vision

In recent years, a considerable amount of effort has been expended on attempting to design an effective, robust vision system for robots. It is not clear quite why this should be so, except that the concepts are attractive and the challenge is academically appealing. It is fair to say that, although there are machine vision systems on the market, none can handle more than a limited range of scenarios: the technology still has a long period of development ahead of it.

Robot vision is seen as the answer to several problems concerning robot intelligence or environmental awareness. When machine vision technology comes of age, there will be a considerable improvement in the capabilities of mobile robots, since the two technologies are closely related.

Machine vision is often cited as the answer to industrial problems such as parts orientation, parts sorting or inspection for quality control. However, engineers should avoid the trap of overlooking the obviously simple answer to such problems.

For parts orientation, there are many successful materials-handling systems which will mechanically orientate component parts rapidly and with complete reliability: it is much more cost-effective to feed pre-orientated parts to a robot than to use a vision system.

Similar solutions apply to parts sorting. One such solution is not to mix dissimilar parts in the first place, but to employ mechanical sorters which select by size or shape, and which are relatively easy to design.

Quality inspection by its very nature implies looking for the unusual or unexpected. Humans are good at this, whereas machines are better at repetitive work, and it is unlikely that machine vision will be as effective as human inspection in the foreseeable future. Robot-like quality inspection systems are already in use, particularly with the accurate measurement of dimensions, but generally these machines do not incorporate vision.

Machine vision systems incorporate three components: lighting, one or more cameras, and a computer for interpretation of the results.

Much can be achieved with clever lighting. A striped light source can be used to define topography, and multiple light sources of different colours can be used to define shadows and surface orientation. Strobed lighting can be used in a variety of ways: to measure velocity in the traditional way, or, when the strobe is synchronised to the interpretation system, to supply identifiable light from different directions.

The camera uses quite straightforward technology. Modern cameras use *charge-coupled-devices (CCDs)*, which are solid-state. A CCD camera can be coupled to an image-capture board in a computer to obtain still pictures, which may be stored in memory or on disc. The pictures are square and resolutions of 128×128 pixels (16 384 picture elements) and 256×256 pixels (65 536 elements) are commonplace. Picture resolutions of 512×512 (262 144 elements) and 1024×1024 pixels (1 048 576 elements) are more costly. The latter gives a very high-quality picture, and is the type of resolution used in aerial photography and photogrammetry.

By far the most difficult aspect of machine vision is the interpretation of the picture. There are two reasons why this is so. Firstly, the task itself is so difficult that sufficiently powerful strategies and algorithms have not yet been developed. Secondly, even for the algorithms which exist, the amount of computation required is so vast that present computers can only interpret the results at a snail's pace. The technology requires better algorithms, faster algorithms and much faster computers. There is a considerable body of literature which covers interpretation methods for machine vision, but it is not appropriate to review it here.

Much research has concentrated on object recognition, which is particularly difficult if the object is randomly orientated in three-dimensional space, or is partially obscured. One technique uses image processing to identifiy the edges of the object, and then attempts to match the shape to templates contained in a library.

Another area of research measures distance from the camera to the surface, to facilitate turning a two-dimensional picture into a three-dimensional surface. This technique was originally developed for photogrammetry, whereby a series of photographs taken by aircraft can be used to generate a map – essentially a drawing of an undulating surface. The photographs are used in pairs, termed *stereopairs*, and the vertical (Z-direction) displacement of a point on the surface is inferred from the X-direction shift of the point, relative to some datum.

The photogrammetry technique involves some pixel matching, which is a rudimentary form of object recognition. For autonomous mobile robot navigation, both object recognition and stereo distance measurement are necessary: the object recognition, however, must be far from rudimentary.

For instance, a robot vehicle must be able to recognise the edge of a road, the shape of which it may never have seen before, and the texture of which changes constantly.

12.5 Servosystems for CNC machines and robots

The hardware associated with servosystems is covered in the companion volume to this book. Position sensors for servosystems are also discussed elsewhere in this chapter. Robot drives can use hydraulic, pneumatic or electric power. In view of the excellent power-to-weight ratio and stall-torque characteristics of hydraulic drives, it is surprising that they are not more often used, or that there is not a wide range of miniature precision hydraulic components on offer. But such has been the dedication of roboticists to electric power that considerable research has been directed towards developing direct drive (gearboxless) electric motors for robots. However, hydraulic power is the most commonplace type for CNC machine tools.

Servosystems differ from regulators in that the set point (desired position) changes frequently and rapidly. There are three variables of interest: position, velocity and acceleration. In addition, it is often necessary to control torque. With regulators, the setpoint is usually kept constant and the control system is only required to compensate for fluctuations in load and parasitic disturbance variables.

Servomotors usually run at constant velocity in response to a steady drive signal, but the measured variable in a servosystem is position, which is the integral of velocity. Since the forward-path transfer function contains an integration, servosystems are always of Type 1, and hence there should be zero position error once the system is at rest. The topics of control system *Type Number* and *steady-state error* are covered fully in Section 14.2.

For both CNC machines and robots, there are two types of machine: point-to-point and continuous-path. With the former, the user is only interested in the final position of the tool axis when it is at rest; there is no interest in the trajectory followed to arrive at this position. There is, accordingly, no interest in the positional accuracy of the servosystem under constant-velocity or constant-acceleration conditions. With continuous-path operation, however, the trajectory accuracy is important at all times, so that any position error occurring under constant-velocity conditions is significant, and special techniques are required to reduce it.

A servosystem would normally be third-order or higher, the three orders typically deriving from the integration, the inertia of the load and the inductance of the motor windings. Similar sources of transfer function poles can be found with hydraulic systems. It is commonplace for servosystems to be quite oscillatory, with considerable overshoot and undershoot, so that some dynamic compensation is necessary in order to minimise the overshoot and yet obtain the fastest rise time (or greatest slew rate) possible.

There are several ways to effect such compensation. It would be possible to increase the friction: this would certainly reduce the overshoot and, to an extent, improve the rise time, but it is an inelegant solution. Alternatively, it would be possible to introduce derivative action into the controller, or a lead compensator into the forward path. The most usual method is to use velocity feedback damping, for which a measure of velocity is obtained either by differentiating the measured position or by means of a tachogenerator either geared to, or mounted on, the motor drive shaft. The velocity signal is fed back to the controller by way of an adjustable gain. By adjusting both controller gain and velocity-feedback gain, the servosystem can be tuned for optimum rise time and minimum overshoot.

Tuning the system requires considerable expertise. It is possible to tune by trial-and-error, but the most professional method is to use the root-locus method. The mathematical background to the method is discussed in most texts on classical control theory, and there are available several computer programs which will plot root-loci. It is necessary to obtain the transfer functions for the various components, which requires a mathematical characterisation of the system. A family of root-loci must be plotted for a range of values of velocity-feedback gain (or of position of the introduced *zero*), and from these the optimum gain values can be found.

A difficulty with servosystems for continuous-path machines is the presence of position error under constant-velocity conditions. If a robot arm or CNC machine tool is moving with constant velocity, a non-zero position error exists because of the need to supply power to the motor, which in turn requires a difference between position setpoint and measured position, in order to drive the power amplifier. Integral action can be used to cure this error, but the addition of another integration in the forward path changes the system to Type 2, which is inherently unstable without compensation. To stabilise a Type 2 system, it is necessary to design a special compensating controller and for that a rigorous application of the Nyquist Stability Criterion is required. The simpler, and thus more widely used, approach is to employ velocity feedforward compensation.

The basic design strategy for a velocity feedforward system is to differentiate the position set point signal, to yield *desired velocity*. This velocity signal is multiplied by an adjustable gain and injected into the forward-path, after the controller gain block. The velocity feedforward gain should be tuned so that, under constant-velocity conditions, the position error (*position setpoint* minus *measured position*) is zero, and the motor drive signal is totally provided by the velocity feedforward.

Such a scheme is ideal for continuous-path robot control, since the computer can calculate velocity and position on a continuous basis, and supply both signals to the controller; just supplying final position is inadequate. Provided that this constraint is respected, the robot can control both its position and velocity simultaneously. The block diagram for the complete servo system is shown in Fig. 12.7.

Other variables which need to be controlled can be handled by feedforward compensation, and acceleration and load torque are the principal

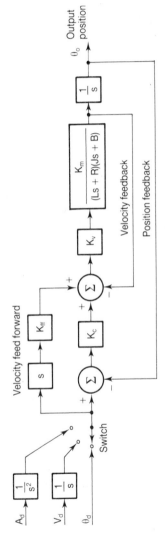

Fig. 12.7 Block diagram for a complete servo system

A_d demanded acceleration; V_d demanded velocity; θ_d demanded position; θ_o output position; K_{ff} feedforward gain; K_c position controller gain; K_v velocity controller gain; K_m motor system gain; L inductance; R resistance; J moment of inertia; B friction

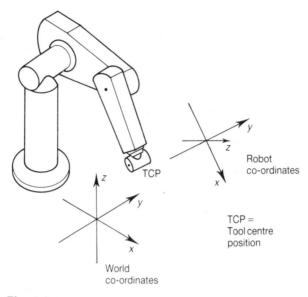

Fig. 12.8   Co-ordinate systems for programming robots

candidates. By coupling these strategies with strain measurements the need for active compliance and gravity compensation outlined in Section 12.4 can be satisfied.

12.6   Programming for CNC machines and robots

It would be fair to say that the lack of popularity of robot arms in the workplace, and the mediocre success (in comparison to that of computers) in the marketplace, are largely due to the extreme difficulty of programming: this is the result of the high number of degrees-of-freedom. A computer program, for example, is *linear* – a set of instructions executed in a time sequence – and has only one degree-of-freedom. An operator of an earth-moving machine can operate two degrees-of-freedom with ease, and a highly-skilled operator can, if only at times, master four degrees-of-freedom. How, then, can one program six degrees-of-freedom to move simultaneously?

There are four principal ways in which robots are programmed:

- discrete logic programming methods
- *step through* methods
- *lead through* methods
- using a programming language.

12.6.1   Discrete logic programming methods

These methods only apply to simple *pick-and-place* robots, which are driven by on/off control servomechanisms. Some pick-and-place robots do not even have intelligent controllers, but are driven from limit switches placed on the arms themselves. Reprogramming the robot in such cases consists of simply repositioning the limit switches. With other pick-and-place machines, the controller is just a sequencer or programmable logic controller, which can be re-programmed by use of ladder logic or of mnemonics as described in Chapter 9.

12.6.2   Step-through programming

This is the principal method used for programming robots. The robot is driven directly by the operator from a control panel, which in most cases is hand-held on the end of an umbilical cable, and is referred to as a *teach pendant*. Teach pendants vary, and can have buttons or a joy-stick for moving the servomotors. Even on the joystick, it would be unusual to be able to move more than three degrees-of-freedom simultaneously. With a six-axis robot, there would normally be a simple choice of having the three movements on the waist, shoulder and elbow, or on the three wrist axes. A switch would be used to change from one mode to the other.

The method consists of moving the robot to a desired position, and then storing the position in memory, as a set of co-ordinates. For a complete program, the sequence must be linked by having a *header* (usually with a sequence name) and a *terminator*.

Programming in step-through mode is exceedingly time-consuming, but, for most robots, it is the only method available. With many robots, it is possible to store short sequences as library routines, which can then be called again later in other, longer, sequences.

One disadvantage with this programming method is that the operator or programmer has very little control over the robot trajectory between two consecutive points in the sequence. This is because each servomotor runs at its allotted speed, irrespective of the speed of the others. If the trajectory is important, intermediate points on the trajectory must be specified, in order to define the path more explicitly.

The more modern robots on the market are available with three alternative co-ordinate systems (Fig 12.8.): namely, joint co-ordinates, cartesian co-ordinates, and robot or tool-orientated co-ordinates. This advance somewhat simplifies the business of developing a routine by the step-through method.

With joint co-ordinates, each movement of the joystick corresponds to movement of one servomotor and hence of one joint only. For instance, rotating the joystick would rotate the waist, left-right movement would move the shoulder, and up-down movement would move the elbow.

With cartesian co-ordinates, one of each of the joystick movements would move X, Y, and Z directions with respect to some datum position on the floor, so that rotating the end of the joystick might cause the robot to move vertically, perpendicularly to the floor.

With robot co-ordinates, the three movements would be X, Y and Z, but with respect to the position of the centre of the robot tool (usually referred to as *Tool Centre Position*, or *TCP*). The Y direction would be along a line parallel to the gripper, towards, and away from, the robot. The X direction would be normal to the centre-line of the wrist and normal to the plane of the gripper. The Z direction would be normal to the centre-line of the wrist and in the plane of the gripper.

12.6.3 Lead-through programming

Step-through programming is clearly unsatisfactory for situations where a smooth path with well-defined accelerations and velocities is required, such as with spray-painting. The answer in this case is to use a *lead-through* method. Only certain robots can be programmed in this way, since the method requires that the robot be totally *relaxed* and be light enough for an operator to be able to move the tool freely. Sometimes, a lightweight mock-up is used: this is just a frame equipped with position transducers, devoid of any other equipment.

For programming, the robot arm or mock-up is held by the operator and led through the entire sequence of moves. For skilled work, such as spray-painting or welding, the best tradesman available would move the robot, so that the robot would be able to repeat the high-quality work consistently.

While the programming takes place, the robot controller runs an internal clock, which determines the sampling frequency: this frequency can be set by the operator. The robot position is sampled (say) once every 300 milliseconds, and the sample values are stored as a sequence of co-ordinates in the same way as with step-through programs.

12.6.4 Robot programming languages

There have been many attempts to use high-level languages for programming robots, but they have been largely unsuccessful. The difficulty arises because there is no natural way (either by language or mathematically) to describe movements in three dimensions; the only practical way is to execute them physically.

Anyone who uses a *Computer-Aided Design and Drafting (CADD)* system will be familiar with the problem. Producing original sketches is difficult enough in two dimensions, using only straight lines, arcs and circles, but irregular shapes are even more difficult. The problem is compounded when a third dimension is added. The great power of CADD comes into play when a library of sketches has been built up and the editing features are used. Multiple copies of a sketch can be added, and they can be stepped and repeated, rotated, mirror-imaged, enlarged, reduced and even distorted.

The power of the better robot languages is similar to that of a CADD system. Routines from a library can be concatenated, and each routine can be similarly modified by stepping and repeating, rotating, mirror-imaging and so on. In this sense, the language is really just an editor; it does not help the user to create the basic library routines. In most cases, the basic routines must still be developed using the tedious but reliable step-through method with the teach pendant.

There are many robots available with standard computer programming languages. The popular languages are BASIC, FORTH, PASCAL and C.

In most cases, the languages are enhanced with special functions for driving the robot, and in some cases come with a library of routines already programmed. These standard languages can be very useful for fluent programmers, experimental technologists and for students, but for serious roboticists in a competitive manufacturing environment they can be quite troublesome, because the factory employees do not always have the appropriate skills.

Some industrial robots on the market also have a programming facility for programming in a high-level language (such as C) from a personal computer. Potential users should be wary of such offerings, because the robot computer may not be of the same type as the personal computer, in which case it will be possible to write and compile on the personal computer and to download the compiled code for execution on the robot only if a cross-compiler is supplied. If a cross-compiler is not available, the language becomes little more than a way to stop and start the robot and to move files around; programming is still undertaken by way of the teach pendant.

Although many industrial robot arms are available only with a teach pendant or a conventional programming language, or both, there are some which are supplied with a special-purpose language.

Most special languages are similar to BASIC and (unfortunately) many are called by the same name. However, such languages must include instructions for driving the robot, and a sample of typical instructions is as follows:

10	T6 = 0. 1.0 9.5	(defines a position T6)
20	SPEED 100	(defines speed of movement)
30	MOVE T3	(move arm to position T3)
40	WAIT + IE2	(arm wait for conveyor to move)
50	OUTPUT+OGI 200	(close gripper)

Other programming languages for robots are similar to PASCAL, again with special functions for driving the robot, and often with libraries of procedures already written. Languages are discussed in more detail in Chapter 9.

12.6.5   Programming CNC machine tools

In many ways, the programming of CNC machine tools is similar to that of robots. The problem is easier though, because there are generally (but not always) fewer degrees-of-freedom and the machine has a more specific purpose with a better-defined workspace.

CNC machines do not normally have a teach pendant but they do have comprehensive control panels which enable the operator to set all the speeds (spindle and feed), drive to any position for any degree-of-freedom, engage clutches, change tools, open and close jaws and so on. Most panels also have a rugged monitor and keyboard for programming, and some form of file input – often a punched paper-tape or mylar-tape reader. Modern machines are often connected to a communications network.

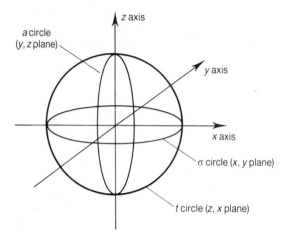

Fig. 12.9   The three cartesian coordinates and the three Euler angles

The usual method of programming is by means of a language which is generally supplied with the machine. Most languages are interpreters like BASIC and many of them are very similar to BASIC. A commonplace language is the so-called G-CODE, which has been in use since before 1970. There are now some translators available which will convert CADD files into G-CODE, to provide the link between CADD and *Computer-Aided Manufacturing (CAM)*: this is still an area of research.

12.7   Co-ordinate transformations

Part of the difficulty with the programming of robots is that the program-mer cannot easily picture the movements of the arm if the robot geometry is spherical or revolute. The mind most easily conceives movements related to cartesian co-ordinates and the conceptual or programming workspace is usually in the cartesian co-ordinates X, Y and Z, as shown in Fig 12.9. But these three variables can only describe a position in three-dimensional space, and to describe tool orientation fully, a further three dimensions are required. The ones most commonly used are the three Euler angles, the a circle, the o circle and the t circle.

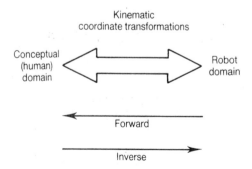

Fig. 12.10   Coordinate transformation

For a six degree-of-freedom robot, the position variables for each of the six servosystems are not the same as the six conceptual variables, so that tool position and orientation information must be converted from one domain (or set of variables) to the other. The process is called *co-ordinate transformation*, or *kinematics.*

Transformations can be carried out in either direction. The transformation from the robot workspace to the conceptual workspace is the *forward transformation* or *forward kinematic problem*, whilst the transformation from the conceptual space to the robot space is the *inverse transformation* or *inverse kinematic problem*, as shown in Fig 12.10.

The forward transformation is arithmetically intensive, but is quite a straightforward calculation. There is one unique solution, since the robot arm occupies an identifiable position in either domain.

The inverse transformation is an order of magnitude more difficult. Some positions in the cartesian domain are unreachable by the robot, some positions have only one solution, but one or more regions have no unique solution: that is, the robot arm is able to employ any one of several possible sets of joint settings to reach that position. For the last situation, the transformation algorithm must employ some kind of qualitative decision as to which is the preferred route to the final position. One simple approach is to move each joint in turn, each time minimising the error between actual position and desired position. The waist is moved first, then the shoulder, elbow, wrist pitch, wrist yaw and wrist roll; this process is repeated until the final position is achieved. Other optimisation methods work equally well.

There has been much research on the topic of the inverse kinematic problem, and there is a considerable body of detailed literature available.

An interesting development has been the use of synchro-resolvers for sensing joint position, for this has enabled much of the tedious transformation calculation work to be carried out in analog form. Because the resolver is used to return the sine and cosine of the measured angle, calculation of a cartesian joint vector is easier, and the calculation of the forward transformation becomes a vector addition. The optimisation method for obtaining the inverse transformation uses an iteration by way of the forward transformation, so that it too is facilitated by the use of resolvers.

12.8   Workcells and robot interfacing

The introduction of robots into the manufacturing environment is usually accompanied by the implementation of *Just-In-Time (JIT)* methods. A full discussion of JIT is not appropriate here, but some aspects are important enough to be covered.

One objective of JIT is to reduce the inventory, by minimising the amount of stock and work-in-progress sitting idle on the factory floor. Ideally, material should always be in the process of having value-added work being applied, and the time between entry of raw material into the factory and shipment of finished goods should be as short as possible.

Another objective is to reduce batch sizes to as low a figure as is theoretically possible – very often one item of finished goods. By reducing batch sizes goods can be made to order, instead of customers having to wait for orders to build up before a batch is made, or stocks of finished but unsold products occupying expensive warehouse space. Reduction of batch sizes implies a very streamlined and flexible operation. Tool changes must be carried out virtually without stopping the machine, and machinery must be capable of handling different products on the same assembly line without interruption.

Robots and CNC machine tools are intrinsically flexible, because of their ability to store several programs and switch freely between them. Rapid tool changes and adaptability to the manufacture of different products simultaneously are simply a matter of programming.

The usual way to attain flexibility is to use a *workcell*, whereby several machine tools performing related operations are grouped together and are serviced by a robot. All the machines are interconnected electronically, so that the operations are synchronised. The robot can service the machines as required or in the correct sequence, so that operations proceed smoothly even though some machines may be faster than others. Changes in the pattern can be accommodated, as can tool changes, product changes, breakdowns and human intervention.

The use of a robot in co-operation with other machines, including CNC machine tools, is only possible if the machines are interconnected electronically. In most cases, the interfacing is by way of logic inputs and outputs, and machines are equipped with buffered inputs and outputs which can be accessed by the program. Outputs can be used to actuate grippers and special tools, or to convey machine status to another machine. Inputs can be fed from on/off sensors or limit switches, or used to sense the status of other machines.

Most machines are equipped with a serial input, usually in the form of RS-232C. Normally, the purpose of this is for down-loading programs and for keyboard control, but sometimes it is possible to use it for parameter-passing between programs running on different machines. Some machines are equipped with networking capability, but different manufacturers have either adopted different standards or have created their own.

To solve the problem of different programmable robots and CNC machines having different networking protocols, the *Manufacturing Automation Protocol (MAP)* and the *Total Office Protocol (TOP)* standards have been developed. The MAP system is designed to permit full functional interconnection of any two intelligent machines in the manufacturing environment. Because of its complexity and expense, however, very few machines are fitted with it.

12.9   Social implications of robotics

There are two principal aspects to the social implications of robotics: the effect of automation on employment prospects generally, and the more specific effects on the relationship between workers and supervisors in the factory.

12.9.1 General effects on employment

There have been many authoritative treatments of this topic, and a very wide range of non-authoritative comments. Since the subject is highly emotive, reference material should be chosen with care, and only used if it is well researched. One early authoritative commentator was Joseph Engelberger, who was the founder of Unimation Inc., the pioneer of robot arm technology.

In general, the press has adopted an attitude that robotics pose a threat to the worker, because anthropomorphic robots will directly replace people, and the catchcry is often taken up by others. Unions are sometimes accused of being implacably opposed to the introduction of robotics.

The truth, as usual, is more complex. More often than not, workers and unions support the introduction of robotics for two main reasons. Firstly, the robots are introduced to carry out work which is either unpleasant, dangerous or both. Employers consistently experience difficulty persuading workers to undertake certain tasks, and workers, generally, whole-heartedly endorse the use of robots to perform them. Typical examples are: spray-painting inside enclosed areas, de-flashing hot castings in a foundry, working in areas of extreme heat, lifting heavy or awkward items, and boring, repetitive, work requiring high concentration, such as the welding of frames for furniture. The second reason is that the introduction of robotics will either help to make a manufacturer more competitive, or at least provide a tangible indication of a commitment by the company management to competitiveness, growth and stability. This, in turn, is reflected in an increased feeling of security and a positive attitude on the part of the workers.

As Engelberger and others have pointed out, the people most likely to be opposed to the introduction of robotics are the factory managers and the members of the boardroom. It is not entirely clear why this should be so, but there are many possible reasons. It is conceivable that the managers have researched the topic very carefully, and found the introduction of robotics not to be cost-effective. They may be afraid of making a firm commitment to the development and competitiveness of the company, being happy to take a complacent and conservative approach. They may be afraid to use a technology which they do not understand and are therefore less able to control. The company, for reasons of taxation structures or cash flow, may be unable to find the necessary capital. The management may genuinely believe that the unions would not accept the proposal.

A spokesman for the Ford Motor Company in Australia has said that the introduction of robots into the Broadmeadows Plant has not displaced a single worker, but workers *are* displaced by the redesign of motor-cars, such as the replacement of an assembled component in steel by a single plastic moulding.

There is, in principle, no reason why robots should replace workers on the factory floor – a robot is, after all, only another machine. Robots and CNC machines tools do, however, change the nature of work. Intelligent machines require well-educated personnel, and a continuous program of retraining is a necessary part of their introduction. There will always be some disruption and workers will be required to change the type of work. Traditional demarcation lines need to change and, unfortunately, there may be some workers who are unable to cope.

12.9.2   Employer–worker relationships

An authoritative researcher in this field is Professor Howard Rosenbrock, who was commissioned by the British Government to prepare policy material on the social aspects of automation. Rosenbrock argues that the use of more efficient manufacturing methods cannot in itself be detrimental to humans. The difficulty is with the way in which the benefits are shared out.

It is a commonplace mistake for managers to use a robot to take over the more interesting parts of a worker's job, leaving the worker to continue the unpleasant or repetitive aspects. In the case of a welder, a welding robot may be used to perform the difficult welds around tricky contours and in tight corners, leaving the person – who previously may have taken pride in that aspect of the work – to lift and carry, remove scale and attend to breakdowns. Such a scenario generally leads to a breakdown of morale and probably an actual reduction in productivity.

It is much better to have a robot perform the fetching and carrying, and have the worker carry on proudly with the work for which he or she was originally trained. It is even better to have a program of training so that the worker may assume responsibility for more challenging work such as robot programming – and with a corresponding increase in remuneration!

Robots are often depicted in textbooks as working alongside people, performing roughly the same sorts of task. Although these illustrations are used to show the versatility of robots, the effect is to reduce the worker to the level of a robot. This is a dangerous misconception, because automation must be introduced with sensitivity and mutual trust between labour and management. Feelings of pride, self-esteem and job security are very important attributes in a workforce, and their loss would outweigh any advantages of automation.

12.10   Mobile robots

One of the first mobile robots to be built was *ELSIE*, a small self-guided tricycle vehicle designed in 1948. The designer, W. Grey Walter, was attempting to simulate animal behaviour, particularly the tendency to be attracted towards light. ELSIE moved pseudo-randomly in arcs until strong light struck a photo-detector on the front of the vehicle, when it adopted a straight-line motion; this motion reverted to arcs again if the light source became hidden or the vehicle struck an obstacle. Six such vehicles were displayed at the 1951 Festival of Britain and at least three survive.

ELSIE was a true autonomous vehicle, but many of its successors are not. Many so-called mobile robots are just remote-controlled vehicles without any on-board intelligence, however rudimentary, and an excellent example is the military bomb disposal 'robot'.

There are several types of robotic vehicle available for factory use, and these are refered to as *Autonomous Guided Vehicles (AGVs)*. AGVs are usually programmed to follow some kind of track, which is usually a wire or steel strip buried in the floor, or a reflective stripe. The vehicles are smart enough to find a destination, and to stop and start and avoid collisions. These vehicles are under the control of a centralised computer, which directs vehicles to certain destinations for various tasks.

Of course, there are very many intelligent autonomous vehicles in the military and aerospace spheres, from self-guided weapons to the experimental autonomous truck. Possibly the most successful of these is Voyager 2, which fulfils all parts of the Robot Institute of America's criterion except one: it is not really a manipulator.

Truly autonomous, unguided, robots are still a long way from reality, although there are several working examples. Both the French and the Japanese have developed cleaning robots, one for sweeping railway station platforms and one for polishing corridor floors in office blocks. But even with such relatively simple tasks, there are many problems to be overcome. Avoiding obstacles, either fixed, temporary or human, is a problem, and the robot must not cause damage or injury. Touch sensors must be sensitive and universal. A wall is easy to detect, but a table is more difficult, since the vicinity of the obstacle consists of 95 percent vacant space.

More complex sensors for detection of obstacles all have their problems. Industrial infra-red detectors are very reliable but are constrained by beam width, reflectivity of the surface, and either fixed or unpredictable distance measurement. Ultrasonic detectors suffer badly from echoes, especially in a changing environment and the results are difficult to interpret, though both infra-red and ultrasonic detectors are constantly being improved in reliability and in value-for-money. Passive infra-red sensors, which will detect human and animal movement, are now produced cheaply for the security market and can be adapted for use on mobile robots. However, robot vision still presents enormous difficulty with interpretation.

Navigation is also a problem, but there are some answers. By measuring distance at the wheels, it is possible to obtain a quite accurate estimate of overall position, and the absolute position can be checked periodically in several ways. Magnetic or optical markers on floor or walls can indicate a datum point; point sources of laser beams, radio, light or sound can provide reference information at a direction-finding detector on the robot. Satellite navigation is also a possibility in some applications, and a digital compass can be fitted to the vehicle.

12.10.1   Walking machines

One aspect of mobile robotics that has attracted considerable research interest is the development of walking machines. Much has been discovered about the mechanics of legged locomotion, and some designs have been quite successful.

Walking machines have quite a long history. Steam-driven walking ploughs were in widespread use in the Fen Country of Britain's East Anglia in the early part of the twentieth century. The machines were larger than a tractor, and were equipped with several rows of spades which dug the earth rather than ploughed it. The driver was seated on the top, and the machine 'walked' across the fields. Walking drag-line cranes are widely used around the world.

Both examples are inherently stable mechanisms, and the most successful design is also inherently stable – it is the six-legged 'insect' crawler. The design represents a stable system because there are at least three legs in contact with the ground at any one time, and the three can be adjusted for levelling. The remaining three legs can be moved through the air simultaneously. To obtain adequate mobility, each leg must have a minimum of three axes, so that the insect crawler has eighteen degrees-of-freedom for the legs alone.

A five-legged crawler can be made stable, but only having four legs will cause instability for some of the time. The walking drag-line is stable with three legs, but the 'feet' are so large that the machine can perch on either the middle one or the outside two alone.

There has also been considerable work undertaken on unstable and dynamically stable walking machines, all of which have one thing in common. While in motion, the unit is in constant danger of falling over, and only the action of placing the next leg within a certain time-constraint can prevent that from happening. The control system must be either constantly balancing it or constantly transferring it from one potentially unstable state to the next.

Other examples of existing mobile robots are a four-legged trotting machine which mimics the horse, two-legged walkers which mimic people, and a one-legged hopping machine. The hopping machine can be made so robust that even a hefty kick will not cause it to lose its balance. All of these machines are experimental, and none is capable of carrying its own power source.

13

Experimental Testing of Plant, System Elements, and Systems

13.1 The need for characterisation

The design of a closed loop system cannot be completed rigorously until the properties of every element in the loop can be specified mathematically. The steady state sensitivity and time- and frequency-dependent characteristics are normally specified in terms of a linear transfer function, and the departure of the steady state sensitivity from a straight line output/input relationship is normally specified in terms of a static characteristic.

Characterisation (or *Identification*) is the name given to the set of alternative experimental procedures available for determining the unknown dynamic and steady-state properties of a system element. In particular, the techniques would be applied to the testing of plant, final control elements and feedback transducers, although this does not preclude their application from the testing of controllers and of completed closed loop systems, in order to verify the design procedures which have subsequently been adopted. Sections 13.2 to 13.8 provide descriptions which assume that DC electrical systems are being tested, whilst Section 13.9 to 13.11 indicate how the techniques could be modified for AC-carrier, digital, and pneumatic systems.

13.2 Experimental procedures for obtaining static characteristics

Figure 13.1 shows a typical arrangement for establishing the static characteristic of a final control element–plant process–transducer combination. If the signal generator is set up to inject a steady signal into the control element, and the value of this signal is measured by meter A, then the steady-state value of the transducer output signal will be measured by meter B, after sufficient time has been allowed to elapse for transient components of response behaviour to decay to zero. A graphical plot of the reading of meter B to a base of the reading of meter A will represent the static characteristic, if the procedure is repeated for a sufficient number of different levels of steady input signal: the steady-state sensitivity will be the gradient of this characteristic.

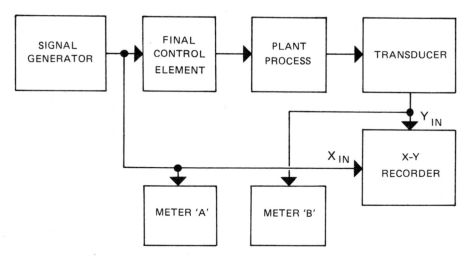

Fig. 13.1 Arrangement of test equipment to measure the static characteristic of a final control element-plant process-transducer combination

An alternative approach is to sweep the input signal, through an appropriate range of values, by causing the signal generator to inject (say) a triangular wave into the control element. The output/input relationship can now best be recorded on either an X–Y plotter or a storage oscilloscope operated in the X–Y mode.

With either technique, the signal should be stepped or swept in each direction, alternately, if there is any possibility of hysteresis being present in the characteristic being plotted. With the latter technique, the presence of dynamic lags can artificially create a resemblance to hysteresis in the experimental plot if the sweep rate is too high; this can be verified by repeating the experimental procedure at a lower rate of sweep: if there is no difference between the two plots, then there is no significant contamination arising from dynamic lags. In any case, it will always be necessary to make sweep rates extremely small: for example, several minutes might be required for sweeping through the full excursion.

Once the form of the static characteristic is known, a suitable procedure for undertaking dynamic testing can be determined. If the static characteristic is a simple straight line relationship, then the plant being tested can normally be regarded as linear, which will mean that dynamic testing will yield the same form of response irrespective of the size of the applied disturbance: that is, the magnitude of the test signal applied during the dynamic test is not critical. (An exception to this occurs in plant possessing the phenomenon of rate limiting: with such a system, the static characteristic may well be linear but the dynamic response for a large disturbance would differ from that for a small disturbance, due to the finite limit on output slew rate occurring in the former case.)

On the other hand, if the static characteristic is nonlinear, then the input signal conditions for the dynamic testing can be critical: an approximation to a linear transfer function may be obtained by applying a small-signal disturbance superimposed on a known quiescent signal level, and the relevant sensitivity will be the gradient of the static characteristic, measured at that quiescent level; this amounts to 'small-signal testing', and the situation is demonstrated by Fig. 13.2. Repeating the dynamic test at a different quiescent level will normally yield different experimental data.

Large-signal disturbances can also be applied when characterising non-linear plant; in these cases, the disturbances would normally be sinusoidal, so that operation is in the frequency domain. With test equipment capable of rejecting harmonic signal components, it is possible to measure experimentally the 'Describing Function', which is used to relate the fundamental component of the periodic output waveform (after sufficient time has been allowed to elapse for transient components to decay to zero) to the applied sinusoidal input waveform.

There is one type of process for which the static characteristic cannot be measured by the alternative techniques so far described, and this occurs when the plant contains an integration process. (An example is when the transducer is measuring liquid level and this is being controlled by the manipulation of liquid inflow, or outflow, rate by the final control element: flow rate is proportional to the time rate of change of level, so that level is proportional to the time integral of flow rate.) In this case, the output will not attain a steady state (in fact, it eventually ramps) when a steady input is applied, so that this will not readily yield data for the static characteristic.

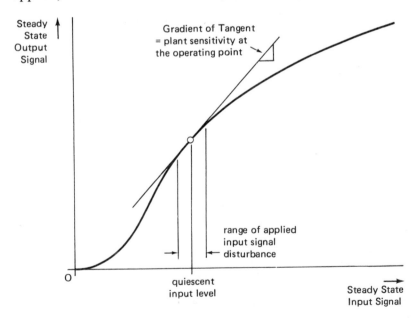

Fig. 13.2 A representative nonlinear characteristic showing how small-signal sensitivity is defined and measured

One solution is either to record the output ramp using a Y–T plotter and to measure the gradient from this, or to record the output rate of change experimentally by attempting to use hardware to generate the time derivative of the output signal. A second alternative is to add a suitable transducer to the plant, in order to measure the rate of change of the output variable, but only very rarely will this technique be practicable. Unfortunately, because it is ramping, the slewing output may well arrive at the limit of available excursion before any meaningful data have been collected! Often a better alternative is to test the hardware in a closed loop configuration in which output ramping will not occur under conditions of steady applied input signal: this type of testing will be dealt with more fully later, in Section 13.6.3

13.3   Experimental procedures for measuring basic plant parameters

In a limited number of instances, not only the form of the transfer function of the plant will be known in advance, but the relationship between the parameters of the transfer function (typically, the time constants) and certain basic parameters of the plant will be known as well. If these basic parameters can be measured, then the transfer function (and possibly the static characteristic and the describing function) can be calculated.

The types of basic parameter which are amenable to this approach include the following:

- mechanics   – mass, moment of inertia, viscous friction, coulomb friction, static fraction, gear backlash
- electric drives   – magnetic saturation level, torque constant, back-EMF constant, armature reaction, armature and field resistance and inductance
- hydraulic drives – degree of lap in valves, orifice flow coefficients, physical dimensions of cylinders, pistons, etc., oil density and viscosity.

This list is by no means complete. As a general rule, parameters such as those listed can be measured using relatively simple experimental equipment and the measurement procedures need not be dealt with here.

On rare occasions, manufacturers of final control elements and feedback transducers may supply sufficient information to enable transfer functions and static characteristics to be formulated quantitatively: a possible hazard is the wide (and possibly unspecified) tolerances which may apply to the quoted parameters.

13.4   Characterisation by step response testing

13.4.1   Experimental techniques for obtaining step responses

Step response testing represents the most simple experimental technique for recording dynamic behaviour, since the signals required can be generated simply, either by making and breaking a steady supply or by stepping between two steady signal levels. It therefore requires a minimal complement of test equipment. Figure 13.3 shows a simple method for manually stepping an input signal between two steady levels: this generates a step superimposed on a quiescent level. The transducer response would be recorded to a base of time, on the Y–T recorder or a storage CRO in its Y–T mode.

Figure 13.4 shows a technique for obtaining repeated step responses, for both directions of applied step, by injecting into the plant a suitable squarewave, superimposed if necessary on a steady quiescent level. The half-period of the squarewave must be greater than the settling time of the step response in order that adjacent responses shall not interfere with one another. The sequence of responses may be recorded to a base of time; alternatively, if the signal generator incorporates a suitable, synchronous, triangularwave auxiliary output, then this output may be used to synthesise a synchronous time base to be used in conjunction with an X–Y recorder or a storage CRO in its X–Y mode: the resulting trace consists of a set of superimposed pairs of step responses.

13.4.2   The order of the response

The degree of ease or difficulty encountered in interpreting experimental step responses (and also impulse responses) depends on the *Order* (or *Degree*) of the plant being tested. Once any dead time and steady-state offset in a linear step response are allowed for, the remaining data can in theory be decomposed into an additive set of simple time-dependent terms, equal in number to the set of terms in the corresponding linear transfer function. The number of such terms is the order of the system and it can readily be shown that, as the order of the system increases, so it becomes

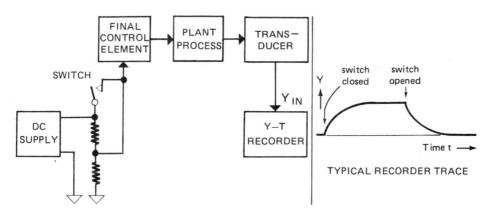

Fig. 13.3   Simple arrangement for manually stepping an electrical input disturbance between two steady levels, and recording the response of the hardware under test

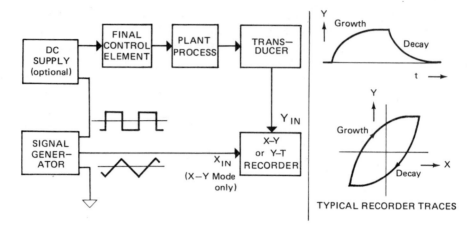

Fig. 13.4   Arrangement of test equipment for generating repeated step responses of a final control element — plant process-transducer combination

increasingly difficult to graphically decompose the response into its constituent parts (which, in effect, represent terms in a partial-fraction expansion): if these components cannot be determined, then the transfer function cannot be computed quantitatively.

In practice, first and second order responses are easily analysed, provided that the transfer function contains no numerator terms in the Laplace operator s. Third and higher order responses are difficult or impossible to decompose, especially when the transfer function contains numerator terms in s. In the latter event, only frequency response testing will yield data amenable to analysis approaching a reasonable degree of reliability.

13.4.3   Examples of simple first order step responses

Figure 13.5 shows typical first order growth and decay responses, with and without dead time, which might be obtained experimentally. Dead time can be measured directly as indicated, provided that the instant of application of the step disturbance is also recorded; time constant is most accurately measured by taking the time for the response to grow to 63 percent of its final excursion, or to decay to 37 percent of its initial value, as the case may be. In the case of all the transient responses being discussed in this chapter, the output is assumed to be stationary when the input step is applied. The presence of numerator terms in s in the transfer function will always result in curves which are more complex than those shown here, and these will necessarily be more difficult to interpret.

13.4.4   Examples of simple second order step responses

The approach to the interpretation of second order responses depends upon the degree of damping present, bearing in mind that lightly damped responses are those exhibiting a relatively large number of significant oscillations before the response settles.

In the case of light and medium damping, the damped natural frequency ω_t can be computed by measuring the period of the oscillation, as shown in Figs. 13.6 and 13.8: thus, $\omega_t = 2\pi/\text{period}$, rad/s. Figures 13.7 and 13.9 give alternative techniques for computing damping factor ζ for light and medium damping cases, respectively. Figure 13.7 involves plotting the natural logarithm of each overshoot and undershoot to a base of the numerical position in the series of overshoots/undershoots, whereas Fig. 13.9 represents a standard curve of peak overshoot versus damping factor, which is most appropriately used in mid-range. Once ζ is known, then the undamped natural frequency ω_n can be computed from the formula

$\omega_n = \omega_t/\sqrt{1 - \zeta^2}$; this will then yield the transfer function

$\omega_n^2/(s^2 + s2\zeta\omega_n + \omega_n^2)$

In the case of heavy damping, the most accurate technique involves constructing a tangent to the point of inflexion, as shown in Fig. 13.10: the orientation of this tangent can be matched to the corresponding tangent for

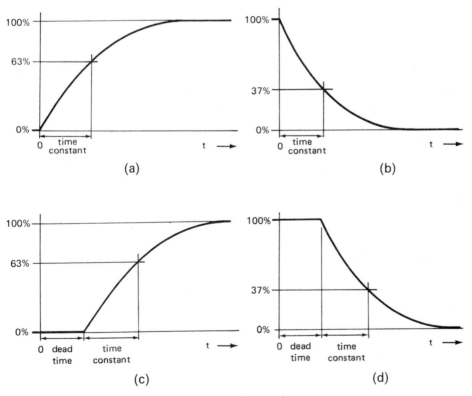

(a) (b)

(c) (d)

Fig. 13.5 Some examples of typical first order responses: (a) exponential growth; (b) exponential decay; (c) dead time plus exponential growth; (d) dead time plus exponential decay

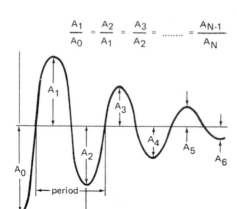

$\frac{A_1}{A_0} = \frac{A_2}{A_1} = \frac{A_3}{A_2} = \ldots\ldots = \frac{A_{N-1}}{A_N}$
A_1
A_2
A_3
A_4
A_5
A_6
A_0
period
period
0
t

Fig. 13.6 Example of a lightly damped second order step response

$\ln\left(\frac{A_N}{A_0}\right)$
N
0
1
2
3
4
5
6
Gradient = $\frac{-\zeta\pi}{\sqrt{1-\zeta^2}}$
hence ζ is evaluated

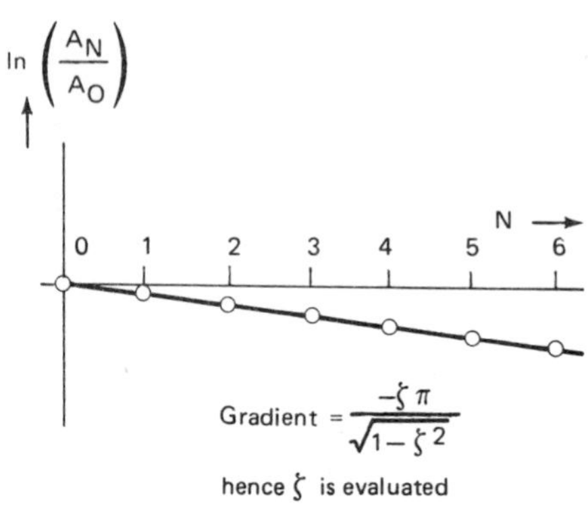

Fig. 13.7 Graph plotted from data measured from a response like that of Fig. 13.6, showing the evaluation of damping factor

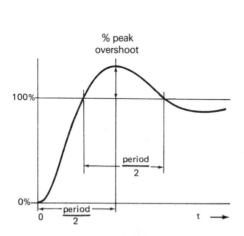

% peak overshoot
100%
0%
period/2
period/2
0
t

Fig. 13.8 Example of a slightly underdamped second order step response

% peak overshoot
100
90
80
70
60
50
40
30
20
10
0
% peak overshoot = 100 e $^{-\zeta\pi/\sqrt{1-\zeta^2}}$
hence ζ is evaluated
0
0.2
0.4
0.6
0.8
1
ζ

Fig. 13.9 Graph of percentage peak overshoot versus damping factor ζ for second order step responses, used to evaluate ζ

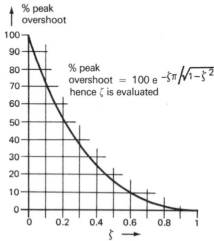

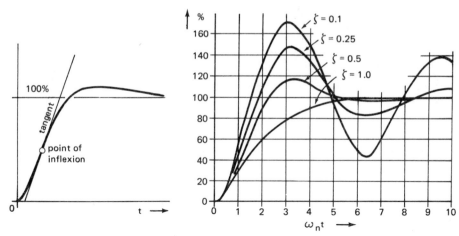

Fig. 13.10   Example of a heavily damped second order step response showing the use of measurements made at the point of inflexion

Fig. 13.11   Representative family of normalised second order step responses, for four different values of damping factor ζ

each of a set of standard quadratic response curves, which are freely available in the literature. Figure 13.11 represents one such set of standard curves. Using interpolation where necessary, the values of ω_n and ζ can be obtained from the curve representing the best match.

 As with the case of first order responses, the presence of numerator terms in s in second order transfer functions will always result in curves which are more complex than those which have been described.

13.4.5   Treatment of third order step responses

In the absence of terms in s in the transfer function numerator, a third order step response can theoretically be decomposed into a constant term (that is, the steady-state level) together with either three first order exponential decays or one first order exponential decay plus an under-damped second order oscillatory decay. Identification of the parameter values and form for the transfer function involves identification of the component parts of the experimentally-obtained composite response curve. Figure 13.12 is an example of a third order response which fairly readily can be decomposed graphically into three terms, which are amenable to being quantified, whereas Fig. 13.13 is an example of a third order response for which graphical decompositon is virtually impossible.

13.4.6   Limitations of step response testing

It can be seen that graphical interpretation may be either extremely unreliable or impossible for third and higher order step responses, plus

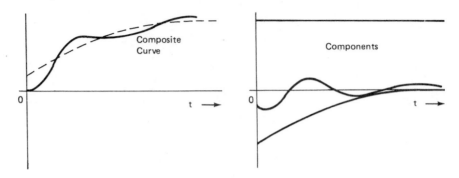

Fig. 13.12 Example of a decomposable third order response, together with its component response curves

those responses influenced by transfer function numerator terms in s. With appropriate software, it may be possible, using numerical analysis and curve fitting, to use a digital computer to identify the transfer function; a more commonly used alternative is to abandon step response testing in favour of frequency response testing, which is much more versatile.

13.5 Characteristisation from the impulse response

The other class of time domain responses would be obtained following the experimental application of an impulse disturbance. (In addition, ramp responses are sometimes generated, usually in repetitive form, by injecting a suitable triangularwave disturbance.) Since the application of an impulse containing sufficient energy to obtain a significant response would be difficult to generate and control, and might be destructive to the plant being

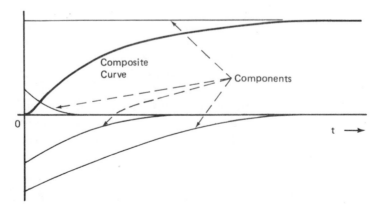

Fig. 13.13 Example of a non-decomposable third order response, showing the actual component response curves

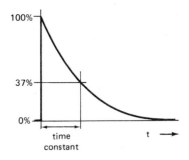

Fig. 13.14 Typical first order impulse response without dead time

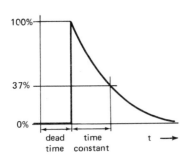

Fig. 13.15 Typical first order impulse response with dead time

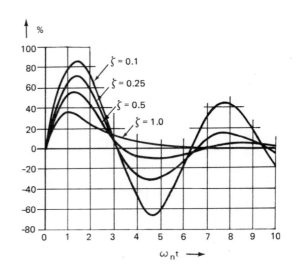

Fig. 13.16 Family of normalised second order impulse responses with dead time, for four different values of damping factor ζ

characterised, the usual procedure is to synthesise the effect of an impulsive disturbance.

13.5.1 Experimental techniques for obtaining impulsive responses

The first technique is suitable when data representing the rate of change of the output variable are available. This comes about from the mathematical relationship

$$\frac{d}{dt}\left\{\mathcal{L}^{-1}\left[\frac{1}{s}G(s)\right]\right\} = \mathcal{L}^{-1}\left[G(s)\right]$$

where $G(s)$ is the transfer function being identified.

This means that the impulse response must be identical to the rate of change of the step response. Data could be obtained by graphical differentiation of step response data but, of course, if the step response data are already to hand, there will be little point in not characterising directly from the step response. However, in those cases where the plant includes a rate of change transducer (such as a tachogenerator in a position servosystem), this transducer will generate the response of interest when a step disturbance is applied to the plant.

Impulse response analysis is of much greater practical application when the plant is subjectd to random signal testing. This is an advanced area of analysis and testing, and will be dealt with separately in Section 13.7.

13.5.2   Examples of simple first and second order impulse responses

Figure 13.14, 13.15 and 13.16 are examples of simple first and second order impulse responses. The techniques used for quantitative interpretation are similar to those employed on comparable step responses, so that no separate explanation is required.

13.6   Characterisation from the frequency response

Frequency response techniques have the merit that their application is not restricted to low order transfer functions. Moreover, the effects of numerator terms are distinguished as readily as those of denominator terms of the transfer function. Dead time and the presence of negative numerator terms are also amenable to identification using these techniques. The experimental method involves the application of suitable test sinewaves of known magnitude and the computation of gain magnitude and phase shift between the output signal and either the applied input signal or a second output signal. The measuring equipment should be capable of operating down to very low frequencies (sometimes lower than 1 mHz) and should preferably be frequency selective in order to reject harmonics. These requirements tend to make the testing procedures very time consuming and also involve a much greater expenditure on test equipment, especially if good harmonic rejection is required; this last property is essential for the measurement of describing functions.

Note that equipment normally used for frequency response testing in the audio frequency range is usually useless for control systems, because it has a lower frequency limit typically in the region of 10 Hz, which is beyond the bandwidth of most control systems.

All practical frequency responses roll off at higher frequencies, so that output signal levels diminish at the higher end of the spectrum, as the frequency is swept upwards. Since all experimental signals will be corrupted by parasitic noise to a greater or lesser extent, it is to be expected that the experimental data may become increasingly erroneous as the signal frequency is progressively increased: the effect can be minimised by employing test equipment having good noise rejection capability.

Frequency response analysis methods may also be applied to frequency response data which have been computed from time domain data obtained experimentally. This is particularly useful in those cases where the time domain responses are too complex for graphical decomposition. In these cases, the generation of the frequency domain data is undertaken either by a dedicated Fourier Transform Analyser or by a Fast Fourier Transform package on a general purpose digital computer.

13.6.1 Experimental techniques for obtaining frequency responses

Figure 13.17 shows an experimental set-up for measuring a frequency response experimentally. The output signal from the transducer is displayed on the Y channel of a recorder or a suitable (preferably storage) CRO. It must be borne in mind that, because the frequencies involved are normally very low in value, all connections must be DC coupled and the recording instrument must be capable of recording very slowly varying signals.

When the DC offset provision is included for small-signal testing, the offset will, of course, be adjusted to give the required quiescent level; only the steady-state alternating components of the output and input signals are of interest as far as the data display is concerned. The amplitude of the output sinewave is given by the height of the displayed curve. Phase may be measured by any of the following four alternative techniques.

- Record output and input to a base of time, on a two-channel Y–T recorder: phase measurement is based on Fig. 13.18.

- Record output and input, plotted against each other, on an X–Y recorder: phase measurement is based on Fig. 13.19.

- If available, use an auxiliary quadrature triangularwave output from the signal generator, in conjunction with an X–Y recorder: the triangularwave forms a synchronous timebase and phase measurement is based on Fig. 13.20.

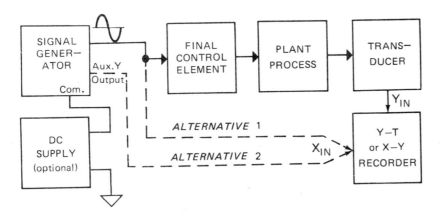

Fig. 13.17 Arrangement of test equipment to measure frequency response data for a final control element–plant process-transducer combination

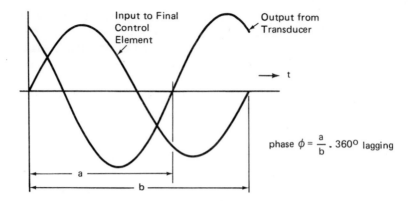

Fig. 13.18   Recordings of input and output sinewaves to a base of time, with measurements for computing phase shift

- If available, use an auxiliary variable-phase sinewave output from the signal generator, in conjunction with an X–Y display. Adjust the variable phase until the displayed ellipse collapses to a straight line: the phase measurement is read directly from the phase shift calibration.

Proprietary Transfer Function Analysers (sometimes called Frequency Response Analysers) are available, although they are expensive. Typically, they compute gain magnitude and phase data, usually using a correlation technique (discussed in Section 13.8), and thereby can provide a high degree of rejection of harmonics, noise and DC offsets; often, they are also autoranging. Usually, the signal generator is incorporated into the analyser, with the sinusoidal output signal being synthesised digitally. The

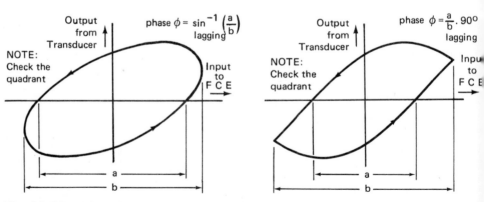

Fig. 13.19   Plot of output versus input sinewaves on an X-Y display, with measurements for computing phase shift

Fig. 13.20   Plot of output sinewave versus auxiliary triangularwave on an X-Y display, with measurements for computing phase shift

more expensive instruments will permit automatic stepping or sweeping of the signal frequency and enable gain magnitude and phase to be measured between two signal output points, neither of which need to be the test input signal point. Because of their harmonic rejection property, they can also be used for measuring describing function data.

By whatever technique the frequency response data are gathered, it is the normal practice to graph the data as Bode gain magnitude and phase plots.

13.6.2   Interpretation of experimental Bode plots

The bulk of the information for identification is contained in the magnitude plot. The phase plot serves as a cross check, and can be shown to be entirely predictable (assuming linear behaviour) from the magnitude plot, except when dead time (which is best identified by time domain testing) and negative numerator terms in the transfer function are present. Negative denominator terms cannot arise, because they result in an unstable system incapable of converging to a steady-state condition following the application of a sinusoidal (or any other) disturbance: frequency domain identification is not possible in this situation without the application of very special techniques.

The objective is to obtain a best fit to the experimental magnitude plot for a suitable set of asymptotes bearing in mind that:

- asymptote gradients can only assume values which are whole number multiples of $\pm 20\,\text{dB/decade}$; i.e. $+40$, $+20$, 0, -20, -40, -60 etc. dB/decade. For this reason, designers often construct templates, resembling set-squares, that have the appropriate gradients for their hypoteneuses
- at each of the intersections of the fitted asymptotes, the vertical dB separation between the asymptote intersection and the experimental curve must equal the predictable figure
- where asymptote gradients change by $+40$ or $-40\,\text{dB/decade}$ at an intersection, the separation between the asymptotes and the experimental curve may be matched to one of a standard set of quadratic lag frequency response curves (these may be inverted as required, for quadratic lead), as presented in Fig. 13.21.

It will be appreciated that some degree of skill and experience is necessary to obtain good asymptote fitting. The transfer function can then be written down by inspecting the asymptote sequence (bearing in mind that the phase plot should be cross-checked for the rare anomalous cases), as follows.

- Note the dB gain at which the low-frequency asymptote (projected if necessary) crosses 1 rad/s (0.16 Hz): this is the transfer function gain constant (in dB).
- The number of integrations in the transfer function will relate to the gradient of the low-frequency asymptote, as follows:

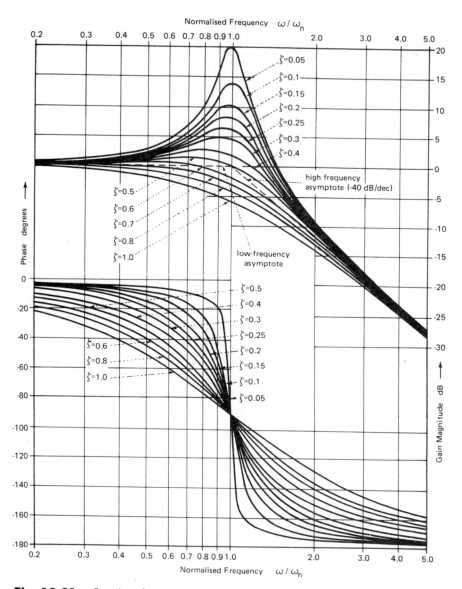

Fig. 13.21   Family of normalised Bode gain magnitude and phase plots for a standard quadratic lag, for a set of values for damping factor ζ; note that, where the gain magnitude peaks, resonant frequency $\omega_r = \omega_n \sqrt{1 - 2\zeta^2}$ and, at this frequency, peak magnitude $M_m = -20 \log (2\zeta\sqrt{1 - \zeta^2})$ dB, $\zeta \leqslant 0.707$

zero integrations – zero gradient
one integration   – – -20 dB/decade gradient
two integrations – – -40 dB/decade gradient.

- Note the break (corner) frequencies at which the asymptotes intersect. Where the gradients change by $+20$ or $-20\,\mathrm{dB/decade}$, the break frequency ω_B yields the value of a time constant T, for $T = 1/\omega_B$. Where the gradients change by $+40$ or $-40\,\mathrm{dB/decade}$, the break frequency ω_B yields the undamped natural frequency ω_n for the corresponding quadratic term, the damping factor ζ for which can be obtained by using Figure 13.21.
- For increasing frequency, if the asymptote gradient changes at an intersection by:

$+20\,\mathrm{dB/decade}$, the corresponding transfer function term is $(1 + sT)$

$-20\,\mathrm{dB/decade}$, the corresponding transfer function term is $1/(1 + sT)$

$+40\,\mathrm{dB/decade}$, the corresponding term is $\left(\dfrac{s^2}{\omega_n^2} + \dfrac{s2\zeta}{\omega_n} + 1\right)$

$-40\,\mathrm{dB/decade}$, the corresponding term is $\left(\dfrac{s^2}{\omega_n^2} + \dfrac{s2\zeta}{\omega_n} + 1\right)$.

Thus, anomalous terms apart, the transfer function can be written down from inspection of the asymptote sequence as the product of sets of first and second order factors. Where break frequencies (i.e. asymptote intersections) appear to be close to one another, it may be found that alternative sets of asymptotes may be fitted to the experimental curve in this region with equal accuracy: in this case, either sequence may be regarded as valid, so that either set of corresponding transfer function terms is also equally valid.

13.6.3 Open loop versus closed loop testing, for characterisation

Figure 13.22 shows a testing configuration for characterising the complete forward path of a control system, the feedback path not being connected. This will yield open loop frequency response data which can be used in the compensation design for the control law, which will subsequently be

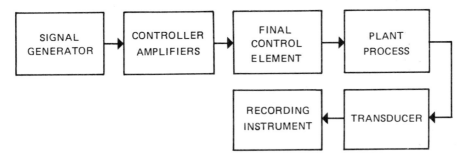

Fig. 13.22 Arrangement of test equipment for open loop testing of a control system

incorporated into the controller. Generally speaking, the following properties will apply to this type of test.

- Because of the high gain often present in the forward path, the input test signal will often need to be considerably smaller than the output signal from the transducer.

- If any integration is present (in the controller or the plant) then an offset, in any of the elements upstream of the site of the integration, will be integrated to produce a ramp component in the transducer signal: test equipment cannot eliminate this ramp and, moreover, the plant process may reach the end of its available excursion before useful data can be obtained.

- In the absence of a physical limit to the travel of the system output, a DC offset can deliberately be added to the signal generator output signal so that, in suitable circumstances, the system output can be given a constant component of motion. This is useful when open loop testing servosystems, for example, because the shaft can be prevented from being reversed by the applied sinewave, with the result that any backlash and static friction in the mechanics can be prevented from corrupting the test data. However, in this case it will be necessary to record output velocity using a tachogenerator because the output displacement signal will not usually be amenable to simple interpretation.

Figure 13.23 shows a testing configuration for characterising a complete closed loop system. It is assumed that the characterisation is being undertaken before the compensation has been designed, so that the controller is not yet implementing the final control law. However, the system must be stable before useful tests can be undertaken.

Fortunately, most practical systems can be made stable if the controller is temporarily arranged to provide simply a reduced degree of gain,

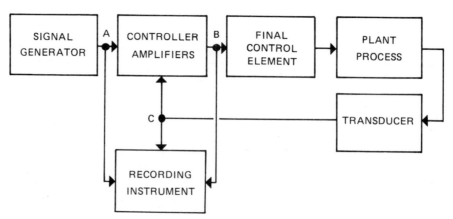

Fig. 13.23   Arrangement of test equipment for closed loop testing of a control system

together with its signal-mixing function. The system performance will obviously not be optimum at this stage, because the compensation has not been designed or implemented. The following properties apply to this type of test.

- The input test signal will be in the same order of magnitude as the transducer output signal for frequencies within the system bandwidth.
- Closed loop data can be measured by comparing signal C with signal A, and 'open loop' data can be measured (despite the fact that the loop is closed) by comparing signal C with signal B.
- Open loop data can also be predicted from closed loop data, using a Nichols chart in reverse: the experimental data are plotted on the M and α contours and the corresponding open loop data can be read off the scales of the diagram axes. (This is not particularly accurate for low frequencies because of the high resolution of the top right hand corner of the chart, as can be seen from Fig. 13.24.) The technique is particularly useful with nonlinear processes, because the linearising effect of the negative feedback, in conjunction with the graphical technique, can yield a notional linearised open loop transfer function.
- Because the configuration is closed loop, offsets cannot generate a ramp component at the output.
- The sinusoidal input signal will cause reversals of the output: this can be disadvantageous in servosystems, for example, because any resulting backlash and static friction in the mechanics will degrade the output data.

13.7   Characterisation by random signal testing

All the characterisation techniques so far described assume that the test input signal is sufficiently large for the resulting output signal strength to be discernible from the ambient noise often present. This will usually mean that a significantly large disturbance must be applied, to the extent that it may be unacceptable if the plant is in service. The alternatives to this are to employ techniques which either make use of the signal variations that occur naturally in normal plant operation, or use a suitable test signal which is too small to embarrassingly disturb the plant operation.

In either case, the techniques adopted involve the use of random signals, and it is then the relationships between the statistical properties of those signals which yield the data for identifying transfer functions. Care is needed when interpreting the data for processes possessing significant nonlinearities.

13.7.1   Correlation functions and power spectral density functions

Figure 13.25 shows a computation process known as *Correlation*. A signal x is delayed by a time τ and subsequently multiplied by another signal y: the

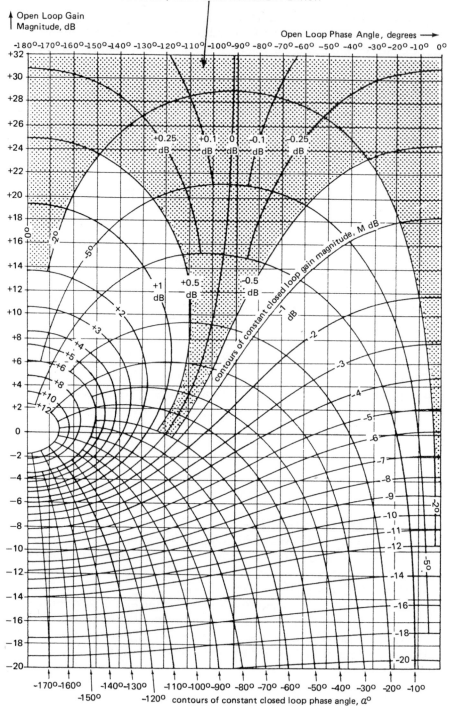

Fig. 13.24 dB gain magnitude versus phase (Nichols) chart, showing the region of high resolution most affected by tolerances in measured data; vertical and horizontal scales represent open loop data, whilst the curved contour scales represent closed loop data

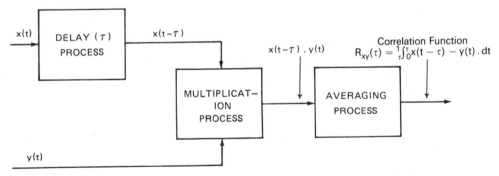

Fig. 13.25 Block diagram representation of a time cross correlation process

product is averaged over a significantly long period of time T to yield an average value $R_{xy}(\tau)$, which is called the 'Correlation Function at delay τ'. The procedure can be repeated for different values of τ, and the resulting variations in $R_{xy}(\tau)$ indicate the relative dependence of the two signals on one another. Thus

$$R_{xy}(\tau) = \frac{1}{T}\int_0^T x(t-\tau).y(t).dt.$$

When x and y are different signals, R_{xy} is called the 'Cross-Correlation Function', and when x and y are the same signal, $R_{xx}(\tau)$ is called the 'Auto-Correlation Function'. If the function is subsequently Fourier transformed, using either a dedicated or a general purpose digital computer, then $R_{xy}(\tau)$ transforms to $S_{xy}(j\omega)$, the 'Cross Power Spectral Density Function', and $R_{xx}(\tau)$ transforms to $S_{xx}(\omega)$, the 'Auto Power Spectral Density Function'. The spectral densities indicate the relative magnitude of the frequency components in the correlation functions: the spectral density functions will be continuous functions if the correlation functions are non-periodic, whereas the spectral density functions will be line spectra if the correlation functions are periodic, which case only occurs when the test signals are periodic.

If x is a White Noise signal, then its Auto Spectral Density $S_{xx}(\omega)$ will be flat (meaning that all frequencies are equally probable) and its Auto Correlation Function $R_{xx}(\tau)$ will be an impulse at $\tau = 0$. Test signals having a flat spectrum over a prescribed frequency range can be generated with suitable hardware, and alternative waveforms are shown in Fig. 13.26.

13.7.2 The application to characterisation

Figure 13.27 illustrates, in a generalised form, the application of the procedure to plant characterisation. The noise generator may be an external noise source connected to inject noise into the plant, or it may be a

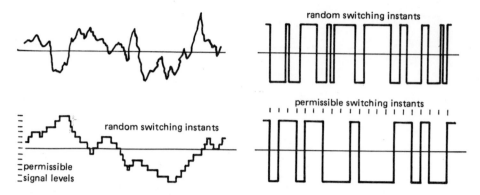

Fig. 13.26   Examples of four different test noise signals, all capable of having a flat density spectrum over a prescribed frequency range

source of noise occurring naturally within the plant. The following relationships, proofs for which may be found in the literature, are relevant in this context.

A $R_{xy}(\tau) \propto g(\tau)$ if $R_{xx}(\tau)$ = an impulse at $\tau = 0$ (that is, if $S_{xx}(\omega)$ is a flat spectrum).

Thus, if the input noise is suitable, the process impulse response, with τ substituted for t, (that is, $g(\tau)$) is proportional to the Cross-Correlation Function (input delayed), $R_{xy}(\tau)$.

B $$G(j\omega) = \frac{S_{xy}(j\omega)}{S_{xx}(\omega)}$$

Thus, gain magnitude and phase data can be obtained for the transfer function by dividing the Cross Density Spectrum (input delayed) by the input Auto Density Spectrum. If the latter is a flat spectrum, this reduces to $G(j\omega) \propto S_{xy}(j\omega)$

C $$\left|G(j\omega)\right| = \sqrt{\frac{S_{yy}(\omega)}{S_{xx}(\omega)}}, \text{ provided that } z(t) = 0.$$

Thus, gain magnitude data can be obtained for the transfer function by dividing the output Auto Density Spectrum by the input Auto Density Spectrum, and square-rooting the result. If the input spectrum is flat, this reduces to $|G(j\omega)| \propto \sqrt{S_{yy}(\omega)}$.

The various relationships provide a duplication of data, so that not all of them would be used for any particular application. The procedure is completed by characterising either impulse response data (method A) or frequency response data (method B or C), using the techniques of Sections 13.5 and 13.6 respectively.

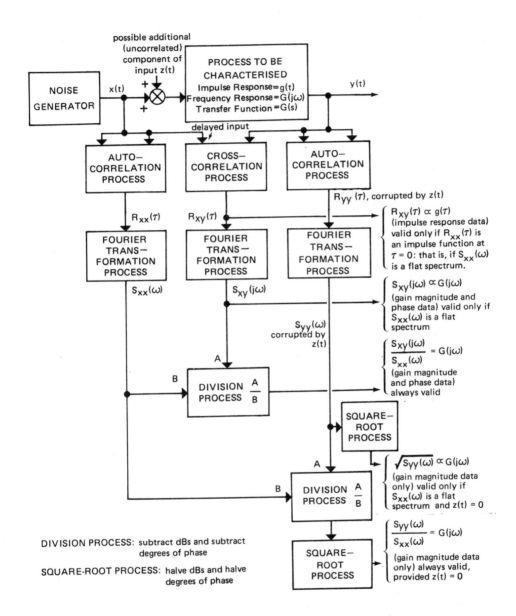

Fig. 13.27 Application of alternative time correlation and Fourier transformation processes to characterise a test process; note that some of the computation can be simplified when x(t) represents a white noise source, and some of the data can be corrupted by the uncorrelated noise signal z(t)

13.7.3   Instrumentation hardware requirements

Random noise generators are commercially available, but only those which generate a noise spectrum of significant magnitude down to DC are suitable: this requirement rules out most commercial noise generators.

Figure 13.28 shows the internal organisation of a 100-point general purpose correlator constructed using hardwired components. The Voltage Range Adjustment stages ensure that the ranges of the two Analog-Digital Converters are fully utilised without overloading. Memory A stores, in scaled digital form, the current and previous 99 sample values of V_x, whilst Memory B stores, again in scaled digital form, the product of this set of 100 scaled sample values of V_x times the current scaled sample value of V_y. Memory C typically stores the accumulated totals of these 100 product values, with summation starting at the instant when the correlation process is enabled by the operator: this is referred to variously as 'Summation Averaging', 'Integration Averaging' and 'Linear Averaging'. This type of averaging is best used when the statistical properties of V_x and V_y are constant. The set of 100 computed average values progressively increases as the duration T of the correlation process is allowed to elapse, where T occurs as the limit in the expression

$$T . R_{xy}(\tau) = \int_0^T x(t - \tau) . y(t) . dt.$$

The value of T occurs as a (changing) scale factor in all output data generated by this time-average, affecting all 100 values in equal proportion.

When the statistical properties of V_x and V_y are changing, $R_{xy}(\tau)$ values will drift accordingly and can be tracked reasonably accurately using an alternative averager, called an 'Exponential Averager'. This is, in fact, a digital synthesis of a simple analog low-pass filter. The output values from the Time Averaging logic now represent the expression

$$R_{xy}(\tau) \cong [x(t - \tau) . y(t)] * e^{-t/T}$$

in which the asterisk $*$ represents the process of convolution. Such an averager weights (exponentially) the recent values of the product $x(t - \tau) . y(t)$ more heavily, but weights the product values less and less heavily as they recede into the past. For this reason, it is sometimes referred to as an 'Exponentially-weighted Past (EWP) Averager'. The choice of time constant τ should reflect the rate at which the statistical properties of V_x and V_y are changing.

All of the logic is clocked synchronously at a rate determined by the setting of the Sampling Interval Clock. This interval Δt will be equal to the increment $\Delta \tau$ in the value of delay τ occurring in the final display, so that it should be chosen so that 100 times this value will span the time scale of interest in the correlation function being generated.

Finally, one hundred point values of $R_{xy}(\tau)$ are generated in the form of a scaled analog voltage, together with another scaled voltage representing each corresponding value of time delay τ, so that an X–Y display of the two voltages will represent one hundred points on the correlation function.

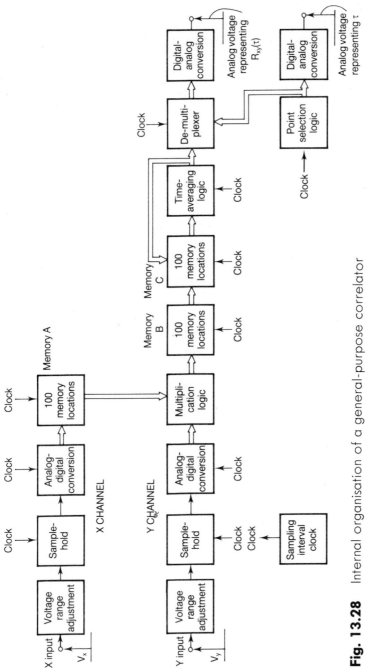

Fig. 13.28 Internal organisation of a general-purpose correlator

Generally, only relative values (and not absolute values) of the points on the $R_{xy}(\tau)$ plot need to be known, in which case the scale factor need not be calculated. Often the 100-point time slice on the correlation function time scale may be repositioned by the insertion of 'pre-computation delay' in the X channel, so that τ no longer commences at zero but at a new value $n.\Delta t$, where n is the value of delay increment chosen (typically an integer multiple of 100) and Δt is the sampling interval (that is, the clock pulse interval).

Such an instrument will output $R_{xy}(\tau)$ and τ data in digital form for processing by a compatible Fourier Transform Analyser, in addition to the analog output representations used for display purposes.

A general purpose digital computer with a suitable realtime interface may also be used to perform the same function as the instrument described above. In this case, the program will need to mechanise the storage, multiplication and averaging processes performed within the instrument.

Low Frequency Spectrum Analysers are available commercially for computing the Fourier transform of voltage signals in realtime; they are constructed using microcomputer technology. On the other hand, Fourier Transform Analysers using similar technology are dedicated to the transformation of Correlation Function data. In both cases, the analyser uses time domain input data samples x(n) to compute the function

$$X(k) = \sum_{n=0}^{N-1} x(n).e^{-j2\pi nkN},$$

where $n = 0, 1, 2 \ldots N - 1$ represents points on a time scale and $k = 0, 1, 2 \ldots N - 1$ represents points on a frequency scale. Interpolation algorithms enable intermediate points to be computed on the frequency scale: for example, the Analyser may compute 1000 frequency spectrum points from 100 point values of correlation function data supplied.

Additional algorithms permit the computation of magnitude, phase, real part, imaginary part, log (magnitude) and log (frequency), thereby facilitating alternative output display formats compatible with Bode, Nyquist and Nichols diagrams.

In addition to the use of dedicated Fourier Transform Analyser instruments, general purpose digital computers with suitable realtime interfaces may be programmed to perform the same task.

13.7.4   PRBN correlation techniques

A highly specialised noise generator-correlator combination may be constructed cheaply, either around integrated circuit devices or using a microcomputer with a realtime interface: the former version will now be described, making reference to Fig. 13.29.

The noise generator is constructed by means of a shift register synchronised to a clock pulse train, with the first bit being set from logic which is gated by the states of certain other bits. The output signal from any element can be shown to switch between the 1 and 0 levels synchronously with the

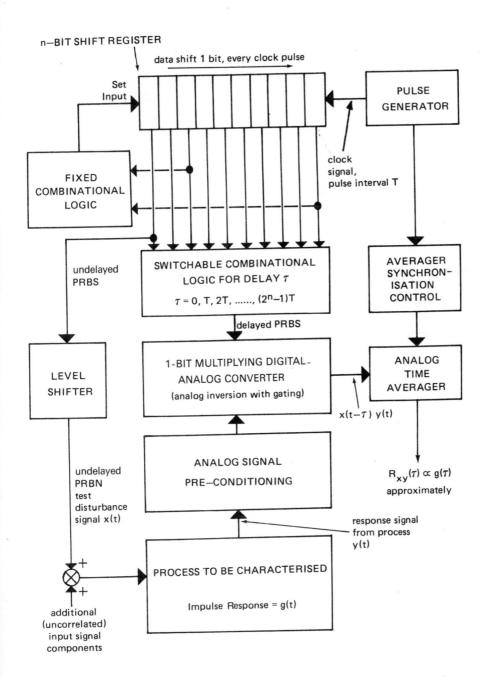

n−BIT SHIFT REGISTER

data shift 1 bit, every clock pulse

Set
Input

PULSE
GENERATOR

clock
signal,
pulse interval T

FIXED
COMBINATIONAL
LOGIC

undelayed
PRBS

SWITCHABLE COMBINATIONAL
LOGIC FOR DELAY τ

$\tau = 0, T, 2T, ..., (2^n - 1)T$

AVERAGER
SYNCHRON-
ISATION
CONTROL

delayed PRBS

LEVEL
SHIFTER

1-BIT MULTIPLYING DIGITAL-
ANALOG CONVERTER

(analog inversion with gating)

ANALOG
TIME
AVERAGER

$x(t-\tau)\ y(t)$

undelayed
PRBN
test
disturbance
signal x(t)

ANALOG SIGNAL
PRE−CONDITIONING

$R_{xy}(\tau) \propto g(\tau)$
approximately

response signal
from process
y(t)

PROCESS TO BE CHARACTERISED

Impulse Response = g(t)

+

+

additional
(uncorrelated)
input signal
components

Fig. 13.29   General arrangement of hardware/software processes for effecting PRBN cross-correlation; note that the output data are valid only if averaging is performed over a whole number of PRBN sequences, but are uncorrupted by the additional input signal components

clock in an apparently random sequence. However, the sequence will normally repeat itself every $(2^n - 1)T$ seconds, where n is the register word length and T is the clock interval: for this reason, it is called a 'Pseudo-Random Binary Sequence' (PRBS) or 'm-Sequence'. This signal can then be level shifted, so that it now switches between $+ V$ and $- V$ volts: such a signal is called 'Pseudo Random Binary Noise' (PRBN), and its auto-correlation function can be shown to approximate an impulse at $\tau = 0$. Such a PRBN generator may be used as a noise source with a general purpose correlator, or may be an integral part of a dedicated PRBN correlator, which constitutes the test equipment processes shown in Fig. 13.29.

Delayed versions of the sequence can be generated, for delay values equal to integer multiples of T, by combining the outputs of the shift register elements in specified combinational logic, each combination yielding a different delay τ.

Multiplication of the plant process output (which is analog) by the delayed PRBS (which is digital) requires a 1-bit Multiplying DAC, and is easily implemented by solid state analog switches and analog inverting and summing amplifiers; averaging over a whole number of sequences can be undertaken, for example, using an analog integrator gated from the shift register clock. Thus, a PRBN correlator is easily, and cheaply constructed. It yields the process cross-correlation function (input delayed) which is, with the test signal used, approximately proportional to a replica of the process impulse response, which can now be characterised. The advantages of this type of correlator include the following:

- it is easily and cheaply constructed
- the test signal bandwidth is proportional to the clock frequency and is therefore readily adjusted
- the spectrum of the test signal extends down to DC
- the sequence repeats exactly, so that the test data should be repeatable
- the test signal is easily arranged to actuate the plant, since it simply switches between two levels.

13.7.5   Using PRBN for frequency response computation

A PRBN generator may also be used with a realtime digital computing system to compute directly the frequency response of a process, provided that a Fast Fourier Transform (FFT) software package is available for the computer. Figure 13.30 shows a typical set-up, which includes both hardware and software elements.

It can be shown that, at any given frequency ω_i rad/s:

$$|G(j\omega_i)| = \left| \frac{P_o(\omega_i) + jQ_o(\omega_i)}{P_i(\omega_i) + jQ_i(\omega_i)} \right| = \left[\frac{P_o^2(\omega_i) + Q_o^2(\omega_i)}{P_i^2(\omega_i) + Q_i^2(\omega_i)} \right]^{1/2}$$

and

$$\underline{/G(j\omega_i)} = \left| \frac{P_o(\omega_i) + jQ_o(\omega_i)}{P_i(\omega_i) + jQ_i(\omega_i)} \right| = \tan^{-1}\left(\frac{Q_o}{P_o}(\omega_i) \right) - \tan^{-1}\left(\frac{Q_1}{P_1}(\omega_i) \right)$$

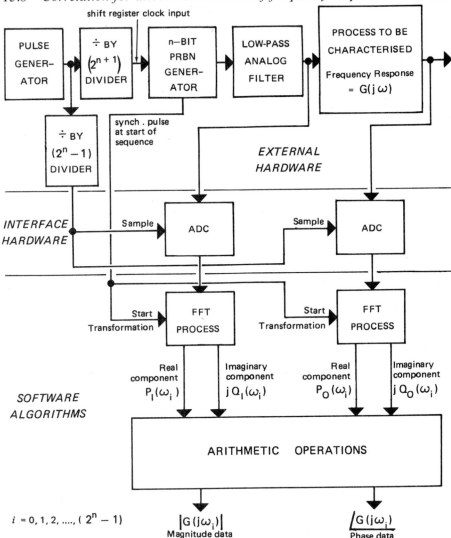

Fig. 13.30   Use of a pseudo-random binary noise source and fast Fourier transform computation to generate frequency response data

The computer is used to generate values for these expressions at each of the set of specified frequencies, corresponding to the set of values of the variable i. The data will be corrupted by the presence of extraneous signals, so that this type of testing is preferably undertaken off-line.

13.8   The use of correlation for direct measurement of frequency response data

A general purpose type of correlator, as referred to in Section 13.7.3, may be adapted to the direct measurement of frequency response data. To do this, it is necessary to synchronise the averaging process to a sinewave signal generator, as shown in Fig. 3.31.

Provided that the correlation averaging is undertaken over a whole number n of cycles of the input sinewave, it is readily shown that

$$R_{xy}(\tau) = \frac{1}{2\pi n} \int_0^{2\pi n} A \sin \omega(t - \tau) B \sin (\omega t - \phi) \, d(\omega t)$$

$$= \frac{AB}{4\pi n} \int_0^{2\pi n} [\cos(\omega \tau - \phi) - \cos(2\omega t - \omega \tau - \phi] \, d(\omega t)$$

$$= \frac{AB}{2} \cos (\omega \tau - \phi), \quad n = 0, 1, 2 \ldots \text{etc}$$

A plot of this function is shown in Fig. 13.32.

From this, it can be seen that

$$|G(j\omega)| = \frac{B}{A} = \frac{2}{A^2} R_{xy}(\tau)_{max} \text{ and } \underline{/G(j\omega)} = -\phi = -\frac{\tau_1}{\tau_2} 2\pi$$

The combination of a dedicated cross-correlator and sinewave signal generator would typically form the basis of the type of proprietary Transfer Function Analyser referred to in Section 13.6.1.

Typically, such an instrument will incorporate digitally-synthesised synchronised sinewave and cosinewave generators, and dual cross-correlators, as shown in Fig. 13.33. The post-computation will compute data such as:

$$\text{gain magnitude } \frac{B}{A} = \frac{2}{A^2} \{[R'_{xy}(0)]^2 + [R''_{xy}(0)]^2\}^{1/2}$$

$$\text{phase } \phi = - \tan^{-1}[R''_{xy}(0)/R'_{xy}(0)]$$

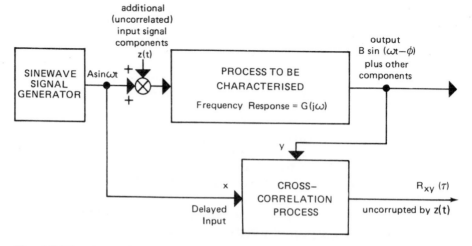

Fig. 13.31   Use of a synchronised sinewave generator and cross-correlation process for direct measurement of frequency response data

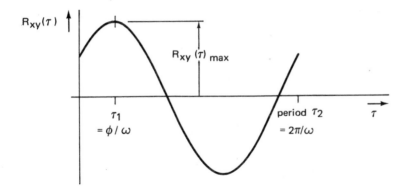

Fig. 13.32   Cross-correlation function obtained with the arrangement of Fig. 13.31

$$\text{real part of gain} = \frac{2}{A^2} \cdot R'_{xy}(0)$$

$$\text{imaginary part of gain} = \frac{2}{A^2} \cdot R''_{xy}(0)$$

Note that cross-correlation values now need to be computed only for zero delay. The availability of the logarithmic function in the post-computation enables the output data to be displayed alternatively in Bode, Nyquist and Nichols formats.

Such an instrument normally has the following features:

- ability to measure gain and phase between any two points in a process, neither of which need be the Asinωt signal injection point, as indicated in Fig. 13.34.
- programmable sweeping of frequency ω upwards and downwards
- programmable test signal amplitude A
- autoranging input signal conditioning
- rejection of harmonics of the fundamental frequency ω
- rejection of uncorrelated noise
- programmable correlation duration
- programmable delay before correlation commences, so that the output data are not corrupted by transient components of response signals
- optional amplitude modulation and demodulation using an external carrier (see Section 13.9).

13.9   Special techniques for testing AC-carrier systems

The types of waveform which characterise the dynamic behaviour of DC control systems are transposed to become the envelopes of amplitude-

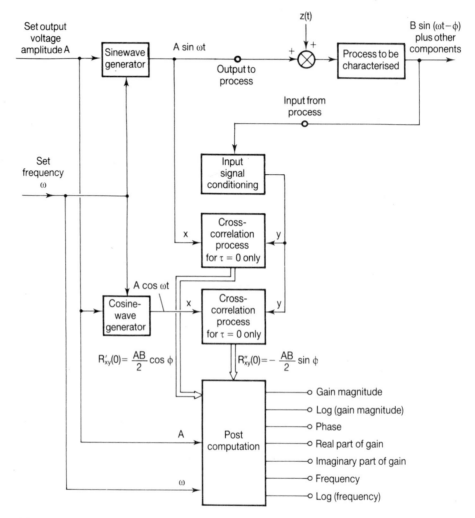

Fig. 13.33   Internal organisation of a frequency response analyser using cross-correlation processes

modulated carrier waveforms, in the corresponding AC-carrier systems. Examples are presented in Fig. 13.35.

Note that, in general, either $y(t) = x(t) \sin \omega_c t$ or $y(t) = \sin [x(t)] \sin \omega_c t$, depending on the nature of the hardware used in the AC system. Note also that the envelope of each modulated waveform cannot be displayed directly.

The data represented by the AC-carrier waveform are thus embodied in the envelope of the waveform, together with the phase of the carrier, measured relative to the carrier reference supply. The types of test signal which the system will require will therefore need to take this form, as will the types of signal representing the system response. Figure 13.36 indicates

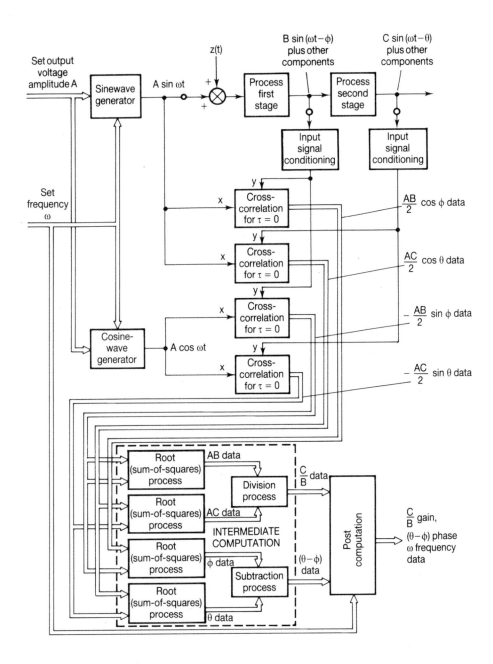

Fig. 13.34   Extension of the frequency response analysis process, to characterise a plant process stage separated from the test signal injection point

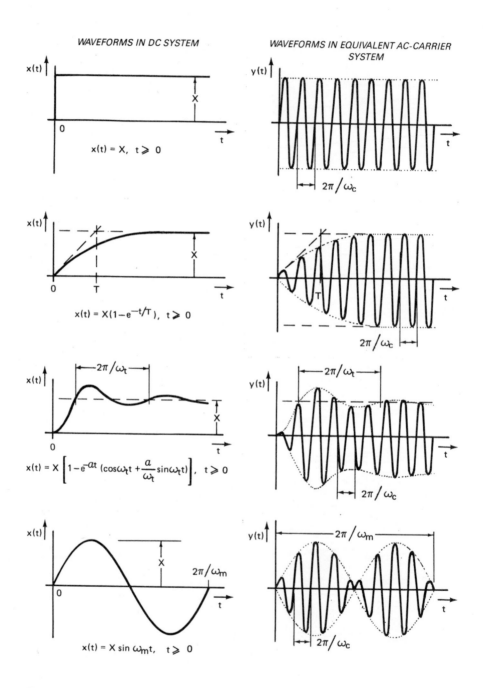

Fig. 13.35 Examples of waveforms occurring under test conditions in DC control systems (in left-hand column) and equivalent waveforms in AC-carrier control systems (in right-hand column)

the hardware which could be required in order to characterise the system.

Whether the AC signals or their DC counterparts are actually measured depends on the nature of the available test equipment. An alternative to using the phase-sensitive demodulator would occasionally be the temporary coupling of a suitable DC transducer to the plant, purely for the purpose of characterisation. Certainly, the waveform from a DC device is more easily interpreted than its amplitude modulated counterpart.

13.10   Special techniques for testing digital systems

In most instances, it will be necessary to use a conventional type of signal generator in order to generate the required disturbance functions. Occasionally, a signal generator may be digitally based, in which case it may have a digital output port: here, it may be feasible to input the digital output word directly into the control system, provided that the word is of the correct format and signal level. If this is not the case, then signal conversion (for

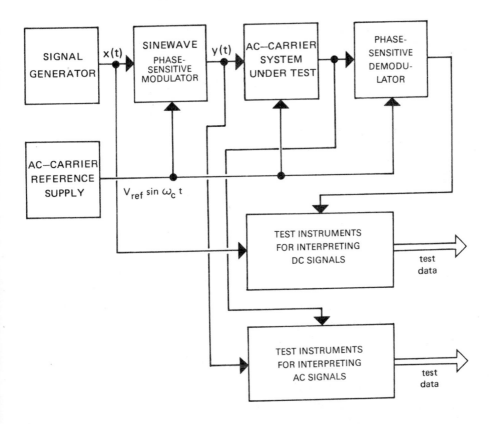

Fig. 13.36   Possible arrangements of test equipment for characterising on AC-carrier control system, showing alternative instruments for analysing counterpart DC and AC-carrier signals

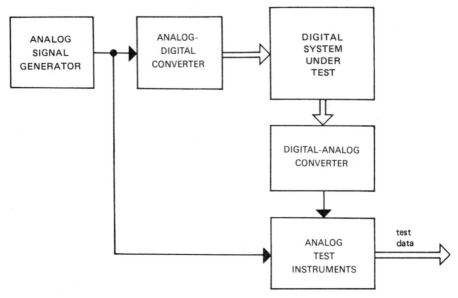

Fig. 13.37   Possible arrangement of test equipment for characterising a digital control system, assuming analog instruments are being used

example, from TTL levels to CMOS levels) and/or code conversion may be necessary properties of the interfacing hardware. Another exceptional case would be where a digital controller is computer based, so that it is conceivable that the required variations in set point value (for example) might be generated by means of suitable program statements.

If the output data from the system are digital, then it would normally be necessary to convert the data to an equivalent DC voltage, assuming that the test equipment requires an analog input signal. Some digitally based test equipment can process digital signals directly, assuming the input word is of the correct format and signal level: if this is not the case, then signal conversion and/or code conversion would be necessary before processing can proceed.

Figure 13.37 represents the situation in which the control system requires digital input and output data and the signal generator and measuring instrument are both analog. The ADC and DAC will need to be chosen for both digital and analog compatibility with the equipment to which they are interfaced, and the sampling frequencies must be high in relation to the bandwidth of the control system by a factor preferably in excess of ten times.

For simple on-off disturbances and for steady-state calibration, the digital input word to the control system might be set up manually using a suitable set of switches connected to an appropriate DC voltage source. Contact bounce can be a problem, although this can be obviated by the use of special 'switch de-bouncer' integrated circuits or bistable elements (see Section 1.14) connected to perform the same task.

13.11   Special techniques for testing pneumatic systems

In most cases, it will be necessary to use an electronic signal generator in order to generate the required disturbance functions: signals may need to be given a DC offset, because they will have to be converted to pressures, which obviously can only be positive. Figure 13.38 shows a typical arrangement for testing a pneumatic system using electronic instrumentation.

Pneumatic strip chart (Y–T) recorders are available for operation from 3 to 15 psi (20 to 100 kPa) signals and these can be used to record the system output response. On these, the pen mechanism is directly actuated by the pneumatic signal, with a bellows or diaphragm acting as the displacement transducer connected to the pen mechanism by means of linkages.

However, the application of electronic test instruments will afford much greater versatility, although these will necessitate the use of an air-to-current converter together with a current-to-voltage converter (for which a resistor of appropriate value should suffice), and optional provision for offsetting the voltage dropped, since this will be unipolar.

It must be understood that, whenever current-to-air and air-to-current converters are used, the relatively slow response times of these converters may have a significant effect upon the dynamic response data collected. It may well be necessary to measure the time constants of the converters and

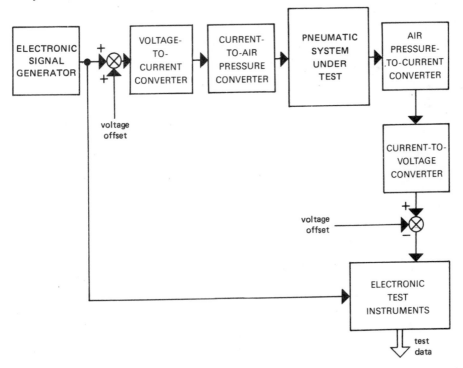

Fig. 13.38   Possible arrangement of test equipment for characterising a pneumatic control system

subsequently to remove the effect of the associated dynamic lag terms from the transfer functions deduced from the characterisation procedure.

For applying pressure step signals to the control system, it would be sufficient to use on-off pressure valves connected to suitable pressure-regulated pneumatic sources. The valves could be operated either manually or electrically by means of solenoids.

Steady-state calibration of pneumatic systems may be undertaken using a manually adjusted pressure-regulated air supply, together with a large-scale precision pressure gauge for indicating system input and output pressures: a small manually operated manifold could be useful as a pneumatic multiplexer for the pressure gauge.

14

Control System Performance
and Commissioning

14.1   The need for specification formulation

The ultimate function of a control system is the control of a property of an item of plant. The final control element and the feedback transducer become an integral part of the plant and must be selected on the basis of the criteria listed in Chapters 1 and 13 of *Basic Control System Technology*; similar criteria also apply to the selection of the reference transducer and its network, if the controller is to be custom designed. Finally, it is necessary for the control engineer to either design or select the controller.

The completion of the design involves the selection and mechanisation of the control law for the controller. In essence, the engineer must be able to predict the performance which the system would achieve in the absence of this control law and then select the law in order that the performance is consequently improved sufficiently to meet the specification. When the completed system is commissioned, the responsible engineer will be required to demonstrate that the system can adequately perform the task for which it has been designed.

In order that the user shall be supplied with a system which will satisfy his needs, it becomes essential for the required performance to be specified very carefully and unambiguously. Inadequate formulation of performance specifications will inevitably lead to much dissatisfaction, recrimination and possibly litigation. The importance of this aspect cannot be overstressed.

The performance of the final control element, plant process and feedback transducer can be tested and characterised, using the techniques described in Chapter 13. This performance will then be defined in terms of static characteristics, frequency (domain) responses, transient (time domain) responses, differential equations, transfer functions etc. In some cases, it may be possible to characterise individually the properties of the final control element, plant process and feedback transducer but, in most instances, only the combination will be identifiable.

The performance of the completed control system is normally defined in terms of a combination of parameters relating to the following four areas of system performance:

- steady-state accuracy in the presence of a steady applied disturbance, after all transient components of response have decayed to insignificance
- frequency domain behaviour, which implies the response to a sequence of specified applied sinusoidal disturbances, after all transient components of response in each case have decayed to insignificance
- time domain behaviour, which implies the transient response to a specified non-periodic disturbance, which will usually take the form of either an impulse, step, ramp or square law (parabolic function)
- noise performance, which implies the effect, on either the controlled variable or the error variable, of specified noise disturbances which typically would be defined in statistical terms.

With all stable systems being operated within a region of linear behaviour, the transient components of response will decay to insignificance if sufficient time is permitted to elapse for this to occur (assuming, of course, that a new disturbance is not applied in the meantime). In some systems containing significant nonlinearities, the transient response following the application of a disturbance may diverge or converge to a bounded oscillatory condition, which implies that the oscillation has a limiting amplitude and frequency. This condition is referred to as 'limit cycling' or, more simply as 'cycling' and it may be acceptable if the amplitude and frequency are within tolerable proportions: such a situation is undesirable but it may prove to be a condition which has to be tolerated. In Sections 14.2 to 14.5, inclusive, the parameters most commonly used to define the four different domains of system performance will be detailed. In any practical case not all of the parameters would be used, because to do so would represent an undesirable duplication of information, which could be conflicting.

14.2 Steady-state accuracy

14.2.1 Factors affecting steady-state accuracy

There are many factors which can contribute to a steady-state error (that is, a steady discrepancy between the actual and the desired values of the controlled variable), and many or all of these will be applicable to a particular control system. Generally speaking, a global limit will be placed upon the tolerable steady-state error, and the apportionment of this figure to the various factors contributing to it is largely at the discretion of the control engineer.

The factors which can contribute to steady-state error are the following:

- reference and feedback transducer imperfections, such as curvature, deadband, quantisation, output signal offset
- computational inaccuracy and output signal offsets in the signal combination processes

- nonlinearities in the loop forward path components, such as deadband, hysteresis, backlash, static friction, coulomb friction
- signal offsets in the loop forward path components
- steady components of load on the plant
- the nature of the reference variable, taken in conjunction with the system Type Number (see Section 14.2.2).

This last factor can be of significant importance but, at the same time, the engineer should not lose sight of the effect of the other factors listed. All texts dealing with the analysis of the behaviour of linear systems will show that the steady-state error arising from a specified type of reference variable disturbance will be related to that disturbance in the manner shown in Table 14.1.

It will immediately be obvious that the number of forward path integrations and the value of the system loop gain are critical parameters when system steady-state error is being considered, and further discussion on these parameters will ensue.

The following are examples of steady component of load on a plant:

- a mass being raised or lowered by a pulley-cable combination, with the pulley drive being controlled by a position control or a speed control loop
- a steady-state load current being drawn from a turbine-alternator set, with loops for voltage control by manipulation of alternator excitation and frequency control by manipulation of turbine steam inflow rate
- a steady outflow rate from a storage vessel subjected to liquid level control by manipulation of liquid inflow rate.

Table 14.1 Relationship between steady-state error and the reference variable disturbance causing that error

Type of reference variation	Steady-state error, for the number of forward path integrations shown		
	0	1	2
step	$\dfrac{\text{(input step size)}}{(1 + \text{system loop gain})}$	0	0
ramp	no steady state: controlled variable cannot track reference variable	$\dfrac{\text{(ramp gradient)}}{\text{(system loop gain)}}$	0
parabola (square law)	no steady state: controlled variable cannot track reference variable	no steady state: controlled variable cannot track reference variable	$\dfrac{\left(\begin{array}{c}\text{parabola second}\\\text{derivative}\end{array}\right)}{\text{(system loop gain)}}$

Essentially, there are nine different techniques available for the minimisation of steady-state error: these are listed below and some will be elaborated upon in the sections which follow. In any particular system, some or most of these methods may be used.

Increasing the loop gain

An increase in system loop gain will reduce the steady-state error in most instances. However, exceptions to this rule will arise where:

- the source of the error is in either the reference transducer, the feedback transducer, or the signal combination process
- the source of the error is in any loop forward path component upstream of the point where the added gain is being incorporated, which suggests that it is preferable to increase the gain within the signal combination process(es).

In most practical cases, an increase in loop gain will result in a reduction in the degree of system stability; it will also tend to result in the transient saturation of forward path elements by reduced levels of reference variation.

Increasing the number of forward path integrations

The effects of inserting an intergration in the sytem forward path will be as follows:

- steady components of load disturbing the loop downstream of the added integration will no longer generate steady-state error
- signal offsets and nonlinearities occurring in forward path components downstream of the added integration will no longer generate steady-state error, but signal offsets and certain nonlinearities occurring upstream of the added integration will will do so, which suggests that integration is best added immediately after, or even incorporated into, the signal combination process(es)
- the addition of integration will always result in a reduction in the degree of system stability, unless suitable transfer function terms are incorporated for compensation purposes
- the addition of integration close to the signal combination network(s) will reduce any tendency towards transient saturation of the forward path elements.

Replacing components with others possessing greater precision

Obviously, there will be a trade-off between the resulting reduction in steady-state error and the capital cost of the components, so that performance considerations need to be weighed against commercial considerations.

Incorporating additional nonlinearities

In some instances, the effect of an unavoidable nonlinearity can be minimised or even nullified by cascading an intentionally introduced nonlinearity with the offending element, so that the combination has an overall static characteristic which, ideally, is linear. Curvature is a particular type of distortion which is amenable to this approach, and the cancellation of the orifice plate square law with a square-root extractor law is a commonplace application of the technique. However, nonlinearities such as saturation, deadband, hysteresis and granularity are not readily amenable to this approach. In some instances, the introduced nonlinearity may be placed in parallel with the offending element. Chapter 2 deals with the alternative techniques available for the construction of nonlinear networks.

Adjusting the calibration of the reference transducer

Where precise control is required, it is preferable that the calibration of the reference transducer should be left until the commissioning stage. In this manner, the calibration can be used to accommodate the effects of such imperfections as single-valued nonlinearities, signal offsets and steady components of load. Inevitably, the calibration will remain precise only if these imperfections are invariable.

Introducing signal offset into the controller

In some controllers, especially general purpose process controllers, provision is included for the adjustment of the offset of the output signal. This facility can be used to cancel out the effects of parasitic signal offsets and steady components of load, but the adjustment will remain precise only if these imperfections remain invariable.

Use of minor negative feedback around nonlinear elements

This application for minor negative feedback is discussed at length in Section 8.5 of *Basic Control System Technology*. The possible results of this technique can be summarised as follows:

● all nonlinearities except saturation can be disguised by negative feedback

● the speed of response of an element can be increased by negative feedback

● the incorporation of minor negative feedback will probably result in a loss in sensitivity, which may need to be restored by additional amplification inserted outside the minor loop.

The design of negative feedback loops around elements possessing significant nonlinearities may require the application of nonlinear system design techniques (see Section 14.6), in order that the minor loop shall have adequate stability margins.

Introducing feedforward control

Where feedback control cannot cope adequately with components of load on the plant, it may be feasible to augment the feedback action with feedforward action. (General-purpose feedforward controllers are discussed in Section 11.3.2 of *Basic Control System Technology*). The mechanisation of feedforward control will necessitate the installation of a load transducer, the controller hardware, and possibly an additional final control element. This technique should cancel the effect of steady components of load and will have the advantage that this cancellation will track long-term changes in these components. In addition, the feedforward control law may be designed to cancel some of the dynamic effects of transient components of load.

Reducing the effect of gear backlash by means of divided reset

If the amplitude of gear backlash in a position servosystem is unacceptably large and the replacement of the gearbox with an improved version proves to be impracticable, a technique which is sometimes used involves mounting displacement transducers on the input and output shafts of the gearbox and mixing, in appropriate proportions, the two position feedback signals generated: this is sometimes known as 'divided reset'.

14.2.2 System type number

The Type Number (or Class Number) of a system is defined as the number of integrations in the loop transfer function or, in the case of systems with minor feedback loops, the effective transfer function of the outer (principal) loop. Loop integrations occur in the forward path of a loop and they can arise as listed in Table 14.2 in relation to the type number shown.

Table 14.2 The contribution of outer loop integrations to system type number

System type number	Total number of forward path integrations in the outer loop	Forward path integrations inherent in the plant transfer function	Forward path integrations introduced by the controller transfer function
0	0	0	0
1	1	0	1
		1	0
2	2	0	2
		1	1
		2	0

Type 2 systems are relatively uncommon, because of the difficulties experienced in achieving reasonable stability margins for these systems (see Section 14.3); such a system is most likely to be created with one integration in the plant and the other in the controller.

An integration exists in a plant process if a steady applied value of the manipulated variable results in a steady-state ramping of the controlled variable (and the measured variable). An example is where the liquid inflow rate to a tank is manipulated in order to vary the liquid level in the tank: a steady inflow rate will result in a steady rate of increase in liquid level, assuming no outflow to occur at the same time and that the surface area remains constant. A second example is where the hydraulic oil flow into, and out from, a cylinder is manipulated in order to vary the displacement of the piston: a steady flow rate will result in a steady rate of change in piston position, provided that leakage past the piston is insignificant. A third example is an electric motor, when the control input voltage (or current) is manipulated in order to vary the angular displacement of the shaft: a steady control signal will ultimately result in a steady rate of change in angular displacement. It is extremely unusual for two integrations to be present in a plant transfer function.

An integration can be introduced into a controller by means of electronic or pneumatic hardware, or software program statements, as appropriate. In the case of the general-purpose process controller the incorporation of integral action (reset action) into the control law will result in the addition of an integral term to the controller transfer function, as shown in Table 14.3. There is no provision made in process controllers for more than a single integration.

In all cases, the s in the denominator of the rationalised transfer function signifies that the control law will increase by 1 the type number of the loop in which the controller is installed.

Minor negative feedback can be applied to eliminate an undesirable integration present in the main system forward path, by making the integration part of the forward path of the minor loop, for which the closed

Table 14.3 Process controller transfer functions involving integral action

Action	Transfer function
I	$\dfrac{1}{sK_pT_I}$
PI	$\dfrac{1}{K_p}\left(1 + \dfrac{1}{sT_I}\right) = \dfrac{1 + sT_I}{sK_pT_I}$
PID	$\dfrac{1}{K_p}\left(1 + \dfrac{1}{sT_I} + sT_D\right) = \dfrac{s^2T_IT_D + sT_I + 1}{sK_pT_I}$

loop transfer function will not include an integration term. This technique establishes the means whereby the type number of a system may be reduced. The stability of the complete system will be enhanced by this additional feedback loop.

14.3 Frequency domain behaviour

In those systems in which it is feasible to characterise the final control element, plant process and feedback transducer in terms of frequency response data, the design of the compensation can be undertaken in the frequency domain. The types of frequency domain parameter which are specified assume linear behaviour, so that they may need to be specified in terms of small (sinusoidal) signal analysis in the event that loop behaviour is significantly affected by nonlinearities. The frequency domain parameters which appear in specifications can be grouped into open loop and closed loop categories.

Parameters applying to open loop frequency response data are:

- gain crossover frequency ω_c
- phase crossover frequency
- gain margin, which typically is specified to lie between 8 and 20 dB
- phase margin, which typically is specified to lie between 30° and 60°
- low-frequency magnitude gradient, which is related directly to the system type number.

These various parameters are defined in Fig. 14.1, using alternative representations of the data.

Parameters applying to closed loop frequency response data are:

- peak gain magnitude M_m (if resonance occurs), measured relative to the magnitude at $\omega = 0$ (a typical specification is 1.3 to 1.5 numeric $= 2.3$ dB to 3.5 dB)
- resonant frequency ω_r, if resonance occurs
- bandwidth ω_b
- high-frequency roll-off gradient, which is related to noise rejection capability.

These various parameters are defined in Fig. 14.2.

Open loop and closed loop frequency domain parameters are interdependent. Thus, ω_c, ω_r, and ω_b will all have the same order of magnitude and usually occur in the sequence $\omega_c < \omega_r < \omega_b$. Similarly, relatively low stability margins will yield a high value for M_m.

14.4 Time domain behaviour

Time domain parameters, when they are used in specifications, normally refer to step response behaviour. The step in question may be applied to the

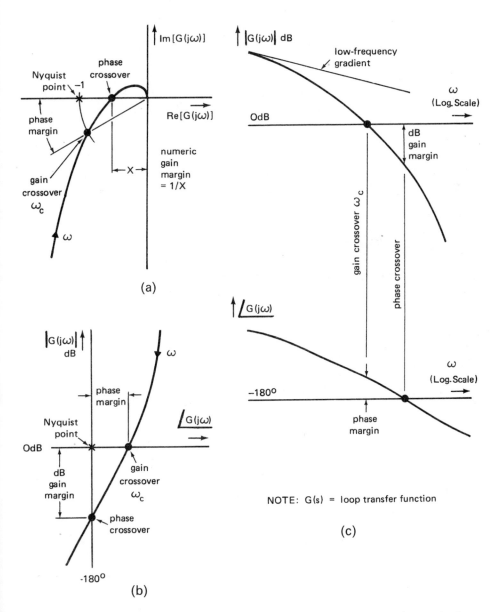

Fig. 14.1 Alternative frequency response representations to demonstrate the definitions of commonly used frequency domain performance parameters: (a) open loop polar diagram; (b) Nichols chart; (c) open loop Bode diagrams

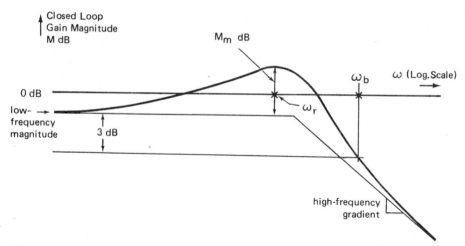

Fig. 14.2 Typical closed loop Bode magnitude plot, demonstrating the definitions of commonly used loop frequency domain performance parameters

reference variable or it may be applied to the load on the plant. Most of these parameters assume linear behaviour, but some (notably rise time, settling time, peak overshoot and decrement) may also be applied to nonlinear behaviour. If behaviour is affected significantly by nonlinearities, it may be preferable to apply the parameters in terms of responses to small steps, for which the transient is likely to approach that for a linear model.

In plant processes involving the physical transportation of material (solid, liquid or gas), the phenomenon of pure time delay (dead time, transport delay) is likely to occur between the point to which the manipulation is applied to the plant and the point at which the effect of the manipulation is instrumented. A notable exception to this occurs in liquid flow processes for which, provided that compressibility effects are minimal, the flow sensor should respond immediatley the flow is manipulated at a different point in the line, Plant dead time can only be reduced by redesign of the plant, because it is not 'disguised' by negative feedback: it will emerge, unaffected in value, in the system closed loop response, except when Smith Predictor (Section 10.9) is introduced into the controller.

The nature of a particular step response largely determines which parameters are most appropriate for specifying that response. Many practical responses conform approximately to the simple linear first order and second order responses discussed in Sections 14.4.3 and 14.4.4 respectively, and when this occurs, the responses are said to exhibit first order or second order 'dominance'. The parameters which are most relevant to these two types of dominance are defined graphically in Fig. 14.3: possible dead time has not been shown in these diagrams.

When an actual response relates to a high system order and does not exhibit low order dominance, the definitions of some of the parameters shown in Fig. 14.3 can easily become obscured. Settling time is still

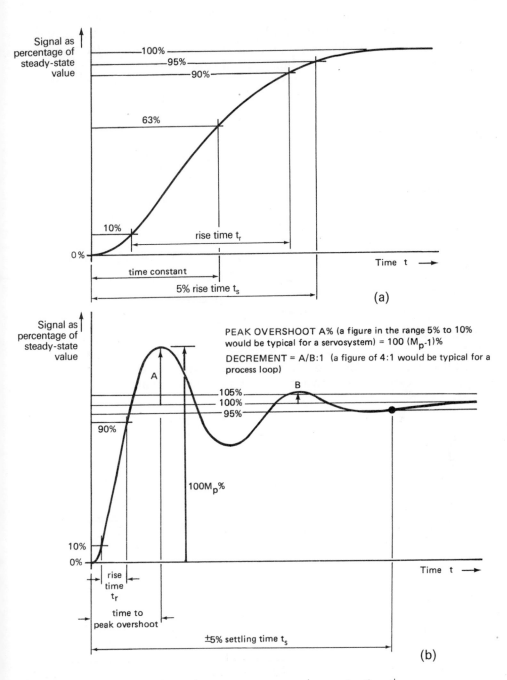

Fig. 14.3   Examples of simple step responses, demonstrating the
definitions of commonly used time domain performance parameters:
(a) dominant first order type of response; (b) dominant second order type
of response

unambiguous, and therefore relevant, but the remaining parameters may need to be applied with caution. Figure 14.4 shows an example of a high order response for which the definitions of peak overshoot, time to peak overshoot, and decrement are not at all obvious. In a case such as this, a limit upon the oscillatory behaviour of the response is best imposed in terms of an upper bound on an appropriate 'performance index'. The index most commonly used for step responses is the 'I T A E Criterion', which is defined as $\lim\limits_{T \to \infty} \int_0^T t|e(t)|\,dt$ and which will penalise both long lived overshoots and long lived undershoots of the steady-state value; $e(t)$ is system error.

Where second order dominance is particularly strong, it may well be appropriate to use those parameters which apply specifically to second order responses:

- damping factor ζ (a figure in the vicinity of 0.707 would be typical)
- damped natural frequency $\omega_t = 2\pi/(\text{period of transient oscillations})$ rad/s
- decay ratio $\alpha = \zeta\omega_n$, where $\omega_n = \omega_t/\sqrt{1 - \zeta^2}$.

Many control systems exhibit a nonlinear phenomenon known as 'rate limiting'. This is manifested as an upper limit on the rate at which the output signal can 'slew' and should be demonstrated clearly by a large-signal step response as shown in Fig. 14.5. Where relevant, this parameter

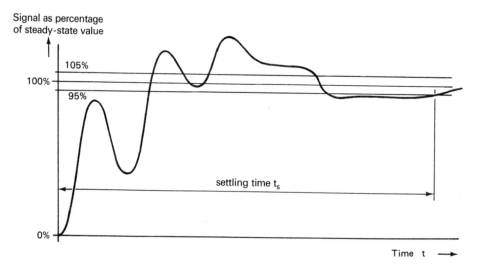

Fig. 14.4   Example of a (fourth order) step response, for the transfer function $\left[\left(\dfrac{s^2}{\omega_{n_1}^2} + \dfrac{s^2\zeta}{\omega_{n_1}} + 1\right)\left(\dfrac{s^2}{\omega_{n_2}^2} + \dfrac{s^2\zeta_2}{\omega_{n_2}} + 1\right)\right]^{-1}$

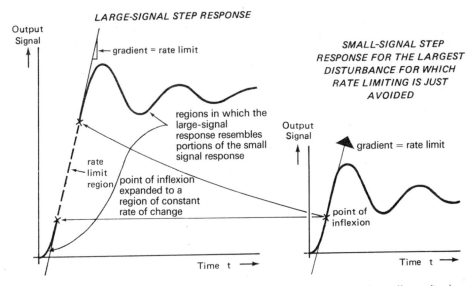

Fig. 14.5   Comparison of transient responses to large and small applied step disturbances for a control system exhibiting rate limiting

will have a marked influence upon large-signal settling time and, in addition, upon large-signal frequency response.

14.5   Noise performance

There are many potential sources of parasitic noise in control systems, amongst which feature the following:

- mains ripple due to imperfect smoothing of electric power supplies
- pressure fluctuations due to imperfect smoothing of hydraulic power supplies
- cutter reaction on machine tool drives
- wind buffeting of antennae and aircraft
- wind and wave buffeting of naval vessels
- air and gas bubbles entering liquid flow lines
- sediment in liquid flow lines
- rumble in continuous belt drives, due to fluctuations in the mass being transported
- imperfect smoothing in demodulation processes
- quantisation and sampling in digital-analog conversion processes
- crosstalk and self-induced high-frequency oscillations.

Where the noise enters the system with a significant amplitude, the system will need to be designed so as to minimise the effect of the noise upon the controlled variable. Some types of noise are random in nature, so that their effect can be specified only in statistical terms; other types of noise are periodic, so that it is practical to specify their effect in frequency domain terms.

The frequency response of that part of the system between the noise source and the controlled variable will have a considerable bearing upon the noise performance and is best specified by parameters such as $-3\,\text{dB}$ bandwidth, resonant gain magnitude and roll-off gradient. If the frequency content of the noise input signal is known, the frequency content of the noise component of the controlled variable can be determined using a knowledge of this frequency response: this will apply to both periodic and nonperiodic noise signals, which exhibit line spectra and continuous spectra, respectively.

A commonly used performance index which can be applied in order to place a statistical amplitude limit upon the effect of noise is the 'Root-Mean-Square Error Criterion', which is defined as

$$\left[\lim_{T \to \infty} \frac{1}{T} \int_0^T e^2(t)\,dt \right]^{1/2}$$

where e(t) is system error: this will penalise both large and long lived noise signal components.

14.6   Analytical design techniques

Clearly, it is not the function of this volume to present the alternative analytical design techniques which are available for the design of control systems: these are covered more than adequately in the many excellent texts devoted to the analytical area. However, because the final stage of designing a control system involves either the design or the selection of a controller, including its control algorithm(s), it is considered relevant to provide a critical summary of the alternative methods which are in common industrial use.

14.6.1   Time domain analysis

Classical time domain analysis is suitable for designing systems which will exhibit first and second order behaviour. Since few practical systems will fall into these categories, it follows that this approach is largely unsuitable as a design method although, of course, the overall time domain performance of the designed system will obviously be of considerable significance.

State variable methods generate time domain performance but their complexity renders them appropriate mainly to high order systems being operated in a linear mode. Solutions are best generated by digital computer and, for low order linear systems, the labour involved will consume more time than classical analysis requires. State variable methods have rarely been used in practice to compute the large-signal behaviour of those high order systems where nonlinearities become significant. For the types of system for which they are most appropriate, state variable methods can be used to optimise simultaneously the values of a set of system parameters such as gain constants and time constants.

Synthesis and analysis of the large-signal behaviour of dominant first and second order systems containing significant nonlinearities can be undertaken using Phase Plane methods, which involve plots, to cartesian co-ordinates, of dx/dt versus $x(t)$, where $x(t)$ is a system variable of interest which typically is equated to system error $e(t)$. Synthesis of phase portraits is not practicable for systems of order greater than two, but the format may sometimes be used to display experimentally-obtained response data as an aid to the interpretation of time domain behaviour.

14.6.2 Frequency domain analysis

The frequency domain is a very powerful medium for the design of systems, although the collection of experimental data can be very time consuming. Experimentally-derived frequency response data can be incorporated directly into the plots, without necessitating the derivation of the associated transfer functions. The effect of dead time is incorporated readily. In contrast, the ultimate derivation of time domain performance data can be very involved when the system order is high.

Frequency domain methods are most appropriate for systems behaving in a linear manner, which may imply small-signal behaviour. However, they can be extended to embrace large-signal nonlinear behaviour if Describing Functions are used although, without resorting to considerable complexity, this technique usually can yield data only on closed loop stability.

The merits of the alternative formats for displaying frequency response data are considered to be as follows:

Bode diagram format
- data values are well distributed
- asymptotes can be used on magnitude plots, for both analysis and synthesis
- is compatible with Nichols chart format
- does not provide open loop–closed loop transformation of data
- multiplication of cascaded transfer functions transforms to the addition of data values
- cannot incorporate describing function data usefully
- can be used for the design of sampled data systems, if a new (dimensionless) frequency variable is defined.

Polar (Nyquist) diagram format
- data values are poorly distributed

- provides open loop–closed loop transformation for linear operation
- simple compensator transfer functions are represented by simple locus shapes – semicircles, circles etc.
- multiplication of cascaded transfer functions transforms to vector multiplication, which is complicated to apply
- can easily incorporate describing function data.

Inverse polar (inverse Nyquist) diagram format
- data values are poorly distributed
- simple compensator transfer functions are represented by simple locus shapes
- when used for feedback compensation, the manipulation of transfer functions transforms to vector addition, which is relatively straightforward to apply
- provides open loop–closed loop transformation for linear operation.

Nichols chart format
- data values are well distributed
- is compatible with Bode diagram format
- provides open loop–closed loop transformation for linear operation
- can easily incorporate describing function data.

Clearly, it can be seen that the alternative formats all have merits and demerits, which explains why they all have a place in design procedures. Frequency response methods are not amenable to rigorous optimisation of any particular parameter: each new trial value will generate a new set of frequency response curves.

14.6.3   s-Domain analysis

The ability to design in the s-domain assumes that the final control element–plant process–feedback transducer combination can be represented accurately by a transfer function, which presupposes that a reliable characterisation procedure has been conducted: see Chapter 13. Analysis is suitable only for linear behaviour, so that it can be used only for small-signal behaviour when significant nonlinearities are present. Moreover, the method cannot incorporate the effect of dead time.

The principal merits of the root locus diagram are:

- it provides a means to optimise rigorously a particular parameter, which need not necessarily be the system loop gain
- it gives a good indication of dominance in the closed loop behaviour
- it readily indicates, in detail, closed loop time domain behaviour and, to a lesser extent, frequency domain behaviour and, moreover, shows how

these behaviours change as the parameter being optimised is changed –
this is particularly useful when designing a control system to be 'robust',
that is, for its closed loop behaviour to be relatively insensitive to
unavoidable changes in a critical parameter occurring within specified
limits

- the method is extended readily to the z-domain for the design of
sampled data systems.

The Routh-Hurwitz stability criterion is a simple test which can be applied
very rapidly in order to predict closed loop stability or instability, but it will
not yield information on the degree of stability such as would be indicated,
for example, by stability margins in the frequency domain.

14.6.4 Transformation of data between time and frequency domains

Closed loop performances in the time and frequency domains are interdependent. Where the behaviour can be regarded as being linear, one
performance is predictable from the other, but the level of difficulty
encountered in undertaking this transformation of data will depend upon
the order of the system.

Where closed loop performance can be regarded as being second order
dominant, the following *approximate* relationships can be used for ball-
park estimates:

$$t_s \cong \pi/\omega_c \quad M_p \cong 0.85 M_m \quad \zeta \cong 1/(2M_m), \text{ where } M_m \text{ is numeric}$$

$$t_s \cong 3\sqrt{1 - \zeta^2}/\zeta\omega_t \quad \omega_t \cong \omega_r \cong 0.75\omega_c \quad t_r\omega_t \cong 1.3$$

Note that all of these parameters have been defined in Sections 14.3 and
14.4.

Certain researchers have compiled performance charts* which enable
closed loop performance parameters to be read off as output data, once
specified open loop performance parameters are applied as input data.
These charts have tended to concentrate either upon particular classes of
servomechanism or upon particular classes of process loop, with the latter
incorporating the effect of dead time: these can be useful for the classes of
system covered, although their application may involve a considerable
amount of graphical interpolation.

14.6.5 Computer simulation

Computer simulation has been used extensively in the aerospace industry,
in nuclear and thermal power generation, and in certain process industries
for the optimisation of system designs. In addition to requiring the

---

*Chestnut, H. and Mayer, F. W.: 'Servomechanisms and Regulating System Design'.
Vol. 1, John Wiley, N.Y., 2/e 1959, pp. 515 to 532.

Wills, D. M. : 'Control Engineering', April 1962, pp. 104 to 108.
       : 'Control Engineering:', August 1962, pp. 93 to 95.

necessary computing equipment, technical expertise and availability of time and finance, computer simulation cannot be successful unless the final control element–plant process–feedback transducer combination can be represented accurately in mathematical terms, using the characterisation procedures of Chapter 13. Because of all of the factors involved, it is hardly surprising that industries with limited resources have not invested heavily in computer simulation.

An interesting and recent phenomenon associated with microprocessor based process controllers is the provision of the capability to allocate temporarily some of the processor hardware to simulation functions during the system commissioning stage. Thus, processor hardware may be 'borrowed' from another loop and used to simulate the plant process in the loop presently being commissioned, so that the controller may first be tuned with the model before ultimately being connected to the plant. In this manner, controller tuning for optimum system performance may be undertaken off-line, in contrast to the more usual on-line tuning procedure. The technique assumes that the plant process can first be characterised and then modelled adequately by the controller algorithms available. Some manufacturers make available simulation software to enable personal computers to be used for modelling plant during off-line tuning procedures.

14.7   Commissioning procedures

The procedures to be outlined here assume that the final control element and feedback transducer have been coupled to the plant and tested for correct functioning. The nature of the next stages of commissioning depends upon whether the controller has been custom designed or is general purpose. In the former case, the nature of the control law (that is, the compensation transfer function) may be subjected to last minute modification; in the latter case, the control law will have been chosen from a small number of alternatives and the parameter values will be subjected to fine tuning as the commissioning proceeds. The first descriptions that follow assume that a single feedback loop is being commissioned: obviously, the procedures will require further development when feedforward action and multiple, interacting, feedback loops are being commissioned.

14.7.1   Commissioning custom designed controllers

When commissioning custom designed controllers, conduct the following operations more or less in the sequence indicated:

- connect the controller to the final control element to complete the forward path, whilst leaving the feedback disconnected
- if not already present, temporarily install means to attenuate the error signal and power up the hardware with minimum signal applied to the final control element

- check that the signal from the feedback transducer has a polarity which is the reverse of that of the signal from the reference transducer, normally indicating that the feedback will be in the negative sense when it is connected – if the sense is incorrect, take suitable steps to cause a reversal (see later)
- if open loop characterisation is to be undertaken, identify the final control element–plant process–feedback transducer combination, using one or more of the procedures described in Chapter 13
- if characterisation is to be undertaken with the loop closed, reduce the reference signal to zero and increase the error attenuation to maximum, close the feedback connection, gently reduce the error attentuation until the system is reasonably responsive to reference disturbances (without approaching instability), and undertake the characterisation by using a suitable procedure (see Chapter 13)
- design and install the compensation hardware – typically, this would involve introducing a compensation transfer function in order to increase the system phase crossover frequency or reduce the gain crossover frequency, or both, together with means to achieve the desired figure for the system loop gain; occasionally, the compensation may be introduced either to the existing feedback path or as part of an additional feedback path ('feedback compensation'), as an alternative to the more usual introduction into the forward path ('forward path compensation' or 'cascade compensation')
- with the loop closed and a high level of error attenuation again present, apply small reference disturbances and monitor the system response
- gently and progressively reduce the error attenuation progressively and check that the system response converges towards the predicted behaviour as the loop gain approaches the design figure
- fine tune the compensator parameters as necessary
- remove the error attenuator if it was to have been temporarily installed
- calibrate the reference transducer, in terms of corresponding steady-state values of controller variable, taking into account relevant loading conditions on the plant
- undertake comprehensive performance tests to confirm compliance with the system performance specification.

If it is initially found that the principal feedback would be in the positive sense, reversal can be effected using one of the following techniques, noting that not all will be practicable in any particular situation:

- installing a sign inverting element in the forward path, within the controller
- interchanging the signal connections from the controller to the final control element
- reversing the action of the final control element

- reversing the sense of the physical coupling of the final control element to the plant
- reversing the sense of the physical coupling of the feedback transducer to the plant
- interchanging the power supply connections to the feedback transducer
- interchanging the signal connections from the feedback transducer to the controller
- installing a sign inverting element in the feedback path within the controller.

Note that, whenever a sign inversion is effected in the principal feedback path, the sense of the calibration of the reference transducer also becomes reversed and this may require correction.

14.7.2   Commissioning of general purpose process controllers in single feedback loops

Two types of empirical approach are commonly used for predicting optimum settings for general purpose process controllers. They both involve time domain testing of the plant: in the first method, the final control element–plant process–feedback transducer combination is subjected to a step disturbance with the feedback signal disconnected; in the second method, a non-optimum feedback loop is completed and subjected to a step disturbance of the set point.

The open loop response method
The system hardware is connected to comply with the block diagram of Fig.14.6. The loop is open and the final control element is subjected to a step disturbance of magnitude X units. The response of the measured

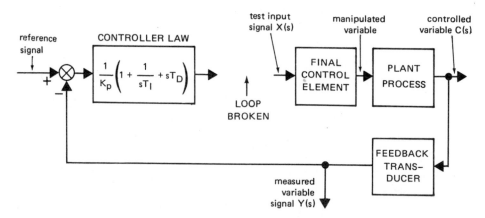

Fig. 14.6   Block diagram of a process control loop suitably broken to facilitate open loop characterisation

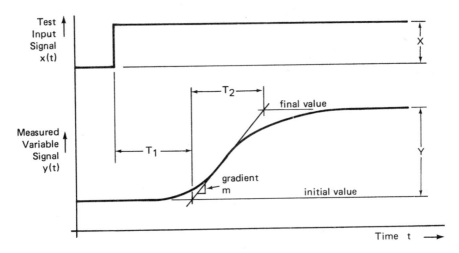

Fig. 14.7 Step input disturbance and representative process reaction curve for the open loop system of Fig. 14.6, showing the construction for quantifying the parameters of the approximate mathematical model

variable y(t) is recorded and analysed on the basis that it will assume the general form shown in Fig. 14.7. The slope m corresponds to the maximum gradient of this 'process reaction curve', T_1 is an equivalent dead time, and T_2 is an equivalent time constant. The response is being modelled mathematically by the transfer function of the form

$$\frac{Y}{X}(s) = \frac{Ke^{-sT_1}}{(1 + sT_2)},$$

corresponding to a gain constant K, a dead time T_1 and a simple lag of time constant T_2. In practice, the actual transfer function is usually somewhat more complicated than this. However, the assumption of the above form is often an acceptable approximation and provides a means for generating reasonable initial settings for the controller. These settings may ultimately be readjusted to obtain a more desirable closed loop response. The values of Y, T_1, T_2 and X are used to determine the recommended values for the controller settings for K_p, T_1, and T_D in accordance with Table 14.4. These settings are intended to yield a 4:1 decrement and minimum settling time for the closed loop step response.

Alternative coefficient values have been proposed by various researchers for the entries in Table 14.4 but, since these formulae are to be used only for the initial controller settings, refinement of the coefficients must be of limited benefit.

In practice, it is necessary to measure the response curve for both positive and negative going input steps and for various magnitudes of input step X in order to determine the most representative model of the above form. If the spread of values obtained for the parameters Y/X, T_1, and T_2 is within

Table 14.4   Ziegler–Nichols recommendations on the basis of process reaction curve data

Type of control law	K_p	T_I	T_D
P	mT_1/X		
PI	$mT_1/0.9X$	$3.33T_1$	
PID	$mT_1/1.2X$	$2T_1$	$T_1/2$

Note Proportional band $= 100K_p$ percent and $m = Y/T_2$

(say) 10 to 15 percent, the expressions for recommended controller settings should yield reasonable values: these settings may ultimately be readjusted to obtain a more desirable response once the loop has been closed.

Reference to Table 14.4, and also to Table 14.5 below shows that the gain $1/K_p$ introduced into the loop has to be reduced when integral action is incorporated, due to the fact that the introduction of an integration term tends to reduce system stability. Conversely, when derivative action is introduced, the gain may be increased, due to the stabilising effect of the derivative term.

Note that the open loop response method will not be suitable for use with plant processes which incorporate an integral term, because then the open loop response will be asymptotic to a ramp.

Note also that the open loop response method will not be suitable for tuning the outer loop of a system in which the (closed) inner loop has itself previously been tuned using this method. This is because the process reaction curve for the outer loop will manifest a (damped) oscillatory component.

In addition, note that the loop should not be closed until it has been ascertained that the sense of the feedback is negative, as discussed in Section 14.7.1

Table 14.5   Ziegler–Nichols recommendations on the basis of limit cycling data

Type of control law	K_p	T_I	T_D
P	$K_{p_m}/0.5$		
PI	$K_{p_m}/0.45$	$T_u/1.2$	
PID	$K_{p_m}/0.6$	$T_u/2$	$T_u/8$

The closed loop response method

This method is based upon the closed loop step response of a non-optimised feedback loop. Having checked that the sense of the feedback will be negative, the system of Fig. 14.6 is restored to closed loop operation with the controller set initially for proportional action only ($T_1 = \infty$, $T_D = 0$) with a very high proportional band K_p. Using the set point control, a small step disturbance of reference input r(t) is applied to the system and the closed loop response is obtained by recording the measured variable y(t). The value of K_p is then progressively reduced until the input disturbance just triggers a continuous limit cycling of the measured variable. This value (K_{pm}) of proportional band and the value T_u of the period of the oscillation are measured. The recommended initial trial values for the controller settings are determined in accordance with Table 14.5, and these again are intended to yield a 4:1 decrement and minimum settling time for the closed loop step response.

The main disadvantage with the Ziegler-Nichols closed loop method is that it can be undesirable to introduce limit cycling into a plant, even for a short duration: cycling may cause practical problems, including danger to plant and personnel. Additionally, some loops may never limit cycle using any of the values of proportional band available.

An alternative approach is initially to follow the above procedure but to stop reducing K_p when a 4:1 (or thereabouts) decrement is obtained for the currently non-optimised closed loop step response. If the decrement actually achieved for this initial response is given by A:1 and the period of the decaying oscillation is P seconds, a reasonable set of new trial values for the controller parameters will be given by

$$K_{p_{new}} = K_{p_{old}}/(0.5 + 2.27\Delta) \text{ percent}$$

$$T_{I_{new}} = P/(1.2\sqrt{1 + \Delta^2}) \quad \text{seconds for P I control}$$

or $\quad\quad\quad = P/(2\sqrt{1 + \Delta^2}) \quad\quad \text{seconds for P I D control}$

and $\quad T_{D_{new}} = P/(8\sqrt{1 + \Delta^2}) \quad\quad \text{seconds for P I D control}$

where $\Delta = \dfrac{1}{2\pi}\ln(A)$.

The advantages of this type of procedure are that the plant is not subjected to cycling. Additionally, the formulae may be re-applied iteratively to implement an optimisation routine which can be undertaken even when the loop is in service: the new values for K_p, T_1 and T_D are set in and the test is repeated, with the current value of $K_{p_{new}}$ being substituted for $K_{p_{old}}$ in the above formula. Normally, the new values for K_p, T_1 and T_D should converge to steady levels as the number of iterations is increased. Figure 14.8 is a flow chart showing the iterative process which, with computer control, could be made the basis for an automatic tuning procedure.

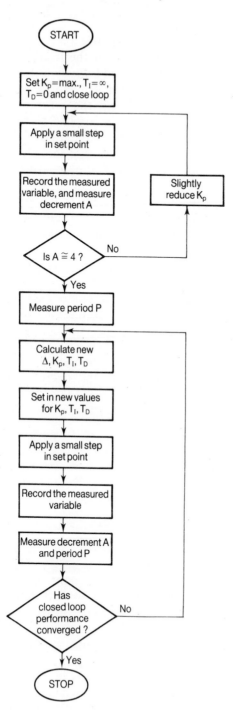

Fig. 14.8   Flow chart showing an interative procedure for optimally tuning a process loop on-line

14.7.3 Commissioning multiple feedback loop systems

A typical procedure will be described for commissioning a system involving two process loops, one enclosing the other, as shown in Fig 14.9. When fully commissioned, process controller PC1 will perform the role of a Feedback Controller, whilst process controller PC2 will take the role of a Cascade Controller. The loops are to be tuned and closed in an orderly sequence, so that at no time is control of the plant lost. It will be assumed that the inner (minor) loop is to be tuned using the Ziegler-Nichols Open Loop Method, whilst the outer (dominant) loop is to be tuned using the Ziegler-Nichols Closed Loop Method.

The following steps represent a typical procedure for orderly commissioning of the plant.

STEP 1.   Set both controllers to Manual action, with K_p at maximum, T_I at infinity and T_D at zero.

STEP 2.   Power up the controllers and the final control element, and start up the plant.

STEP 3.   Using the manual control for adjusting the output O_2 of controller PC2, set O_2 to a quiescent level so that the plant process Stage 1 is running at about its mid-level of operation.

STEP 4.   Manually apply a small step in output O_2 and record the process reaction curve generated by the signal PV_2. Using measurements taken from the curve, calculate the 'optimum' values for the K_p, T_I and T_D settings for controller PC2 and set them in.

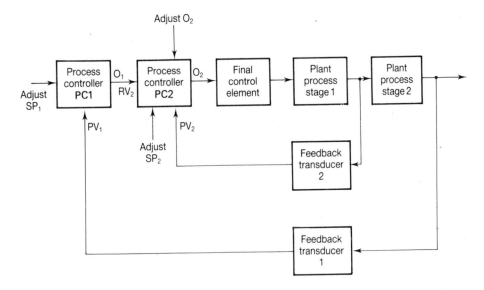

Fig. 14.9   Block diagram for a process control system with two feedback loops

STEP 5.   Controller PC2 is to be switched from Manual to Auto action, with bumpless transfer: that is, there will be no jump in the value of output O_2 when switching occurs. With a modern controller, this facility will be inherent in the design whereby, in the Manual mode, set point signal SP_2 will automatically track process variable signal PV_2, so that the error (deviation) signal is zero at the instant of changeover. In addition, the output from the integral action term will store, as a constant of integration, the value of output signal O_2 which has been set in manually. Note that the O_1–RV_2 connection is disabled automatically at this stage.

STEP 6.   Controller PC2 is switched manually to the Auto mode, so that the inner loop now functions as an isolated single closed loop, with Controller PC2 temporarily acting as a Feedback Controller. Using manual adjustments of set point SP_2, the inner loop may be fine-tuned for optimum closed loop performance and the plant process Stage 1 set to a new quiescent level of operation if desired.

STEP 7.   With modern hardware, output O_1 will automatically track whatever level of set point SP_2 is being set in manually, so that remote variable RV_2 is automatically identical in value to set point SP_2 at the instant at which Controller PC2 is (next) switched from Auto to Cascade mode, so that there is no discontinuity in output signal O_2.

STEP 8.   Controller PC1 now has the potential to influence the behaviour of the system. As in Step 5, this controller will exhibit bumpless transfer when it is switched manually from Manual to Auto mode.

STEP 9.   The outer loop can now be tuned, with the K_p setting of Controller PC1 being reduced until the whole system cycles about the quiescent operating condition for plant process Stage 2 determined by the (manual) setting of set point SP_1.

STEP 10.   Process variable PV_1 is recorded and measurements made from the record enable the optimum settings to be calculated for K_p, T_I and T_D for Controller PC1. These are set into the controller.

STEP 11.   The outer loop can now be fine-tuned, using small adjustments of the Controller PC1 settings and by applying small steps manually to the set point SP_1. The system has now been completely commissioned and is running in its fully operational state.

14.7.4   Other considerations with general purpose process controllers

For the dual feedback loop system just described, it will be seen that the procedure involves setting up the innermost loop first, and then working progressively outwards to the outermost loop: that is, to the loop having greatest authority when the system is fully operational. This general procedure will ensure that, should the outermost loop(s) in any multiple feedback loop system need to be disabled in an abnormal operating situation, the remaining inner loop(s) will continue to function satisfactorily, albeit with a lesser authority.

The converse situation may not necessarily apply: should the innermost loop(s) need to be disabled, it is quite probable that the outermost loop(s) would not be able to continue to control satisfactorily on their own.

With computer control, potential exists for automating start-up and shut-down procedures for plant. Assuming that the tuning procedures do not need to be repeated once commissioning has been completed, the other procedures described in Steps 1 to 11 of Section 14.7.3 can be automated using computer control, with programmed manipulation of offset and set point adjustments and programmed changeover of controller modes. This will yield an automatic start-up procedure, which can then be reversed in order to achieve an orderly shut-down procedure. The use of auto-tuning controllers will add the ability to automatically re-tune controller settings in order to track and counteract the effects of changing operating conditions.

Feedforward controllers are normally tuned after first commissioning the feedback loops into which they are injecting feedforward data. Figure 14.10 illustrates the general situation. Assuming the feedback loop to be

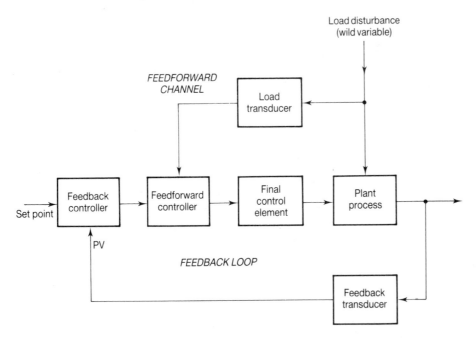

Fig. 14.10   General diagram of a general process feedback loop with feedforward action added

operating satisfactorily (but not optimally) with the set point set at a representative level, the Feedforward Controller may be tuned by monitoring either the error (deviation) signal generated by the Feedback Controller or the process variable signal generated by the feedback transducer.

Static feedforward may be adjusted by changing the gain setting of the Feedforward Controller so as to minimise the effect of steady levels of applied load upon the steady-state values of the error variable (which ideally should be zero). Dynamic feedforward may be tuned by characterising recordings of process variable responses following the application of steps in load disturbance; these responses need to be modelled approximately by a lag-lead transfer function, which one then attempts to cancel by tuning the lead-lag dynamic terms of the Feedforward Controller control law.

Ratio Controllers may be commissioned using the same techniques as for Feedback Controllers. In addition, allowance will need to be made for any differences in sensitivity of measurement of the two (that is, controlled and wild) variables, the ratio of which is to be controlled by the loop.

Bibliography

Signal and Data Processing

Analog Devices Inc. (1988). *Data conversion products databook*. Analog Devices Incorporated, Norwood, Mass.

Arbel, A. F. (1980). *Analog signal processing and instrumentation*. Cambridge University Press, Cambridge, Eng.

Baher, H. (1990). *Analog and digital signal processing*. Wiley, Chichester (Eng), New York.

Bateman, A. and Yates, W. (1989). *Digital signal processing design*. Pitman, London.

Berlin, H. M. (1978). *Design of phase-locked circuits with experiments. (The phase-locked loop bugbook)*. H. W. Sams, Indianapolis, Ind.

Byrnes, C. I., Martin, C. F. and Saeks, R. E. (1988). *Linear circuits, systems and signal processing*. Elsevier, Amsterdam.

Cadzow, J. A. (1987). *Foundations of digital signal processing and data analysis*. Collier-Macmillan, London, New York.

Carter, R. C. and Bruntjen, S. (1983). *Data conversion*. Knowledge Industry Publications, White Plains, N.Y.

Clay, R. (1976). *Nonlinear networks and systems*. Bks Demand UMI, Ann Arbor, Mich.

Clayton, G. B. (1982). *Data converters*. Macmillan, London.

Cluley, J. C. (1983). *Interfacing to microprocessors*. McGraw-Hill, New York.

Connor, F. R. (1986). *Networks*. 2nd ed. Edward Arnold, London.

Coughlin, R. F. and Driscoll, F. F. (1987). *Operational amplifiers and linear integrated circuits*. 3rd ed. Prentice-Hall, Englewood Cliffs, N.J.

Daryanany, G. (1976). *Principles of active network synthesis and design*. Wiley, New York.

Dolezal, V. (1977). *Nonlinear networks*. Elsevier Scientific, Amsterdam.

Dooley, D. J. (1980). *Data conversion integrated circuits*. Wiley, New York.

Enden, A. W. M. van den and Verhoeckx, N. A. M. (1989). *Discrete time signal processing – an introduction*. Prentice-Hall, Englewood Cliffs, N.J.

Godfrey, K. and Jones, P. (1986). *Signal processing for control*. Springer-Verlag, Berlin, New York.

Graeme, J. G. (1987). *Applications of opertional amplifiers: third generation techniques*. McGraw-Hill, New York.

Graeme, J. G. (1977). *Designing with operational amplifiers: applications, alternatives.* McGraw-Hill, New York.

Hnatek, E. R. (1988). *A user's handbook of D/A and A/D converters.* Wiley, New York.

Huelsman, L. P. (1976). *Active RC filters: theory and aplication.* Academic Press, London.

Huelsman, L. P. (1980). *Introduction to the theory and design of active filters.* McGraw-Hill, New York.

Jackson, L. B. (1989). *Digital filters and signal processing.* 2nd ed. Kluwer Academic Publishers, Boston, Mass.

Lindsey, W. C. and Simon, M. K. (1978). *Phase-locked loops and their application.* IEEE Press, Piscataway, N.J.

Lynn, P. A. (1989). *An introduction to the analysis and processing of signals.* Macmillan Education, Basingstoke, Eng.

Nagai, N. ed. (1990). *Linear circuits, systems and signal processing.* M. Dekker, New York.

Oppenheim, A. V. and Schafer, R. W. (1989). *Discrete-time signal processing.* Prentice-Hall, Englewood Cliffs, N.J.

Rutklowski, G. B. (1984). *Integrated-circuit operational amplifiers.* Prentice-Hall. Englewood Cliffs, N.J.

Yuen, C. K., Beauchamp, K. G. and Fraser, D. (1989). *Microprocessor systems in signal processing.* 2nd ed. Academic Press, London.

Microprocessors and Microcontrollers

Inmos Limited (1986). *Transputer reference manual.* Inmos Limited, Bristol, Eng.

Intel Corporation (1987). *Embedded controller application handbooks.* Intel Corporation, Santa Clara, Calif.

Intel Corporation (1989). *Microprocessor and peripherals handbooks.* Intel Corporation, Santa Clara, Calif.

Motorola (1988). *Microprocessor, microcontroller and peripheral databook.* Motorola, Oak Hill, Texas.

Motorola (1989). *M68000 family reference databook.* Motorola, Phoenix, Arizona.

Motorola (1989). *MC68000/08/HC000 8/16/32-bit microprocessor user's manual.* Motorola, Phoenix, Arizona.

National Semiconductor Corporation (1989). *Microprocessor databook.* National Semiconductor Corporation, Santa Clara, Calif.

Texas Instruments Incorporated (1989). *TMS320 third generation user's guide.* Texas Instruments Incorporated, Houston, Texas.

Backplane Buses

Del Corso, D., Kirrman, H. and Nicoud, J. D. (1986). *Microcomputer buses and links.* Academic Press, London.

Data Communications

Inglis, A. F. (1988). *Electronic communication handbook.* McGraw-Hill, New York.

Roden, M. S. (1988). *Digital communication systems design.* Prentice-Hall, Englewood Cliffs, N.J.

Computer Operating Systems

Barron, D. W. (1984). *Computer operating systems for micros, minis and main-frames.* 2nd ed. Chapman and Hall, London, New York.
Blackburn L. and Taylor, M. (1985). *Introduction to operating systems.* Pitman, London.
Comer, D. (1987), *Operating system design.* Prentice-Hall, Englewood Cliffs, N.J.
Deteil, H. M. (1984). *An introduction to operating systems.* Addison-Wesley, Reading, Mass.
Kaisler, S. H. (1983). *The design of operating systems for small computer systems.* Wiley, New York.
Keller, L. S. (1988). *Operating systems communicating with and controlling the computer.* Prentice-Hall, New York.
Kurzban, S. A., Heines, T. S. and Sayers, A. P. *Operating systems principles.* 2nd ed. Van Nostrand Reinhold, New York.
Lister, A. M. and Eager, R. D. (1988). *Fundamentals of operating systems.* 4th ed. Macmillan Education, Basingstoke, Eng.
Lubomir, B. (1988). *The logical design of operating systems.* 2nd ed. Prentice-Hall, Englewood Cliffs, N.J.
Maekawa, M. M. (1987). *Operating systems advanced concepts.* Benjamin/Cummins Publishing Co., Menlo Park, Calif.
Milenovic, M. (1987). *Operating system concepts and design.* McGraw-Hill, New York.
Northcutt, J. D. (1987). *Mechanisms for reliable distributed real-time operating systems – the alpha kernel.* Academic Press, Boston, Mass.
Tanenbaum, A. S. (1987). *Operating system design and implementation.* Prentice-Hall, Englewood Cliffs, N.J.
Turner, R. W. (1986). *Operating system design and implementation.,* Collier-Macmillan, London, New York.

Computer Languages

Brodie, L. (19881). *Starting Forth.* Prentice Hall, Englewood Cliffs, N.J.
Cluley, J. C. (1987). *Introduction to low level programming for microprocessors.* Macmillan, London.
Derick, M. and Baker, L. (1982). *Forth encyclopedia.* Mountain View Press, Mountain View, Calif.
Downes, V. A. and Goldsac, S. J. (1982). *Programming embedded systems with ada.* Prentice-Hall, Englewood Cliffs, N.J.
Dowsing, R. (1988). *Introduction to concurrency using occam.* Van Nostrand Reinhold, London.
Etter, D. M. (1983). *Structured fortran 77 for engineers and scientists.* Benjamin/Cummins Publishing Co., Menlo Park, Calif.
Hoare, C. A. R. (1988). *Occam 2 reference manual.* Prentice-Hall, Englewood Cliffs, N.J.
Kernighan, B. W. and Ritchie, D. M. (1978). *The c programming language.* Prentice-Hall, Englewood Cliffs, N.J.
Ledgard, H. F. (1980). *Ada, an introduction by Henry Ledgard. Ada reference manual.* Springer-Verlag, Berlin, New York.
Merchant, M. J. (1981). *Fortran 77: language and style: a structured guide to fortran.* Wadsworth Pub. Co., Belmont, Calif.

Wirth, N. (1976). *Algorithms + data structures = programs.* Prentice-Hall, Englewood Cliffs, N.J.
Wirth, N. (1988). *Programming in modula-2.* Springer-Verlag, Berlin, New York.

Control Algorithms

Åstrom, K. J. (1987). *Proceedings of the IEEE.* IEEE, Piscataway, N.J.
Rosko, J. S. (1972). *Digital simulation of physical systems.* Addison-Wesley,
Reading, Mass.
Shannon, R. E. (1975). *System simulation: the art and science.* Prentice-Hall,
Englewood Cliffs, N.J.
Smith, J. M. (1977). *Mathematical modelling and digital simulation for engineers
and scientists.* Wiley, New York.
Stephenson, R. E. (1971). *Computer simulation for engineers.* Harcourt Brace
Jovanovoch Inc., New York.
Tsyphin, Y. Z. (1971). *Adaption and learning in automatic systems.* Academic Press,
New York.

Microcomputer Process Control

Moore, J. A. (1986). *Digital control devices, equipment and applications.* ISA Press,
Englewood Cliffs, N.J.
Murrill, P. W. (1988). *Application concepts of process control.* ISA Press, Engle-
wood Cliffs, N.J.
Harrington, J. (1989). *Automated process control electronics.* Delmar Publishers,
Albany, N.Y.
Thompson, L. M. (1988). *Electronic Controllers.* ISA Press, Englewood Cliffs, N.J.

Robotics

Engelberger, J. F. (1980). *Robotics in practice: management and applications of
industrial robots.* Kogan Page (in association with Avebury), London.
Malcolm, D. R. (jr.) (1988). *Robotics: an introduction.* 2nd ed. PWS-Kent Pub. Co.,
Boston, Mass.
Rosenbrock, H. H. (1981). *New technology society, employment and skill: report of
a working party.* London Council for Science and Society, London.
Rosenbrock, H. H. (1989). *Designing human-centred technology: a cross-
disciplinary project in computer-aided manufacture.* Springer-Verlag, Berlin,
New York.
Spong, M. W. and Vidyasagar, M. (1989). *Robot dynamics and control.* Wiley, New
York.
Wolovich, W. A. (1986). *Robotics: basic analysis and design.* Holt, Rinehart and
Winston, New York.

Control System Testing and Commissioning

Bendat, J. S. and Piersol, A. G. (1986). *Random data: analysis and measurement
procedures.* 2nd ed. Wiley, New York.
Clarke, A. B. and Disney, R. L. (1985). *Probability and random processes: a first
course with applications.* 2nd ed. Wiley, New York.

Cooper, G. R. and McGillan, C. D. (1986). *Probabilistic methods of signal and system analysis*. 2nd ed. Holt, Rinehard and Winston, New York.
Davenport, W. B. and Root, W. L. (1987). *An introduction to the theory of random signals and noise*. IEEE Press, Piscataway, N.J.
Hollier, R. H. ed. (1988). *Tracking, identification and control*. IFS Publications.
Jackson, J. B. (1991). *Signals, systems and transforms*. Addison-Wesley, Reading, Mass.

Index